New Biology
for Engineers
and Computer Scientists

New Biology
for Engineers
and Computer Scientists

Aydın Tözeren
Drexel University, Philadelphia, PA

Stephen W. Byers
Georgetown University, Washington, DC

PEARSON

Prentice
Hall

Pearson Education, Inc.
Upper Saddle River, New Jersey 07458

Library of Congress Cataloging-in-Publication Data

Tözeren, Aydın.
　New biology for engineers and computer scientists / Aydın Tozeren, Stephen W. Byers.
　　p. cm.
　Includes bibliographical references and index.
　ISBN 0-13-066463-4
　1. Molecular biology. 2. Cytology. 3. Bioinformatics. 4. Biomedical engineering. I. Byers, Stephen W. II. Title.

QH506.T68 2003
572.8—dc21   2003040485

Vice President and Editorial Director, ECS: *Marcia J. Horton*
Acquisitions Editor: *Dorothy Marrero*
Vice President and Director of Production and Manufacturing, ESM: *David W. Riccardi*
Executive Managing Editor: *Vince O'Brien*
Managing Editor: *David A. George*
Production Editor: *Scott Disanno*
Director of Creative Services: *Paul Belfanti*
Creative Director: *Jayne Conte*
Art Editor: *Greg Dulles*
Manufacturing Manager: *Trudy Pisciotti*
Manufacturing Buyer: *Lisa McDowell*
Marketing Manager: *Holly Stark*

© 2004 by Pearson Education, Inc.
Upper Saddle River, New Jersey 07458

10 9 8 7 6 5 4 3 2

ISBN 0-13-066463-4

Pearson Education Ltd., *London*
Pearson Education Australia Pty. Ltd. *Sydney*
Pearson Education Singapore, Pte. Ltd.
Pearson Education North Asia Ltd., *Hong Kong*
Pearson Education Canada, Inc., *Toronto*
Pearson Educación de Mexico, S.A. de C.V.
Pearson Education—Japan, *Tokyo*
Pearson Education Malaysia, Pte. Ltd.
Pearson Education, Inc., *Upper Saddle River, New Jersey*

Dedicated to:

LNMP—"Nothing compares to you"

Steve

Lynnette Platt Byers, my beautiful, vivacious,
and courageous friend

Aydın

Contents

4 Gene Circuits 114

5 Genomics: The Technology behind the Human Genome Project 136

6 Cell Adhesion and Communication 168

7 Cell Division and Its Regulation 192

8   Development of Multicellular Organisms   215

9   Large-Scale Biology   240

Preface

New Biology for Engineers and Computer Scientists is an easy-to-read modern biology book specifically targeting engineers, engineering students, and computer scientists. Biology is probably the most fascinating science of our time and is extraordinarily ambitious because it seeks to uncover the mysteries of life itself. Recent advances in biotechnology have already enabled scientists to decode the human genome. Considerable progress has also been made in understanding the complex world of proteins, and the networks they form within and across living cells. The systemwide analyses used in the New Biology era represent a paradigm shift for scientists accustomed to investigating discrete pathways or simple biological phenomena. The contribution of computer scientists and engineers to New Biology is undeniable. Genome sequencing and gene prediction are based on sophisticated pattern recognition analyses and the recent advances in nanotechnology and microrobotics have transformed proteomics from science fiction into reality. The exciting new integration of biology, physics, and computational sciences brings to light the need for a new type of engineer, one with a grasp of modern biology. Educating engineers in molecular cell biology has always been a challenge. Biology is perceived by many engineering students as a memorization, not a learning, class. However, New Biology is intellectually stimulating and can be mathematically rigorous; therefore, it has great appeal for engineers and computer scientists.

New Biology is terminology rich, because researchers have identified many of the genes and proteins more or less arbitrarily. In addition, the procedures biologists use originate from a large variety of disciplines and are difficult to capture in a one-year course. As a result, most biology books are large, complex texts covering many topics such as the structure and function of cells, heredity, evolution, biology of plants and animals, and ecology and biogeography. As such, they are extremely useful learning tools for graduate students with a background in biology. On the other hand, *New Biology for Engineers and Computer Scientists* focuses narrowly on what we perceive to be the essentials of New Biology, namely, genes and proteins, cells as the basic units of life, cell division, and animal development.

The contents of *New Biology for Engineers and Computer Scientists* reflect the opinions voiced in an extensive survey among the biomedical engineering faculty in the United States. The survey also attested to the strong need for such a book. *New Biology for Engineers and Computer Scientists* introduces cells as robust complex networks of genes and proteins and adopts a systems view to discuss communication of cells with other cells and with the external environment. Some of the assignments listed at the end of each chapter illustrate the link between biology and engineering. In keeping with the "hands on" approach common in engineering classes, these assignment sections are a particularly important aspect of the learning experience of *New Biology for Engineers and Computer Scientists*. By completing the assignments, the student can both test his or her knowledge and expand it to include areas not covered directly in the text. *New Biology for Engineers and Computer*

Scientists integrates the tools of bioinformatics throughout the text and illustrates their effective use: The assignments that follow are used to further emphasize the important themes of New Biology. These assignments typically have more specifics than the main text and refer the reader for further bioinformatics research to Web sites such as www.ncbi.nlm.nih.gov/ and http://www.genome.ad.jp/dbget.html. Students will learn how to read nucleotide sequences from the Gene Bank, search for similarities among proteins or genes, and learn how to read molecular pathway diagrams. *New Biology for Engineers and Computer Scientists* introduces the reader to advances in functional genomics and protein sciences and to the emerging tools of biotechnology such as microarrays, microfluidic chips, and proteomics. Our experience in teaching biology to engineering students indicates that bioinformatics tools become very powerful in the hands of those engineering students who are eager to uncover the language and content of the decoded genomes. We have refrained from adding extensive sets of numerical or quantitative examples into the book, as this book is in our view a primer for engineers and computer scientists, not a computational biology textbook.

New Biology for Engineers and Computer Scientists is designed as textbook for a course for engineering and computer science undergraduates. The book will also be useful in teaching systems biology to starting bioinformatics or biomedical engineering graduate students with little background in biology. Physicists, engineers, and computer scientists interested in learning about biology and biotechnology will also find *New Biology for Engineers and Computer Scientists* useful. The demand for engineers in the biomedical industry is growing rapidly. Engineering skills, from building microrobots to pattern recognition and large-scale data analysis, are of crucial importance to the biotechnology industry. We believe our book provides an effective tool in teaching basic biology to those engineers and scientists wanting to join the biotechnology workforce.

We wish to express our gratitude to the many authors on whose work we have drawn. We are deeply indebted to Drs. Banu Onaral, Mike Mullins, George Zahalak, Robert Lechleider, Tracey Rowlands, Christopher Avvisato, Orest Blaschuk, and Becky Hoxter for reviewing and contributing to different sections of the book. We are grateful to Nathalie Boyd, Neil Weston, and Zihang Ou for pointing out many errors and inconsistencies. Zihang Ou created many of the figures in the text. The suggestions of Christopher Batich (University of Florida), Holly V. Goodson (University of Notre Dame), Ali Shokoufandeh, (Drexel University), and Nikolaos V. Sahinidis (Univ of Illinois at Urbana) greatly improved the final version of this book. Finally, our thanks go to editors Eric Frank, Dorothy Marrero, Scott Disanno, and other members of the Prentice Hall community for bringing the idea behind this book to life.

AYDIN TÖZEREN
STEPHEN W. BYERS

The Chemistry of Life

1.1 | Introduction

Biology is the study of life, a self-replicating and evolving system of enormous complexity. The large, complex molecules of living cells facilitate their own reproduction and allow for genetic continuity through generations. Nonliving systems can also exhibit forms of self-replication, as exemplified by the growth of a crystal lattice, but such systems do not evolve in ways to meet new challenges. Another important property of life is the capacity of living organisms to capture energy. In a cell, thousands of chemical reactions occur simultaneously at any instant of time. To maintain life, an organism repairs or replaces its structures. These and other life processes require a continuous supply of energy. The primary source of energy comes from sunlight. Plants and some microorganisms capture the energy of light and transform it into heat and chemical energy. Chemical energy derived from sunlight is then used to sustain tissues, reproduction, and growth. Many of the processes involved in living systems are highly complex, even in the simplest organisms.

Radioisotope studies suggest that the molecules that form living beings appeared gradually in our planet after it cooled billions of years ago. At some point during the hundreds of millions of years since then, an appropriate environment might have come into being for a particular "soup" of these molecules to form units able to reproduce themselves. Today there exist over 30 million types of organisms. Despite this diversity, common molecular patterns and principles underlie all expressions of life. Organisms as different as bacteria and humans use the same building blocks to construct proteins, lipids, and carbohydrates. The flow of information from genes to proteins is essentially the same in all organisms. Moreover, living systems on earth use a common currency of energy called ATP. These observations provide support to the thesis that life on our planet evolved from a common origin.

Living systems have many levels of hierarchy, beginning with nonliving particles—atoms and molecules—and moving on to living cells, multicellular organisms,

TABLE 1.1

Atomic Composition of Living Organisms* and the Earth's Crust[†]				
Element	Human	Alfalfa	Bacteria	Earth's Crust
Oxygen (O)	62.8%	77.9%	73.7%	50%
Carbon (C)	19.4%	11.34%	12.14%	0.2%
Hydrogen (H)	9.3%	8.7%	9.9%	0.9%
Nitrogen (N)	5.1%	0.8%	3.0%	1.3%
Phosphorus (P)	0.6%	0.7%	0.6%	0.12%
Sulfur (S)	0.6%	0.1%	0.3%	0.12%
Total	97.9%	99.6%	99.72%	52.6%

*Living organisms also contain a number of mineral ions such as iron, calcium, and magnesium. These minerals account for as much as 2 percent of the human body weight.
[†]The Earth's crust includes the atmosphere, oceans and lakes, and the first 10 miles of solid matter.

families, and societies. Studies of fossils indicate that life on Earth has always consisted of a relatively small number of elements. Six elements out of more than a hundred make up 98 percent of the mass of any living organism: hydrogen (H), carbon (C), nitrogen (N), oxygen (O), phosphorus (P), and sulfur (S) (see Table 1.1). Moreover, with the exception of oxygen and calcium, the biologically most abundant elements are but minor components of the Earth's crust. The elements that constitute life, when combined with each other, produce an astonishing number of different molecules with diverse structures and functions. The capacity to achieve complexity out of a simple set of building blocks is one of the most intriguing properties of living systems.

1.2 | Atoms

1.2.1 Physical Properties of Atoms

All matter, living and nonliving, is composed of atoms, which are the smallest units of matter. In the following we present a brief review of atomic structure and its impact on the chemical properties of elements. Atoms consist of a dense, positively charged nucleus and a cloud of one or more negatively charged electrons (Fig. 1.1). The nucleus is made of two different types of particles: protons and neutrons. These particles have the same mass of 1.67×10^{-24} g. Protons have a positive charge of 1.60×10^{-19} coulomb and neutrons have no charge. Electrons have negligible mass compared to protons and neutrons (9.1×10^{-28} g). The number of electrons in an atom is equal to the number of protons in the atom, and because the negative charge of an electron is equal in magnitude to the positive charge of a proton, an atom has no net charge. By convention, the charge of a proton is shown as $+1$ and that of an electron as -1.

Atoms vary in mass and volumetric size. An *element* is matter that is made up of only one kind of atom and cannot be decomposed by chemical reaction into substances of simpler composition. The number of protons in an atom (the *atomic number*) determines the identity of the element. The atomic number increases by one in the sequence of elements in the Periodic Table (Fig. 1.2). There are over 100 elements

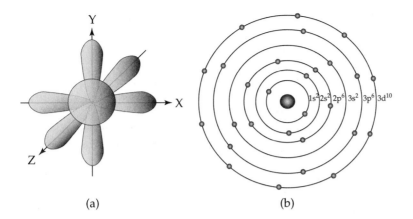

(a)

$1s^2 2s^2 2p^6 3s^2 3p^6 3d^{10}$

(b)

FIGURE 1.1 Outer-shell atomic orbitals of carbon (a). The nucleus of the carbon atom is at the origin of the Cartesian coordinate system (x, y, z). The nucleus is surrounded by the spherical 1s and 2s orbitals and the three mutually orthogonal dumbbell shaped 2p orbitals $(2p_x, 2p_y, 2p_z)$. Schematic diagram of orbitals around a nucleus (b). The orbitals are shown as circles around the spherical nucleus. The symbols identifying various orbitals are defined in the text.

on earth. The most abundant molecules of living systems (hydrogen, carbon, nitrogen, and oxygen) have low atomic numbers.

Atomic mass number refers to the number of protons and neutrons in the nucleus of an atom. Not all atoms of an element have the same atomic mass number. For example, different atomic forms of a single element, *isotopes*, differ in the number of neutrons in the atomic nucleus. The dominant isotope of oxygen has eight protons and eight neutrons. This isotope is assigned an *atomic mass unit* (amu) of 16. By definition, then, an atomic mass unit is 1/16 of the mass of

Group	1	2		3	4	5	6	7	8	9	10	11	12	13	14	15	16	17	18
Period																			
1	1 H																		2 He
2	3 Li	4 Be												5 B	6 C	7 N	8 O	9 F	10 Ne
3	11 Na	12 Mg												13 Al	14 Si	15 P	16 S	17 Cl	18 Ar
4	19 K	20 Ca		21 Sc	22 Ti	23 V	24 Cr	25 Mn	26 Fe	27 Co	28 Ni	29 Cu	30 Zn	31 Ga	32 Ge	33 As	34 Se	35 Br	36 Kr
5	37 Rb	38 Sr		39 Y	40 Zr	41 Nb	42 Mo	43 Tc	44 Ru	45 Rh	46 Pd	47 Ag	48 Cd	49 In	50 Sn	51 Sb	52 Te	53 I	54 Xe
6	55 Cs	56 Ba	*	71 Lu	72 Hf	73 Ta	74 W	75 Re	76 Os	77 Ir	78 Pt	79 Au	80 Hg	81 Tl	82 Pb	83 Bi	84 Po	85 At	86 Rn
7	87 Fr	88 Ra	*	103 Lr	104 Rf	105 Db	106 Sg	107 Bh	108 Hs	109 Mt	110 Uun	111 Uuu	112 Uub	113 Uut	114 Uuq	115 Uup	116 Uuh	117 Uus	118 Uuo
*Lanthanides			*	57 La	58 Ce	59 Pr	60 Nd	61 Pm	62 Sm	63 Eu	64 Gd	65 Tb	66 Dy	67 Ho	68 Er	69 Tm	70 Yb		
**Actinides			*	89 Ac	90 Th	91 Pa	92 U	93 Np	94 Pu	95 Am	96 Cm	97 Bk	98 Cf	99 Es	100 Fm	101 Md	102 No		

FIGURE 1.2   Periodic table indicating the atomic properties of all elements found on Earth. (From http://www.genome.ad.jp/kegg/catalog/elements.html.)

the most common isotope of an oxygen atom. The choice of oxygen as the standard is arbitrary. Nevertheless, it can be justified by the fact that oxygen forms compounds with most elements. *Atomic mass* is the average atomic mass number in a large collection of atoms of an element. Because there may be multiple isotopes of an element in the collection of atoms, the element's atomic mass is not necessarily an integer number. The atomic mass of hydrogen is 1.008 amu and that of sulfur is 32.006 amu.

Since atoms are extremely small, any laboratory experiment dealing with chemicals involves large numbers of atoms. One *gram-atom* is defined as the collection of atoms whose total mass in grams is numerically equal to the atomic mass of the atom. For example, 32.066 g of sulfur is equal to 1 gram-atom of sulfur. The number of atoms in one gram-atom of an element is always equal to 6.0235×10^{23}. This number is called the *Avogadro number*.

1.2.2 Electron Configurations

The extent with which an atom interacts with other atoms depends on the configuration of the cloud of electrons surrounding its nucleus. The precise location of a given electron in an atom at any time is impossible to determine, but it is possible to specify a region where the electron is likely to be at least 90 percent of the time. Such a region is called an *orbital* (Fig. 1.1), which can be occupied by at most, two electrons. The two electrons in an orbital spin about their own axes with equal speed, but in opposite directions.

An electron in an atom can only have certain specific levels of energy. Furthermore, the energy of an electron in a given orbital remains constant. Thus, orbitals constitute a series of constant-energy electron shells around the nucleus (Fig. 1.1b). The innermost electron shell is associated with the lowest energy, and the energy levels increase with increasing orbital numbers. The only way an electron can change its energy is to shift from one discrete energy level (orbital) to another. When an electron jumps to a lower energy level, it radiates energy, and if no lower energy level is available, the electron cannot get any closer to nucleus. An electron will jump to a higher energy orbit when its energy is increased due to increase in temperature or radiation.

Chemists classify the electron orbitals around the nucleus into shells and subshells according to their energy levels (Fig. 1.1b). Because electrons in an orbital have constant energy, schematic diagrams show orbitals as circles of varying radii around a centrally located nucleus. The larger the radius of the orbital, the greater is the energy of electrons in that orbital. The *principal quantum number* identifies the shell to which an electron belongs. This parameter is represented by the letter n. Its lowest value is one, and it increases by one for the next energy level. The number of subshells in a shell is always equal to n. These subshells are identified with small letters s, p, and d in the order of increasing energy levels. The s orbital forms a spherical surface around the nucleus. The p orbitals resemble that of two spheres, one on each side of the nucleus (Fig. 1.1a). An electron in a p orbital has an equal probability of being found in either half of the orbital. A p subshell is constructed of three p orbitals perpendicular to each other. The spatial distribution of electrons in the higher order subshells are considerably more complicated.

1.2.3 Chemical Reactivity and the Octet Rule

The electrons of an atom fill the orbitals from the lowest to the highest energy levels. The shell with the lowest overall energy ($n = 1$) has only one subshell identified as 1s. The integer in front of the subshell symbol s identifies the principal quantum number n. The 1s subshell can contain up to 2 electrons. The L shell ($n = 2$) has the next lowest energy level. It is composed of two subshells: 2s (containing up to two electrons) and 2p (which has 3, 2p orbitals and can hold six electrons). Electrons in the 2p orbital have higher energy than the 2s orbital. In the M shell ($n = 3$), there are three subshells, 3s, 3p, and 3d, each of which can hold 2, 6, and 10 electrons, respectively. Because the energy level of the 4s orbital is slightly smaller than the energy level of orbital 3d, atoms with atomic numbers greater than 18 begin to fill 4s before filling 3d. The six atoms most commonly found in biomolecules have atomic numbers less than 18. Their electronic configurations are shown in Table 1.2.

The terms 1s, 2s, 2p, 3s, and 3p refer to electron orbits written in the order of increasing energy levels. The integers 1, 2, and 3 identify the orbit and the letters s and p the orbital belonging to the orbit; s orbitals can accommodate up to two electrons and p orbitals up to six electrons. In general, atoms (with atomic numbers greater than 2) form molecules that have eight electrons in their outermost orbitals. Therefore, the number of electrons in the outermost shell is an important determinant of the chemical properties of an atom. These electrons are called *valence electrons*. They are responsible for the *combining capacity* of atoms. In the periodic table presented in Fig. 1.2, elements are grouped into vertical columns according to the number of valence electrons. The tendency for eight electrons in the outermost orbitals is called the *octet rule*. Atoms that already have eight electrons in their outermost shell such as argon do not interact with other elements. They are said to be *inert*. In *electron dot formulas*, *valence* electrons are shown with dots surrounding the symbol of the atom:

$$\dot{H} \qquad \cdot \dot{\underset{\cdot}{C}} \cdot \qquad \cdot \overset{\cdot\cdot}{\underset{\cdot}{N}} \cdot \qquad \cdot \overset{\cdot\cdot}{\underset{\cdot\cdot}{O}} \cdot \qquad \cdot \overset{\cdot\cdot}{P} \cdot \qquad \cdot \overset{\cdot\cdot}{\underset{\cdot\cdot}{S}} \cdot$$

The symbols H, C, N, O, P, and S denote, respectively, hydrogen, carbon, nitrogen, oxygen, phosphorus, and sulfur. *Hydrogen* is the simplest element on earth. It forms the bulk of the matter of the sun, but exists only in small amounts in an uncombined form on Earth, suggesting that the Earth's gravity is not strong enough to

TABLE 1.2

Electronic Composition of Bioelements							
Z*	Element	1s	2s	2p	3s	3p	4s
1	Hydrogen (H)	1					
6	Carbon (C)	2	2	2			
7	Nitrogen (N)	2	2	3			
8	Oxygen (O)	2	2	4			
15	Phosphorus (P)	2	2	6	2	3	
16	Sulfur (S)	2	2	6	2	4	
20	Calcium (Ca)	2	2	6	2	6	2

*Z: Atomic Number [number of protons (electrons) in an element].

keep large amounts of hydrogen gas in the atmosphere. Hydrogen has a single valence electron. In forming a stable compound, hydrogen must either empty its electron orbital by giving up its electron to another atom (ionic bond) or fill the orbital by sharing electrons with another atom (covalent bond).

The organic element with the second smallest atomic number is *carbon*. There are six protons in each carbon atom and most carbon atoms have six neutrons. About 1.1 percent of carbon atoms have seven neutrons and a tiny fraction have eight. Carbon has four valence electrons: therefore, its outermost orbitals are exactly half filled. The possession of a half-filled outermost orbit allows carbon to engage in the formation of chemical bonds with many other atoms. The resulting chemical versatility is essential for the reactions of biological metabolism and propagation. Silicon also has four valence electrons and also serves as the backbone of molecules that are large enough to carry biological information. On the other hand, silicon interacts with only a few other atoms, and the large silicon molecules are monotonous compared with the combinatorial universe of carbon-based organic molecules. Carbon is present as carbon dioxide in the atmosphere and is found dissolved in natural waters. It is also abundant in living organisms and in organic matter such as coal and petroleum. Carbon is also one of the most abundant of the higher elements in our solar system.

Nitrogen occupies 78 percent of the volume of the atmosphere. In plants and animals, nitrogen is found in proteins and in the genetic material DNA and RNA. Approximately, 16 percent of the weight of proteins is due to nitrogen. Nitrogen atoms have seven protons and five valence electrons. This corresponds to an outer electron orbit, which is only slightly more than half filled. The compounds of nitrogen, though not as numerous as those of carbon, are just as varied in function. *Phosphorus* has 15 protons and five valence electrons. It is abundant in rock formations in the form of calcium phosphate. This inorganic salt also constitutes 60 percent of bones in the human body.

Oxygen and *sulfur* both have six valence electrons. Oxygen is the most abundant element in the earth's crust. It is found in water, rocks, living organisms, and in the atmosphere. Oxygen has eight protons and six valence electrons in its outermost shell. It can gain two electrons to achieve the stable *octet* configuration. In so doing, it combines with all other elements except inert helium, neon, and argon. *Sulfur* is found in the free-element form in large beds several hundred feet underground. As discussed in the next chapter, sulfur atoms, when present in the amino acid cysteine, play a fundamental role in determining the three-dimensional shapes of many proteins.

1.3 | Molecules and Covalent Bonds

Atoms interact with each other by forming bonds of different types such as covalent bonds, ionic bonds, and hydrogen bonds. These various bond configurations will be discussed in this section and the following sections. Two atoms form *covalent bonds* by sharing one or more pairs of electrons in order to have a complete set in their outer shell at least some of the time. For most elements found in living systems this means eight electrons in the outermost orbitals (octet rule). The only exception is hydrogen, which has one valence electron in a single outermost orbital. In a hydrogen

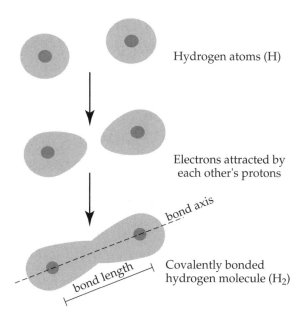

Hydrogen atoms (H)

Electrons attracted by
each other's protons

bond axis

bond length

Covalently bonded
hydrogen molecule (H_2)

FIGURE 1.3 Sharing of electron pairs in covalent bonds. Two hydrogen atoms form a covalent bond when they come close enough that the electron orbitals in the $n = 1$ (M) shell overlap. Each electron is attracted to both protons, but the two protons of the opposing hydrogen atoms repulse each other and the balance of these opposing forces are the determinants of the resulting molecular structure.

molecule (H_2), the two hydrogen nuclei share the two electrons equally and completely (Fig. 1.3).

A *covalent bond* is formed when an outer orbital of one atom containing an electron overlaps an outer orbital of another atom also containing a single electron. These two atoms begin sharing the electron pair (Fig. 1.3). The overlapping orbitals deform as a result of the interaction. The nuclei of the interacting atoms are electrically attracted to the mutually shared electron pair. This attraction results in what is called the *covalent bond*. The line joining the nuclei of two covalently bonded atoms is called the *bond axis* and the distance between the nuclei is called the *bond length* (Fig. 1.3). The bond length is not fixed, and the bond acts much as if it were a stiff spring. Figure 1.4 shows the orbitals of the shared electrons for three commonly known molecules: methane (CH_4), ammonia (NH_3), and water (H_2O).

A combination of atoms held together by covalent bonds is called a *molecule*. A molecule is identified by the symbols of elements that constitute it. The subscript associated with each element refers to the number of atoms of that element in the molecule. For example, the oxygen molecule (O_2) is composed of two oxygen atoms. Carbon dioxide is formed when two oxygen atoms share some of their electrons with a carbon atom; and thus the symbol CO_2. The quantity of molecules in a given volume is typically expressed in terms of moles, or M. The unit M represents one mole per liter. Like the number of atoms in one gram-atom, the number of molecules in one *mole* of matter is equal to the *Avogadro* number, 6.0235×10^{23}.

1.3.1 Double and Triple Bonds

Bonds in which a single pair of electrons is shared between two atoms are called *single bonds*. A single bond is represented by a straight-line segment connecting two atoms.

FIGURE 1.4 Structural representation of simple molecules. Top row: Methane (CH_4) molecules assume the shape of a tetrahedron with bond angles equal to 109.5° (a). Ammonia (NH_3) also takes the shape of a tetrahedron, with the unshared electron pair occupying one corner of the tetrahedron (b). Water (H_2O) is triangular in shape, with the bond angle between the two hydrogen–oxygen covalent bonds equal to 104.5° (c). Note that the clouds of unshared electrons are larger than those of shared electrons. The charge clouds of nonbonded electrons push the bonded pairs toward each other. The middle and bottom rows show the ball-and-stick and space-filling models of these three molecules.

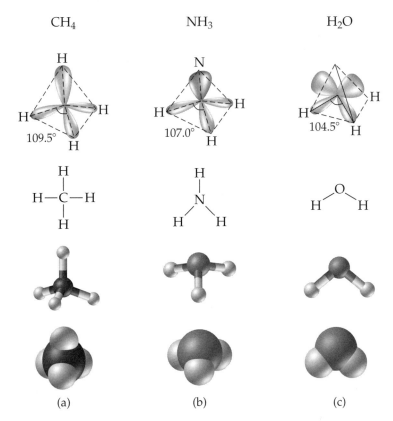

When four electrons are shared between two atoms, the covalent link is called a *double bond*. A double bond is represented by a double-line segment. This is illustrated for ethylene (C_2H_4):

Shown in the left is the standard structural formula of ethylene. On the right is the simplified diagram that masks the presence of C—H bonds. Since carbon will form four bonds with neighboring atoms, it is implicitly assumed that the bonds not shown are C—H bonds. Another example of a six-carbon hydrocarbon is as follows:

Double bonds between carbon atoms occur often in biological compounds (Fig. 1.5).

Class of Molecules	Functional Group	Example	Ball - Stick
Ethanol (Alcohols)	Hydroxyl —OH	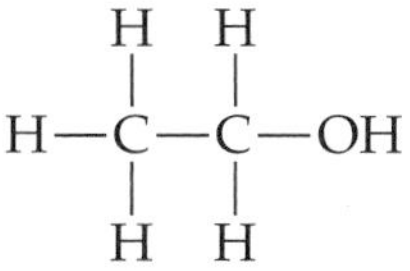	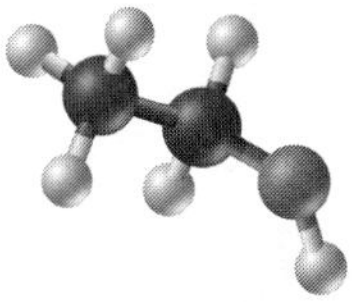
Acetaldehyde (Aldehydes)	Carbonyl —CHO	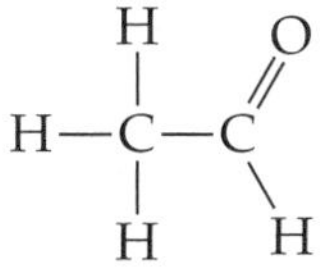	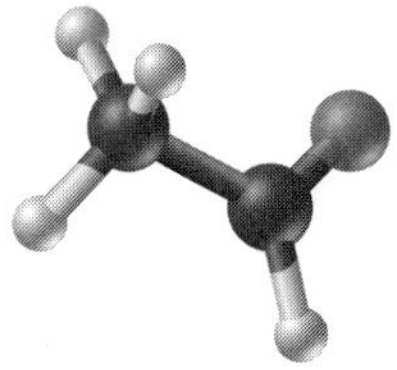
Acetone (Ketones)	Carborryl >CO	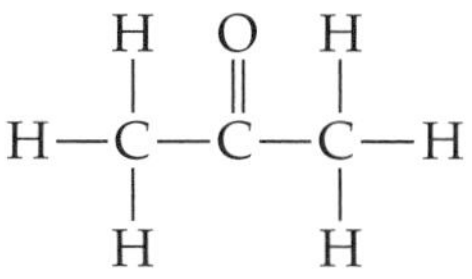	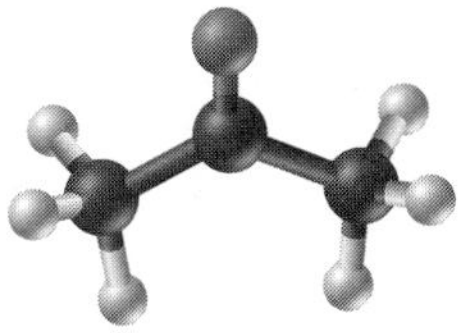
Acetic Acid (Carboxylacids)	Carboxyl —COOH	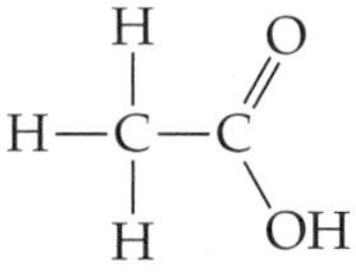	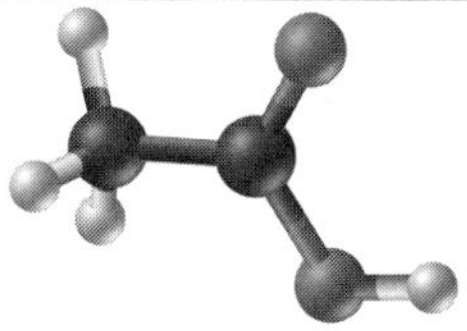
Methylamine (Amines)	Amino —NH$_2$	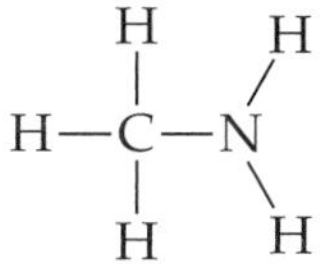	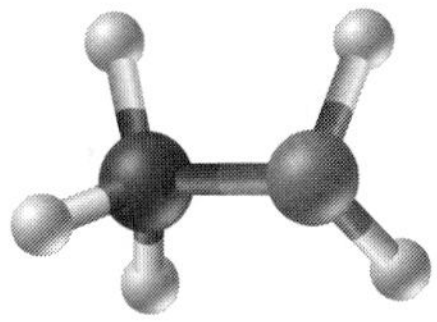
3-Phosphoglycenic Acid (Organic Phosphates)	Phosphate —OPO$_3^{2-}$	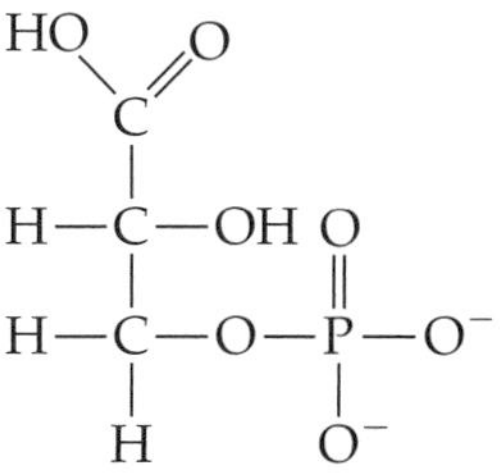	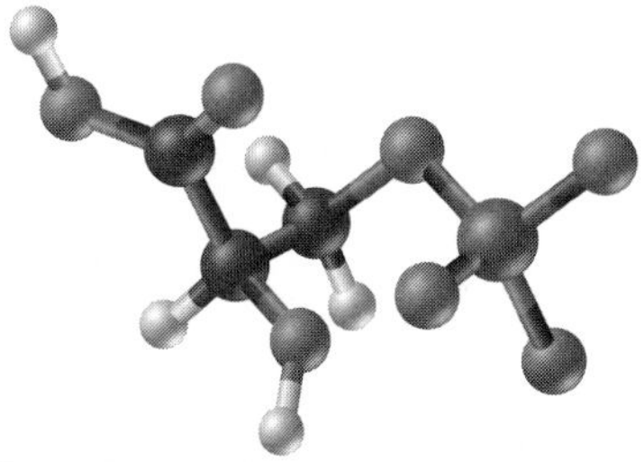
Mercaptoethonol (Thiols)	Sulfhydryl —SH	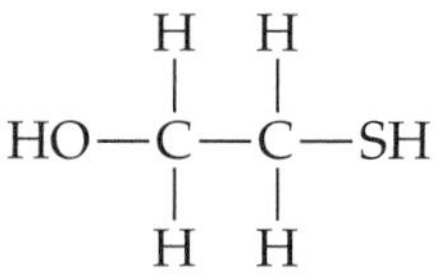	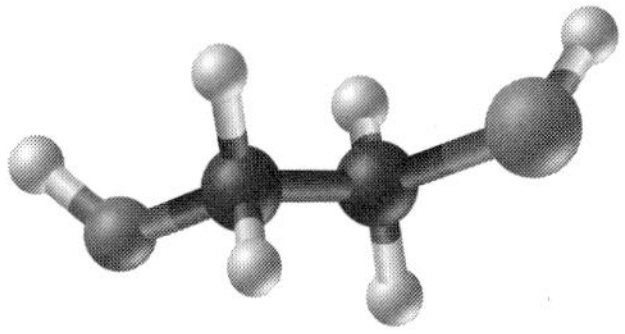

FIGURE 1.5 Examples of small carbon molecules observed frequently in organic compounds. Double bonds are shown with two line segments and single bonds with a single connecting line segment.

Atoms connected by a single bond can rotate freely about the bond axis; however, atoms connected by double bond cannot rotate about the bond axis. Because a double bond occupies more space than a single bond, it is more repulsive. As a result, the H—C—H bond angle is 116° whereas H—C=O angle is 122°.

A *triple bond* involves the sharing of six electrons between two atoms. An important example of a triple bond is the covalent bond between two nitrogen atoms in a nitrogen molecule (N_2). This is a very stable bond. Triple bonds are rarely observed in organic materials. Most nitrogen compounds involve single bonds and are relatively unstable. The precursers of these molecules are produced from atmospheric nitrogen by the nitrogen-fixing bacteria present in the roots of nitrogenous plants such as legumes (peas and beans). These bacteria convert molecular nitrogen to a form that can be used readily by plants.

1.3.2 Three-Dimensional Structure of an Organic Molecule

The pair of electrons involved in covalent bonding is called a *shared pair*. The pairs of outermost electrons not involved in bonding are called *unshared pairs*. Each bond and each unshared pair in the outer level of an atom form a charge cloud that repels all other charge clouds. The repulsions between the charge clouds determine the three-dimensional shape of the molecules.

The physical rules that dictate the three-dimensional shape of a molecule can be listed as follows:

1. In a molecule, atoms are oriented so that repulsion between electron pairs (either bonding or unshared) around an atom is minimized. In other words, electron pairs spread as far apart as possible to minimize repulsive forces.

2. Repulsive forces between various electron pairs in a molecule are not equal in strength. An unshared pair is acted upon by only one nucleus, and its cloud occupies more space compared to that of a shared pair. The repulsive forces between the unshared pairs of electrons are greater than the repulsive forces between two shared pairs.

The geometry of a molecule is a fundamental determinant of its chemical properties. Biologists use various representations to capture the three-dimensional shapes of molecules. Figure 1.4 shows four of these representations for the molecules methane (CH_4), ammonia (NH_3), and water (H_2O). The top row indicates the three-dimensional shapes of the shared and unshared outer orbitals. The second row presents the planar sketches of the molecules. The third row represents molecules as if they were composed of balls and sticks. Balls represent the locations of the nuclei of atoms in the molecule and sticks represent the covalent bonds along the bond axes. The bottom row shows the space-filling models of these three molecules. In methane, the central carbon atom forms covalent bonds with four hydrogen atoms in order to have eight electrons in its outermost shell. Because this results in four pairs of electrons equally repelling each other, the shape of the methane molecule is a tetrahedron. The angle the two adjoining bonds make (bond angle) is 109.5° in this case. In the ammonia molecule (NH_3), the three pairs of bonded electrons and one unshared pair must be positioned to reduce repulsion to a minimum. The resulting structure is again a tetrahedron, but the unshared pair occupies more space than the other three electron pairs. The angle between the bond pairs is 107° in ammonia.

Two unshared pairs of outer electrons of oxygen are present in a water molecule: Both of these clouds are larger than the clouds of bonded electrons. This additional cloud size results in a reduction in the bond angle (104.5°) between the bond axes connecting O to H atoms. Again the electron clouds are tetrahedral, the molecule (lines connecting the nuclei) is V-shaped.

The three-dimensional structure of thousands of small organic molecules and information on their chemical properties can be found by using the compound search engine of the Kyoto University KEGG Web site http://www.genome.ad.jp/ dbget/ligand.html. Another highly used site is that of the Protein Data Bank or PDB, which is the single worldwide repository for the processing and distribution of three-dimensional biological macromolecular structure data (http://www.rcsb.org/pdb/).

One can search for the molecular structure and other information by typing the name of the compound in the ligand search engine. KEGG provides structural formulas in their simplified form. For example, the benzene molecule (C_6H_6) that forms a carbon ring is shown in KEGG simply as a hexagonal ring with no symbols attached, as shown earlier.

In general, KEGG masks the symbols for carbon and hydrogen atoms. All corners with no symbol attached represent the locations of carbon atoms. The presence of hydrogen bonds is implied when a carbon atom has less than four explicitly drawn bonds. A more detailed description of the KEGG representation is presented in Chapter 2.

1.3.3 Physical Strength of a Covalent Bond

Covalent bonds constitute stable links between atoms and are the strongest of bonds connecting molecules. The strength of a covalent bond is measured by the energy required to break it. Rupture of covalent bonds can occur in two ways: The linkage between two atoms may be broken symmetrically to provide a pair of free radicals whose reactivity is derived from the unpaired spins of their electrons. This mode of cleavage is often observed in the breaking of identical or similar atoms. Alternatively, the bond may be broken asymmetrically to produce a pair of ions, one of which is electron deficient and the other, electron rich. The bond strengths of some covalent bonds important in biological systems are shown in Table 1.3.

In general, the energy required to break a covalent bond is much greater than the internal (thermal) energy available at body temperature (0.6 kcal/mol). Chemical changes between molecules take place when the energy needed to break bonds is supplied by some other source, such as the energy released by the formation of new bonds.

As Table 1.3 indicates, covalent bonds between carbon atoms are highly stable. This property enables carbon to form more than seven million chemical compounds. Although silicon also has four valence electrons, silicon's larger atomic radius prevents two silicon atoms from approaching each other closely enough to form stable bonds with each other. Both nitrogen and phosphorus are electron-rich atoms having five valence electrons. The repulsive forces between the unshared pairs of electrons make the N—N bonds less stable than C—C bonds. As a result, the bond energy of an N—N bond is about half the bond energy of a C—C bond. This is why extended chains of bonded N atoms are unstable compared to carbon compounds. Since phosphorus has a larger diameter than nitrogen, it forms even less stable chains of covalent bonds.

TABLE 1.3

Energy Required to Break Covalent Bonds (kcal/M)			
Bond	Energy	Bond	Energy
C=O	170	C—H	99
C=N	147	C—O	84
C=C	146	C—C	83
P=O	120	S—H	81
O—H	110	C—N	70
H—H	104	C—S	62
P—O	100	N—O	53

Next on the periodic table, with six valence electrons, are oxygen and sulfur. Oxygen can make only two covalent bonds; it is involved in most important biochemical reactions, but does not form the backbone of biomolecules. The concentration of sulfur is much lower than oxygen in living systems. Sulfur is also less electronegative than oxygen. Sulfur bonds play a fundamental role in the determination of the three-dimensional shape of a protein. Some of the frequently observed covalent bond groups in organic compounds are illustrated in Fig. 1.5.

1.3.4 Electronegativity and Polar Bonds

The capacity of an atom to attract electrons from a neighboring atom is called *electronegativity* and is measured on a scale from 4 (fluorine, the most electronegative element) to a hypothetical 0 (Table 1.4). Table 1.4 shows oxygen as highly electronegative in comparison to carbon and hydrogen.

Covalent bonds between two atoms with comparable electronegativity (such as that between C and H) correspond to an equal sharing of electrons between the two nuclei. When highly electronegative atoms form covalent bonds with atoms of weaker electronegativity, shared electrons spend more of their time closer to the atom with stronger electronegativity. The resulting bond is called a *polar covalent bond* or simply a *polar bond*. Thus, covalent bonds can have partial charges when the atoms involved have different electronegativity. The symbols δ^+ and δ^- are used to indicate partial charges located at different parts of the molecule. For example, the water molecule can be shown in this notation as

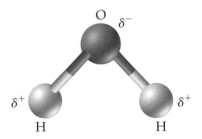

TABLE 1.4

Electronegativity of Bioelements		
Element	Number	Electronegativity
Hydrogen (H)	1	2.1
Carbon (C)	6	2.5
Nitrogen (N)	7	3.0
Oxygen (O)	8	3.5
Phosphorus (P)	15	2.1
Sulfur (S)	16	2.5
Calcium (Ca)	20	1.0

In the water molecule, the oxygen atom exerts a stronger attraction to the bonding electrons than the hydrogen atoms do, resulting in unequal sharing of bonded electrons. When oxygen binds to hydrogen, the bonding electrons spend much more time near the oxygen nucleus than the hydrogen nucleus. Consequently, the oxygen end of the molecule is slightly negative and the hydrogen end slightly positive. Unequal sharing of electrons between two atoms in a molecule results in charge polarity. As discussed later in this chapter, *polar* molecules dissolve easily in water and interact with other polar molecules. In contrast, *nonpolar* molecules dissolve better in a nonpolar environment such as acetone or ethanol.

1.4 | Ionic Compounds and Electrostatic Bonds

Consider a compound in which one of the two interacting atoms is much more electronegative than the other. In such a compound, one or more electrons from the outer shell of the less electronegative atom are transferred to the more electronegative atom. The result is two electrically charged particles, which are called *ions*. An ion with a positive charge such as Ca^{2+} or H^+ is called a *cation*. An ion with negative charge, such as the hydroxyl ion OH^-, is called an *anion*. The net charge of an ion is shown at the superscript immediately to the right of the name of the atom from which it was derived. An electrostatic force holds two ions together due to their differing charges. This force is called the *ionic bond*. The electrostatic force between two ions is large enough to make ionic compounds stable substances. Ionic compounds have high melting points, conduct electricity in the molten state, and tend to be soluble in water.

An example of ionic bond is found in table salt. A sodium atom (Na) has only one electron in its outermost shell of electrons and a chlorine atom (Cl) has seven electrons in its outermost orbits. When the two atoms meet, the highly electronegative chlorine atom takes the single unstable electron from the sodium, producing two ions, Na^+ and Cl^-. Another example of an ionic bond is found in calcium phosphate $[Ca_3^{2+} (PO_4)_2^{3-}]$, major component of bone. The calcium atom has two electrons in its outermost shell and becomes an anion (Ca^{2+}) by losing two electrons. The phosphate ion PO_4^{3-} is a complex ion, meaning that it is a group of covalently bonded atoms that carry an electrical charge.

An example of a covalent bond between a molecule and an ion is exhibited in the ammonium ion (NH_4^+):

$$
\begin{array}{ccc}
H & & H \\
| & & | \\
H-N-H & & H-N^{\pm}-H \\
& & | \\
& & H
\end{array}
$$

In ammonia (NH_3), the nitrogen forms three covalent bonds with three hydrogen atoms. One of the outermost electron pairs of nitrogen is not involved in a covalent bond; thus, this electron pair is not shared. When dissolved in water, ammonia picks up a hydrogen ion (H^+) to become an ammonium ion (NH_4^+), which has a net positive charge of 1. The recruited hydrogen ion shares a previously unshared bond. Again the outermost shell of nitrogen is filled with eight electrons. Because ammonia is highly reactive when dissolved in water, it is used as a main ingredient in many cleaning reagents. Other molecules also become ions when immersed in certain solvents. Consider for example the case of ethanol. Ethanol has no charge, whereas ionized ethanol has a charge of −1. This transformation depends on the acidity of the medium, a topic discussed later in the chapter.

1.5 | Water and Hydrogen Bonds

At room temperature, oxygen and hydrogen exist in gaseous form. The water molecules they form, however, are mostly in the liquid state under the same physical conditions. This is in contrast with gaseous carbon dioxide, another compound formed by two elements. Water has important physical properties that make life possible on Earth. It has the ability to dissolve many other substances. As a result, it serves as a medium in which a great variety of chemical changes occur. These special properties of water are due to the *hydrogen bonding* between water molecules (Fig. 1.6). If hydrogen bonds did not exist, water would be a gas at room temperature, very much like other gases such as ammonia and carbon dioxide.

Hydrogen bonding is caused by the polar nature of covalent H—O bonds that hold together water molecules. In liquid water, the hydrogen atoms of water molecules are attracted to the unshared electrons of oxygen atoms of adjacent water molecules. In chemical diagrams, hydrogen bonds are represented by a series of three dots. The polarity of water molecules leads to extensive interactions between them. The mutual attraction between water molecules causes water to have melting and boiling points at least 100° higher than they would be if water molecules were nonpolar.

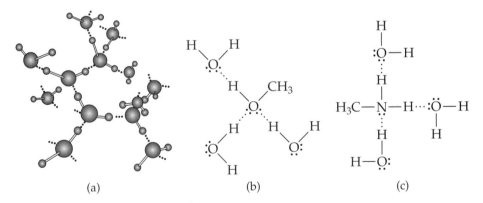

(a)   (b)   (c)

FIGURE 1.6 Hydrogen bonding. Bonding is maximal when two oxygen atoms and a hydrogen atom fall into a straight line (a). In liquid water, water molecules form transient hydrogen bonds with several others, creating a fluid network (b). Hydrogen bonds between methanol (CH_3OH) and water, and methylamine (CH_3NH_2) and water (c). Each of the two pairs of nonbonding oxygen electrons can accept a hydrogen atom in a hydrogen bond. The single pair of unshared electrons in the nitrogen outer shell is also capable of becoming an acceptor in a hydrogen bond.

Water molecules form hydrogen bonds with other *polar* molecules including those of nitrogen and oxygen compounds. Both nitrogen and oxygen are considerably more electronegative than hydrogen. Because all covalent N—H and O—H bonds are polar bonds, the H atoms in these bonds can participate in hydrogen bonding. The two atomic groups that often engage in hydrogen bonding in living systems are the amino (—NH₂) and hydroxyl (—OH) groups (Fig. 1.5). Thus, the presence of amino or hydroxyl groups makes many molecules soluble in water. When table salt dissolves in water, the negative (oxygen) ends of the water molecules surround the sodium ions (Na⁺). On the other hand, the chloride ions (Cl⁻) are surrounded by the H atoms of the polar water molecules. Thus, water molecules cluster around cations and anions in solutions, blocking their association into a solid. Hydrogen bonds also form between different parts of large polar molecules such as proteins (Fig. 1.6). A hydrogen bond is much weaker than a covalent bond in physical strength and duration. However, because many hydrogen bonds can form between different parts of a molecule, they can greatly influence the physical properties of biological substances.

1.6 | Lipids and van der Waals Attraction

Carbon and hydrogen share bonding electrons equally. Hydrocarbons, compounds of carbon and hydrogen, are therefore nonpolar. When hydrocarbons are dispersed in water, they slowly come together to form aggregates (Fig. 1.7).

These aggregates take the form of droplets that join to form larger drops. The separate grouping of nonpolar substances in an aqueous environment is due to the inability of nonpolar hydrocarbon molecules to form hydrogen bonds with water. Their presence distorts the usual water structure. The disruption decreases upon formation of a hydrogen-bonded cage surrounding nonpolar hydrocarbons. The cage compresses nonpolar molecules and separates them from the water molecules. In other words, nonpolar solute molecules are driven together in water not primarily because they have high affinity for each other, but because water bonds strongly to itself. Because they do not dissolve in water, nonpolar substances such as olive oil are also called *hydrophobic* or water fearing. Nonpolar molecules do not interact with ionic particles or polar molecules of any type.

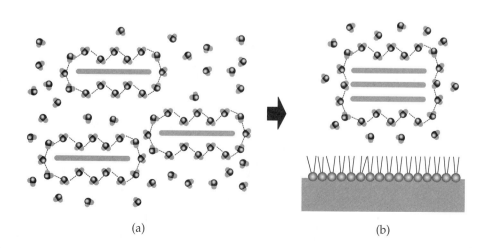

(a)   (b)

FIGURE 1.7 Nonpolar molecules disturb the hydrogen bonds between water molecules, forcing water to form a cage around them (a). More stable clusters of nonpolar molecules form as the surface area of cage is minimized. As a result, these molecules are pushed together to form larger and larger droplets. Oil forms a single thin layer on the air–water interface (b). The hydrophobic tails of oil molecules extend into the air to avoid contact with water.

The hydrophobic effect drives a number of very important biological phenomena. For example, the formation of cell membranes by lipid bilayers as discussed in Chapter 2 is driven by the hydrophobic nature of the hydrocarbon chains of the lipids that make up the bilayer. The hydrocarbon chains interact with one another, but not with the polar environment. Consequently, lipid bilayers have a hydrophobic core with the hydrophilic heads facing the polar environments outside or inside the cell. Even though the formation of an organized bilayer would appear to impose more order, the entropy (disorder) of the system actually increases because restrictions on the movement of neighboring water molecules decreases with the formation of a lipid bilayer.

Similar hydrophobic effects drive the folding of proteins, the molecular machines of life. In this case, regions rich in hydrophobic amino acids (building blocks of proteins) tend to fold away from more polar regions and from the outer surface of the protein in contact with the polar environment. In some proteins, this leads to the formation of hydrophobic pockets or clefts, which often provide a binding site for small hydrophobic molecules such as steroid hormones.

When molecules come very close to each other, whether they are pushed toward each other by water or any other mechanism, the random variations in the electron distribution of one molecule creates an opposite charge distribution in the adjacent molecule. The result is a brief, weak attraction called *van der Waals* interactions. Although each such interaction is brief and weak at any one site, the summation of many such interactions over the entire surface of a molecule can produce substantial interaction. Van der Waals forces act on nonpolar molecules brought together by a polar solvent.

1.7 | Acids and Bases

An *acid* is a substance that can donate protons (H^+), and a base is a substance that can accept protons. An example of an acid is hydrochloric acid (HCl), which dissolves in water to generate H^+ and Cl^- ions. Molecules having a carboxyl group (—COOH) are also acidic because COOH tends to dissociate in water to form COO^- and H^+. Ammonia (NH_3) is a base because it readily picks up a hydrogen ion to become an ammonium ion (NH_4^+). Many important biological molecules have an amino group (—NH_2) associated with them. Like ammonia (NH_3), —NH_2 is also a base because it can take up a hydrogen ion to become —NH_3^+. Acids and bases have distinct physical features. Acids in water have sour taste and cause the purple dye litmus to turn red. *Bases* taste bitter, feel slippery, and turn the purple litmus paper blue.

Biological molecules bear functional groups that can undergo acid–base reactions in living systems. Important properties of such molecules vary with the acidities of the solutions in which they are immersed. The main ingredient of the liquid phase of biological organisms is water. A small fraction of water molecules dissociate into positively charged hydrogen ions (H^+) and negatively charged hydroxyl ions (OH^-). When an acid is added to water, the concentration of H^+ increases, and conversely when a base is added the concentration of H^+ decreases. The hydrogen ion concentration affects the rates of chemical reactions that occur in aqueous medium. In pure water at 25°C, the concentration of hydrogen ions [H^+] is equal to 10^{-7} mol/L. Since there are 6.0235×10^{16} (*Avogadro* number) ions in one mole of H^+, this concentration amounts to the same number of hydrogen ions per liter. The concentration of hydrogen ions in a solution is expressed as

$$pH = -\log[H^+],$$

where log denotes 10-based logarithm and $[H^+]$ is the concentration of $[H^+]$ in moles per liter (or M). The pH of water at 25°C is 7. The pH of cell fluid (cytoplasm) is normally about 7.2. Values of pH below 7 indicate acidic solutions and values above 7 indicate basic solutions. Gastric fluids, vinegar, soda, lemon juice, and contracting muscle cell fluid are acidic whereas sweat, human blood ($[H^+] = 4 \times 10^{-8}$ M) and household ammonia are basic.

Small shifts in cellular pH play an important role in the regulation of a number of cellular phenomena including cell division and cell growth. The relative number of acidic and basic amino acids (building blocks of proteins) in different proteins varies significantly. Although many proteins are denatured by acidic conditions, some proteins, such as gastric enzymes and lysosomal enzymes, are in their most active state at low pH. In contrast, enzymes in the small intestine are most active in somewhat basic conditions. Some cellular membranes have the ability to pump protons from one side to the other and create a pH gradient. As we discuss later, such a proton gradient serves an important function in mitochondria where it is an important component in the generation and storage of energy.

1.8 | Chemical Reactions

A chemical reaction occurs when atoms combine or change binding partners. In a chemical reaction, one form of matter is changed into another. Consider the combustion reaction that takes place when propane (C_3H_8) reacts with O_2 in a stove, the resulting equation can be written as

$$C_3H_8 + 5\,O_2 \rightarrow 3\,CO_2 + 4\,H_2O.$$

In this equation, propane and oxygen are the reactants, and carbon dioxide and water are the products. The arrow shows the direction of the chemical reaction. Numbers preceding the molecular formulas indicate the proportions of the number of molecules reacting or produced. Subscripts indicate the number of a certain atom type in a molecule. Since matter is neither created nor destroyed in a chemical reaction, the number of atoms of a given type on the left-hand side of the equation equals that on the right-hand side.

1.8.1 Energetics of Chemical Reactions

The forces of thermodynamics were instrumental in the evolutionary development of living cells. In thermodynamics, a "system" is defined as the part of the universe that is of interest. Examples of a system include an isolated tumor, a cell, or even a single protein molecule. The rest of the universe is then called the "surroundings." A system is said to be open if it can exchange matter and energy with its surrounding. Living cells are open systems as they take up nutrients, release waste products, and generate work and heat. On the other hand, an organism that was frozen for research purposes can be idealized as a closed system.

The first law of thermodynamics is also referred to as the conservation of energy law and states that energy can be neither destroyed nor created. There are many forms of energy such as mechanical energy, electrical energy, and chemical energy. In a chemical reaction, energy is merely transformed from one form to another. Let ΔU, q, and w denote, respectively, the energy gained by an open system, the heat absorbed by the surroundings, and the work done by the system on the surroundings. Then the first law can be written as follows:

$$\Delta U = q - w. \tag{1.1}$$

Heat is a consequence of random molecular motion, whereas work is associated with organized motion (force) and can be mechanical, electrical, or chemical in nature. The energy of a system depends only on its current properties (state) and not on how it reached that state, whereas both heat and work are path dependent. The processes in which the system releases heat ($q < 0$) are called *exothermic* processes, and those in which the system gains heat are known as *endothermic* processes.

The *second law of thermodynamics* states that in all processes, some of the energy involved irreversibly loses its ability to do work. The disorder of the universe (its entropy) increases with each ongoing process. For a system 1 and its surroundings, the second law of thermodynamics can be written in the form

$$\Delta S_1 + \Delta S_2 = \Delta S_u > 0. \tag{1.2}$$

In this equation, S refers to entropy, which is the measure of disorder in a system. The terms S_1, S_2, and S_u represent the entropies of the system, surroundings, and the universe, respectively. The inequality (1.2) indicates that the change in entropy of the system (ΔS_1) could be positive or negative, but the overall change of the entropy of the universe must always be negative. For simple systems such as a gas in a chamber, entropy can be estimated by counting the number of different ways gas particles can exist. However, estimation of entropy using combinational probability is much more difficult in more complex media such as the cytoplasm of a cell.

In closed systems, the entropy of the system must increase with time. Therefore, any spontaneous process must lead to an increase in the entropy of the system. Such a spontaneous process occurs when oil drops in water merge to form a separate phase from that of water. Nonpolar molecules composing oil drops cannot form hydrogen bonds with water and therefore distort the usual water structure, forcing the water to make a cage of hydrogen bonds around it. Formation of such cages restricts the motion of adjacent water molecules and thereby increases the structural organization of water. When nonpolar molecules cluster together, the structural organization of the water is reduced and the overall entropy of the system increases.

The first and the second laws of thermodynamics can be combined in the following form when chemical reactions are considered:

$$\Delta H = \Delta G + T\,\Delta S. \tag{1.3}$$

In this equation, ΔH is the overall change in bond energy due to reaction (kcal/mol), ΔS is the change in entropy (a measure of the energy lost to disorder in the system), and ΔG is defined as the change in free energy, first introduced into the thermodynamics literature by J. Willard Gibbs in 1878. Of the energy that is released by a chemical reaction, only a certain percentage can be used to do work, no matter how efficient the process. The energy that can be used to do work is called the free energy. For spontaneous chemical processes that occur at constant temperature, the following relation holds:

$$\Delta S_1 \geq q/T. \tag{1.4}$$

In this relation, q refers to heat imported into the system and T is the absolute temperature. Equality holds only at equilibrium, where changes could occur infinitesimally. For systems that can only do pressure–volume work at constant pressure and temperature, the criterion of spontaneity can then be written as

$$\Delta G \leq 0. \tag{1.5}$$

Again, ΔG denotes the change in free energy. To gain insights into the preferred direction of chemical reactions, consider a chemical reaction in which A and B react to produce C and D:

$$A + B \leftrightarrow C + D. \tag{1.6}$$

In this equation, the double arrow indicates that the reaction proceeds in both directions. Whether the reaction will proceed overall in the forward or reverse direction depends on the change in free energy due to the reaction.

The parameter ΔG is an indicator of spontaneity for constant-temperature and pressure processes (Fig. 1.8):

1. If $\Delta G < 0$, then the free energy of the products is less than the free energy of the reactants, and the forward reaction will tend to occur spontaneously. This type of reaction is said to be *exergonic* or energetically favorable.

2. If $\Delta G > 0$, then the reverse reaction tends to occur. Such a reaction is said to be *endergonic*.

3. If $\Delta G = 0$, both forward and reverse reactions occur at equal rates; the reaction is at steady state. If the reaction occurs in an isolated chamber, then the steady state corresponds to *chemical equilibrium*. However, living systems are not in chemical equilibrium. Cells are not isolated chambers, and there is continuous exchange of mass and energy across the cell membrane between the cell interior and the external environment.

There are two parts to the free-energy change that occurs in a chemical reaction. One part is called the *change in standard free energy*, and it is represented by the symbol $\Delta G°$. The other part depends on the concentrations of reactants and products. For a chemical reaction of the form

$$A + B \rightarrow C + D, \tag{1.7}$$

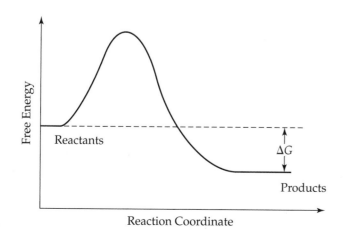

FIGURE 1.8 Free-energy change during an energy-releasing reaction expressed as a function of the progress of reaction. The conversion of substrate to product will only occur if it results in the loss of potential energy.

the change in free energy is given by the equation

$$\Delta G = \Delta G^\circ + R T \ln \{([C][D])/([A][B])\}, \tag{1.8}$$

where $\ln$ denotes natural logarithm, $R = 1.987$ cal/mol is the gas constant and $T = 273.15^\circ + C^\circ$ is the temperature measured in degrees Kelvin. The terms in brackets represent the volumetric concentrations of the reactants and the products in M. The standard free energy ΔG° is the free-energy change of the reaction when all reactants and products are in a standard state. For biochemical reactions, the standard state means restricting initial concentrations of the chemicals to unit values (1 mol/L), the temperature to 25°C, and the pH to 7.0. Standard free energy is an intrinsic property of the chemical reaction itself and can be determined experimentally by measuring the equilibrium concentrations of the products and reactants under standard conditions. Standard free-energy changes have been measured and tabulated for many biochemical reactions. For the oxidation reaction of glucose, the standard free-energy change $\Delta G^\circ = -686$ kcal/mol. Thus, when 1 mol of glucose reacts with oxygen under standard conditions, 686 kcal of energy is released.

The second part of the free-energy change is dependent on temperature and the concentrations of reactants and products. The free-energy change (ΔG) increases with increasing temperature and increasing concentrations of reactants. The change in free energy decreases with increasing concentration of the products. The net reaction proceeds in the forward direction when the sum of the two components of free-energy change is less than zero.

Many chemical reactions in biological systems are *endergonic* reactions where the change in free energy $\Delta G > 0$ and therefore the forward reaction is not favored. How do such reactions proceed in the forward direction? The answer lies in the additive property of free energy. A chemical reaction with a positive ΔG may be coupled to a reaction that has a negative ΔG of a larger magnitude, so that the sum of the two reactions has a negative ΔG. Consider, for example, the following two-step reaction process:

(1) $A \rightarrow B + D$ $\Delta G_1^\circ = 12$ kcal/mol;

(2) $D \rightarrow X + Y$ $\Delta G_2^\circ = -16$ kcal/mol.

Reaction (1) would not occur spontaneously under standard conditions because $\Delta G_1^\circ > 0$. When reaction (1) is coupled to reaction (2), however, the net change in standard free energy becomes negative ($\Delta G_1^\circ + \Delta G_2^\circ = -4$ kcal/mol). As Reaction (2) converts D into products, the concentration of D is reduced to values that make the free energy change negative [see Eq. (1.1).] Reaction (1) will operate in the forward direction in order to replenish the equilibrium concentration of D. As a rule, as long as the overall pathway is *exergonic* ($\Delta G_3 = \Delta G_1 + \Delta G_1 < 0$), the sequence of reactions will operate in the forward direction.

The standard free-energy change $\Delta G^{\circ\prime}$ is related to the equilibrium constant K_{eq} of a reaction by the equation

$$\Delta G^{\circ\prime} = -RT \ln K_{eq},$$

in which $\ln$ denotes natural logarithm. A chemical reaction having a positive $\Delta G^{\circ\prime}$ can proceed spontaneously if it is coupled with a reaction having a negative $\Delta G^{\circ\prime}$ larger value. This exergonic reaction is most often the hydrolysis of ATP. Endergonic organic reactions are coupled to the energy-releasing properties of ATP, a molecule that is considered the universal energy currency of living systems (Fig. 1.9).

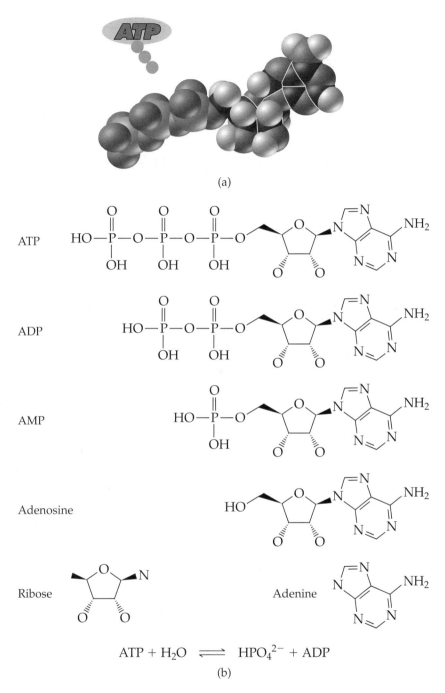

(a)

(b)

FIGURE 1.9 Structures of ATP, ADP, and AMP (a). The hydrolysis of ATP releases free energy (b). An ATP molecule consists of the base adenine bonded by three phosphate groups. In the chemical conditions in a living cell, the phosphate groups are ionized and bear negative charges that repel one another. When ATP hydrolyzes and the phosphate group breaks away, the charges are spread over two molecules that are free to move apart from each other. Hydrolysis of ADP to AMP also releases energy, but this reaction does not occur often in cells. Hydrolysis of the last phosphate group does not spread the negative charges apart, so the change in free energy is small.

1.8.2 ATP: The Standard Energy Carrier

Most food products would readily burn in the presence of oxygen to produce carbon dioxide, water, heat, and light. The metabolic machinery of the cell traps a significant portion of the chemical energy of foodstuffs into a more readily usable form stored in ATP. ATP consists of the nucleotide, adenine, the sugar ribose, and a triphosphate unit and participates in most cellular reactions and processes that require energy. The energy-extracting processes of cells are extremely complex, involving large numbers of enzymes and ligands. Nevertheless, this multistep process can be divided into distinct pathways than can be independently analyzed by explicitly listing inputs and outputs.

ATP is a very stable molecule: One will find bottles of ATP stored at room temperature in any biochemistry laboratory. However, when catalyzed by the appropriate enzyme (an ATPase), ATP can undergo the so-called hydrolysis reaction $ATP \rightarrow ADP + Pi + H_2O$, where ADP and Pi represent respectively, adenosine diphosphate and PO_4^{3-}. ATP hydolysis has very high activation energy and releases 7.2 kcal/mol of energy that can be used to drive energy requiring cellular processes. Endergonic reactions coupled to ATP hydrolysis include the synthesis of large macromolecules, contraction of muscle cells, and the transport of molecules in and out of a cell against a concentration gradient.

The hydrogen atoms shown in the figure as bound to phosphate groups represent protons in the cytosol that associate with the electronegative oxygens of the phosphate groups. The active form of ATP is a complex of ATP with Mg^{2+} or Mn^{2+}. In this complex, the positively charged divalent cation interacts with oxygen by replacing two of the hydrogen atoms in the outer phosphate groups. The chemical reaction

$$ATP + H_2O \rightarrow ADP + Pi + H^+$$

is called ATP hydrolysis.

The change in free energy as ATP is converted to ADP represents that portion of the total energy, which is available for doing work. The free energy G is a quantity (always a positive scalar) similar to gravitational potential energy and the elastic energy of a spring. In ATP hydrolysis, the term change in free energy (ΔG) refers to the free energy of ADP + Pi minus the free energy of ATP. The hydrolysis of ATP results in a change of free energy (ΔG) of -12 kcal/mol under physiological conditions. The standard free energy $\Delta G^{\circ\prime}$ is -7.3 kcal/mol for hydrolysis of ATP to ADP.

The four hydrogen ions associated with the triphosphate of ATP dissociate easily from the molecule, leaving the negative charges on the oxygen ions unmasked. Electrostatic repulsive forces arising from these negative charges stretch the molecule and increase its potential energy. These repulsive forces are reduced when the outer inorganic phosphate group detaches from the rest of the molecule. This is one reason why conversion of ATP to ADP results in a reduction of free energy. One other factor contributing to the energy-carrying capacity of ATP is that ATP–ADP turnover increases entropy. ADP and the orthophosphate have many more possible positional configurations when compared with the structural configurations that are possible for ATP.

Let us next illustrate how ATP hydrolysis could drive a reaction that requires energy input. Suppose that the free energy of the conversion of A into B is +4 kcal/mol, as in

$$A \rightarrow B \qquad \Delta G^{\circ\prime} = 5 \text{ kcal/mol.}$$

Because standard free energy is positive, conversion from A to B does not go forward under the standard conditions. Let us now couple this reaction with ATP hydrolysis:

$$C + ATP \rightarrow D + ADP + Pi + H^+$$
$$\Delta G^{\circ\prime} = (+5 - 7.3) \text{ kcal/mol} = -2.3 \text{ kcal/mol}$$

The standard free-energy change for the coupled reactions is equal to the standard free-energy change due to ATP hydrolysis plus the change due to the conversion of A into B. The resulting standard free-energy change is negative and the conversion of A to B goes forward.

In biological systems, the concentration of ATP is maintained at a much higher level than that of ADP. The ATP-generating system of cells maintains the ratio $\{[ATP]/[ADP][Pi]\}$ of the order of 500. In a typical cell, an ATP molecule is consumed within a minute following its formation. Even though ATP is inherently stable, the turnover of the ATP–ADP cycle is very high: A resting human converts about 40 kg of ATP to ADP in 24 hours. This massive hydrolysis of ATP is catalyzed by enzymes known as ATPases, which are involved in almost all energy-requiring reactions. Some reactions in the living systems are driven by nucleotides other than ATP. Guanosine triphosphate (GTP) drives synthesis of peptides and also plays an important role in supplying the energy needed for signal transduction. The structure of GTP is very similar to that of ATP, and GTP hydrolysis is catalyzed by enzymes known as GTPases.

Enzymes catalyze the transfer of the terminal phosphoryl group from one nucleotide to another:

$$ATP + GDP \rightarrow ADP + GDP.$$

The standard free-energy change from GTP to GDP is comparable to that of ATP to ADP. Other nucleotides such as uridine triphosphate (UTP) and cytidine triphosphate (CTP) also drive energy-requiring biosynthetic processes. The reader could refer to *Kyoto Encyclopedia of Genes and Genomes* (KEGG) database by the Institute of Chemical Research, Kyoto University and utilize the search engine at http://www.genome.ad.jp/dbget/ligand.html to learn about the pathways that involve the hydrolysis of UTP and CTP. It is interesting to note, however, that most of the central molecules involved in energy transfer in living systems are nucleotides that contain the five-carbon sugar ribose rather than deoxyribose. A likely explanation for this phenomenon is that RNA evolved before proteins and DNA. When proteins replaced RNA as the major catalysts to achieve greater versatility, these enzymes utilized ribonucleotides because they were already well adapted to their metabolic roles. The role of energy-carrying ribonucleotides in metabolism is further illustrated in Chapter 3.

Besides ATP and GTP, a large number of biological molecules contain high-energy bonds. Why is it then that ATP and similar molecules such as GTP are used as the standard energy source in biological reactions? The answer to this question is that the free energy of hydrolysis of ATP (also GTP) is sufficient to drive most of the coupled reactions in living systems. Higher release of free energy would be wasted as heat. The situation is comparable to using currency to buy goods. If commerce were restricted in a way that no change is returned after a purchase, one would be tempted to use smaller bills rather than a hundred-dollar bill in purchasing an item that costs four dollars. Otherwise, much of the effort spent in earning a hundred-dollar bill would have been wasted.

1.9 | Enzymes

Burning of propane gas ($C_3H_8O_2$), although energetically favorable, does not occur spontaneously. One has to use a flame to ignite propane and initiate burning. Most chemical reactions require an initial input of energy to get started. The added energy increases the kinetic energy of the molecules and thereby increases the frequency of collision. Collision is needed to overcome the repulsive forces between the electrons surrounding molecules and to break existing chemical bonds with the molecules. *Activation energy* is defined as the energy barrier that needs to be overcome for a reaction to proceed (Fig. 1.10). Burning of propane involves breaking of covalent bonds between carbon and hydrogen and between two oxygen atoms. As presented in Table 1.3, the dissociation of bonds requires infusion of energy. The activation energy for this reaction is supplied by the flame used to initiate burning. The energy released in the forward reaction in the form of heat is sufficient to overcome the activation energy barrier at later times. The importance of activation energy on the progression of the reaction is illustrated graphically in Fig. 1.10. Consider two molecules A and B reacting to form P and Q: $A + B \rightarrow P + Q$. For this reaction to occur, A and B must approach closely enough so that an unstable (high-energy) complex (say, X*) forms. The intermediate product X* is called the transition state. This product is so high in free energy that it is unstable and eventually breaks down to give P and Q. When the free energy of the reactants and products are shown as a function of the reaction coordinate (measuring the extent in which the reaction proceeded), the transition state lies at the crest of the energy profile curve in Fig. 1.10. The rate of reaction is considerably enhanced when heat is supplied to overcome the activation barrier.

Chemical substances called catalysts provide an alternative way of overcoming activation energy. A catalyst is a substance, present in small amounts relative to the reactants, that accelerates a reaction without being consumed in the process. Catalysts form complexes with reactants and lower the activation energy barrier for the reaction being catalyzed (Fig. 1.10). Because the forward and reverse reactions are catalyzed to the same extent, the reaction equilibrium is not disturbed. Catalysts have no effect on the ΔG of the reaction.

Biological catalysts are called enzymes. In a living cell, thousands of reactions can occur at constant temperature at any instant of time. Any extraction or infusion of heat would affect these reactions indiscriminately. A flux of heat also breaks the

FIGURE 1.10 Enzymes overcome the energy of activation and facilitate a chemical reaction to go forward. Not all reactant molecules have the same levels of kinetic energy. Some molecules surmount the energy barrier and react, thereby forming products (a). An enzyme reduces the activation energy barrier of a chemical reaction (b).

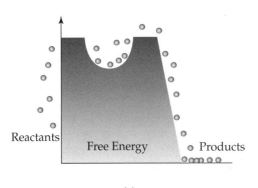

(a)

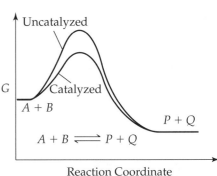

(b)

hydrogen bonds between various parts of proteins—molecular machines of life—and has other destructive effects.

Even if a biological reaction is *exergonic* (negative free-energy change), it does not proceed in the absence of enzymes. For example, digestion of proteins is an energetically favorable process in which the peptide bonds between the monomers (amino acids) of a protein molecule are broken. Nevertheless, in a sterile aqueous solution at room temperature, hardly any degradation would occur over days or even months. This is because the activation energy for breaking the peptide bonds is high enough to prevent all but a few protein molecules from engaging in the forward reaction of degradation. In biological systems, in reactions where covalent bonds are made and broken, differences in energy between reactants and products range from 1 to 20 kcal/mol. On the other hand, the activation energy associated with a bioreaction can be on the order of 100 kcal/mol. As a result, the covalent bonds of structural molecules remain intact over very long periods of time.

Let us now illustrate the effect of an enzyme on a simple chemical reaction in which a covalent group G is transferred from donor D to acceptor A:

$$D\text{–}G + A \rightarrow D + A\text{–}G.$$

This overall reaction proceeds via two half reactions: (1) the formation of the transition state and (2) the decay of the transition state to form products. These reactions proceed as follows:

$$D\text{–}G + A \rightarrow D\text{–}G\text{–}A;$$
$$D\text{–}G\text{–}A \rightarrow D + A\text{–}G.$$

In these equations, $D\text{–}G\text{–}A$ represents the transition state where D, G, and A are bound to each other for a brief period. The free energy of the transition state is typically higher than the free energy of the reactants, creating an *energy barrier* against the forward reaction. An enzyme alters the transition state through which a reaction must proceed by temporarily binding to one or more of the reactants (substrates) through multiple weak interactions. Remarkably, many enzymes have evolved to bind (and stabilize) the transition-state conformation of the reactant rather than the starting substrate. Consequently, the transition states that occur in the presence of an enzyme have considerably lower free energy, thus allowing the reaction to proceed. For example, the reaction

$$D\text{–}G + A \rightarrow D + A\text{–}G$$

is transformed into a set of intermediate reactions in the presence of enzyme E:

$$D\text{–}G + A + E \rightarrow D\text{–}G\text{–}E + A \rightarrow D + G\text{–}E + A$$
$$\rightarrow D + G\text{–}E\text{–}A \rightarrow D + A\text{–}G + E.$$

In the first reaction, enzyme E combines with $D\text{–}G$. In the second reaction, D detaches from $E\text{–}G$. Another transition product, $E\text{–}G\text{–}A$, is subsequently produced, and in the last step, enzyme detaches from the acceptor molecule A. Thus, the concentration of enzyme E is not changed by the resultant chemical reaction. As illustrated in this example, much of the catalytic power of enzymes comes from their bringing substrates together in favorable orientations in *enzyme–substrate* complexes. In doing so, enzymes may utilize a lock-and-key type mechanism where the substrate binds to a site of exact complementary fit (Fig. 1.11a). Alternatively, an enzyme

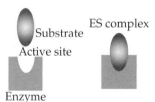

(a) Lock-and-key

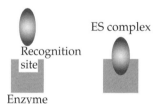

(b) Induced fit

FIGURE 1.11   Enzyme–substrate interactions. In the lock-and-key mechanism, binding of substrate to the complementary active site positions it for catalysis (a). In enzymes with induced fit action, binding of substrate induces a conformational change in the enzyme that brings the part of the enzyme site involved in catalysis to the proximity of the substrate (b).

may induce strain or distortion of the substrate to promote catalysis. In such cases, enzyme–substrate interaction may occur according to induced fit action where the binding of substrate induces a conformational change in the enzyme that brings part of the enzyme site involved in catalysis to the proximity of the substrate (Fig. 1.11b). Studies of protein structure by X-ray crystallography have indicated that the active sites of most enzymes are largely preformed; however, in most cases, conformational changes are induced by substrate binding.

Essentially all biochemical reactions are enzyme catalyzed. Enzymes are reaction-specific catalysts. Each enzyme binds only a small number of reactants, and frequently just one reactant. Many diseases are due to genetically determined abnormalities in the synthesis of a few enzymes. Quite often, the rate of an enzyme-catalyzed reaction is 10^6 to 10^{12} times that of an uncatalyzed reaction under otherwise similar conditions.

1.9.1 Factors Affecting Enzyme Activity

Concentrations of substrates and enzymes in an enzyme-mediated reaction, the temperature, pH, and the presence of enzyme inhibitors are among the major factors affecting enzymatic activity. Rates of biological reactions double with a 10°C increase in temperature, hence the increase in heart rate with increasing room temperature. For most enzymes, maximum activity is observed between pH values of 5 and 9, and many enzymes denature at extremely low and high pH values. Moderate changes in pH may also have significant effects on the ionic state of the enzyme and the substrate. Changes in pH could alter the charge distribution on the binding site, resulting in a change in the efficiency of catalytic activity.

As mentioned earlier, enzyme concentrations are typically much lower than the concentrations of reactants. In enzyme-mediated reactions, the rate of reaction increases in proportion to enzyme concentration. The rate of reaction shows hyperbolic behavior, on the other hand, with respect to substrate concentration. This is illustrated in Fig. 1.12 for the reaction $S \rightarrow P$. The figure shows that the rate of reaction is proportional to substrate concentration at low substrate concentrations, but becomes concentration independent (zero order) at high concentrations. The hyperbolic dependence of rate of reaction on substrate concentration reflects the fact that when all the enzyme molecules available are already combined with substrate molecules, increasing substrate concentration does not increase the rate of reaction. The equation for the rate of product formation $V = d[P]/dt$ can be written as

$$V = k_{cat}[E_t][S]/\{[S] + K_m\}, \tag{1.9}$$

where V is typically identified as the velocity of reaction and $[E_t]$ and $[S]$ are the concentrations of the enzyme (bound or unbound) and the substrate, respectively. The parameters k_{cat} and K_m refer to time constants involving the enzyme-mediated reaction. In particular, the dimensionless rate constant K_m is equal to the half maximal velocity of the reaction. This equation is called the Michealis–Menten equation. Linear double reciprocal plots using this equation present a straight line where slope and intercept can be deduced by computing the rate of reactions at alternate S values for constant $[E]$. It is important to note that even in the face of marked changes in the rates at which reactions occur, enzyme-mediated catalysis does not change reaction equilibria, both forward and reverse reactions are catalyzed

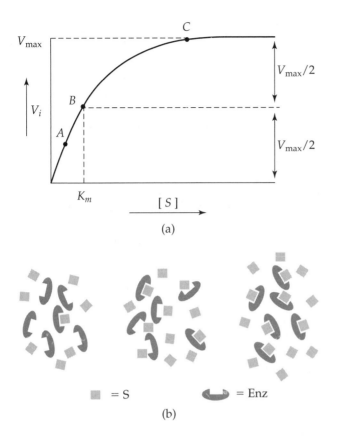

FIGURE 1.12   Effect of substrate concentration on the velocity of an enzyme-catalyzed reaction (a). The velocity of reaction $S \rightarrow P$ is defined as the time rate of formation of product P. The points A, B, and C in graph (a) represent low, medium, and high concentrations of the enzyme relative to the substrate concentration (b).

equally well. Enzyme concentration or activity is often the point of regulation of chemical reaction pathways in biological systems. Examples of control of cellular activity with modification of enzyme concentration will be discussed in Chapter 4.

1.10 | More on Bioelements

1. *Discuss the possible reasons why elements with atomic numbers 2 to 5 do not constitute important components of living systems.* The element with atomic number 2 is helium with an electron configuration of $1s^2$. This electron configuration is stable, making the helium atom inert. Helium is not capable of bonding with other atoms and therefore would not be a good building block in a living system, which requires the building of complex molecules. The element with atomic number 3 is lithium with an electron configuration $1s^2\,2s^1$. Lithium has one electron in its outer shell and is more likely to lose the lone electron rather than attract electrons to achieve a full outer shell. Because lithium has lower electronegativity (0.98) than hydrogen (2.1), it would give the electron rather than share it with other atoms. Hydrogen, on the other hand, can form covalent bonds with organic elements such as carbon, nitrogen, and oxygen and in solution. Compounds of lithium such as lithium carbonate and lithium sulfate are used as antidepressants. Because lithium has similar electron configuration

as sodium and it is a smaller atom, it can use the sodium/potassium channels to enter living cells from blood flow. Precisely how lithium affects brain cells to alleviate mental disorders is not known. However, the effect of lithium can be attributed to its action on receptor-mediated neuronal pathways.

The element with atomic number 4 is beryllium with an electronic configuration of $1s^2\,2s^2$. Because beryllium's s subshells are filled, this element is relatively inert compared to other alkali elements. Beryllium is not a building block of complex molecules; it does not react with water even at very high temperatures. The element with atomic number 5 is boron, with the electron configuration $1s^2\,2s^2\,2s^1$. Boron has three valence electrons. However, its electronegativity (2.04) is less than that of hydrogen (2.1) and of nitrogen (3.0). The atomic radius of boron (1.17 A) is larger than carbon (0.91 A) and nitrogen (0.75 A). A large atomic radius indicates that the outer subshell electrons are relatively far from the radius. Thus, in a living system, molecules containing boron would be more loosely held than those molecules containing carbon or nitrogen. The weak bonding capacity would not favor the longevity of complex molecules in a living system. Nevertheless, boron is a trace mineral that is essential for plants. In minute amounts, boron may also be essential for humans and animals in energy utilization and the development and maintenance of bone. Animal studies show that boron improves the production of antibodies that help fight infection and markedly decreases peak secretion of insulin from pancreas.

2. *Fluorine is the most electronegative element and yet it is not part of life. Why?* The extremely high electronegativity of fluorine makes it a very reactive element. It partakes in ionic bonding, but not covalent bonds. In a living system, fluorine would strip electrons from atoms, thus destroying bonds that keep complex molecules intact.

3. *Discuss the differences between covalent bonds and metallic bonds.* In metallic bonds, electrons in the outer shell form a cloud around the lattice of nuclei. A covalent bond, on the other hand, is the sharing of one or more electrons between two atoms. In metallic bonding, each atom is bound to several neighboring atoms. The bound electrons are relatively free to move throughout the three-dimensional structure of the lattice of atoms, a property that makes metals good conductors of electricity.

4. *Why do heavy elements form covalent bonds much less frequently than lighter elements?* The electrostatic force by which electrons are attracted to the nucleus is inversely proportional to the square of the distance between the nucleus and the electron. Typically, heavy elements have a larger atomic radius than lighter elements. Therefore, the outer shells of heavy elements are held less tightly than those shells of lighter elements. The more easily the electrons in the outer shell part from the rest of the atom, the less likely the atom is to have covalent bonding.

5. *Describe the various isotopes of oxygen and carbon and their significance for living systems.* Isotopes of an element differ in the number of neutrons in nucleus. There are four naturally occurring isotopes of oxygen. These isotopes are O^{15} with a half-life equal to 122.2 s and the three stable isomers O^{16}, O^{17}, and O^{18}. The most abundant isotope is O^{16}. The naturally occurring oxygen in air and water is a combination of the three stable isotopes.

The isotopes of carbon include C^{11} with half-life of 20.3 min, C^{12} (stable), C^{13} (stable), C^{14} with half-life equal to 5730 years, and C^{15} with half-life of 2.5 s. The three principal isotopes of carbon that occur naturally are C^{12} (98.89%), C^{13} (1.11%), and C^{14}. Plants and animals utilize carbon within the biological food chain and as a consequence take up C^{14} as well as the other two isotopes. Since there is no incorporation of radioactive C^{14} after an organism dies, measuring the concentration of C^{14} in a fossil helps determine its age.

6. *Provide examples of nutrients rich in (a) nitrogen, (b) phosphorus, and (c) sulfur.* Nitrogen is found in all proteins, most vitamins, and hormones. Nitrogen accounts for approximately 16 percent of all proteins by weight. It is also found in fertilizer products in the form of ammonia, nitric acid, and urea. Phosphorus is found in some proteins in milk, eggs, fish, poultry, legumes, nuts, and whole grains. Similarly, meat, milk, eggs, and legumes are rich in sulfur-containing amino acids.

7. *Estimate the number of different types of enzymes in a typical bacteria.* One of the most heavily studied bacterial genomes is that of *Escheria coli* (*E. coli*). The *E. coli* genome has 4290 genes. Another bacterium, Haemophilus influenza, has 1743 genes. Not all of these genes encode enzymes. *E. coli* has approximately 2000 enzymes. The list of these enzymes can be found in the EcoCyc database.

8. *Why is ammonia a good cleaning agent?* Ammonia has the chemical formula NH_3 and is a good cleaning agent because of its ability to attract protons associated with dirt. When aqueous ammonia comes into contact with dirt, it takes a proton and is converted to the ammonium ion NH_4^+. Once the dirt particle loses a proton, it becomes negatively charged and becomes soluble in water. The dirt then detaches from what it was initially attached to and is swept away by the drained water.

1.11   ASSIGNMENTS

1.1   Brief Quiz: (a) What are the elements that make up organic materials? Discuss their chemical properties. (b) What is chemical reactivity? What is the octet rule? (c) Discuss the four types of bonds between atoms and/or ions. (d) Why do fatty acids cluster when immersed in water? (e) What are enzymes, and what role do they play in living systems? (f) Why is sulfur less electronegative than oxygen? (g) What significant role does sulfur play in living organisms?

1.2   Determine the chemical and structural formulas of the organic compounds *glyceraldehyde 3-phosphate*, *succinate*, and *glutamine* using the ligand/compound database called the *Kyoto Encyclopedia of Genes and Genomes (KEGG)* by the Institute of Chemical Research of the Kyoto University. Specifically, go to the Web site http://www.genome.ad.jp/dbget/ligand.html, and press

"Search enzymes and compounds." In the next page, input your search keywords (the name of the compound) into the search box, and then press "Submit." The next page will present you with a list of enzymes that interact with the compound (those lines that begin with *ec:*) and the different forms and variations of the compound itself (*cpd:*). When this search was conducted in October 2000, there were 36 entries for glutamine, 21 of which were enzymes. The compound name glutamine appeared 23rd on the list. When this line was entered, the next page provided the structural as well as the chemical formula of the glutamine ($C_5H_{10}N_2O_3$).

1.3   The energy required to break a covalent bond must be equal to the work done against forces of interaction between the two atoms. The atomic force microscope is being increasingly used in the quantification of repulsive and attractive forces between atoms (Fig. 1.6). Force resolution of the instrument ranges from about 10^{-12} N to about 0.5×10^{-11} N. Conduct a literature survey on atomic microscopes, and describe their elementary design.

Ans: The microscope has an elastic beam whose tip interacts with a target atom on a plane of atoms. If the force between the leading atom of the tip and the target atom is attractive, the beam bends toward the target atom and vice versa. The deflection of the cantilever beam at its tip is proportional to the force of interaction between the interacting atoms. The constant of proportionality is a linear function of the cantilever beam's force constant ($k = 0.2$ N/m). The force between the tip of the cantilever and the target atom in the sample is dependent on the distance between the atoms. The equilibrium distance corresponds to zero force. By convention, negative values correspond to attractive overall forces, and positive values correspond to repulsive forces acting on the tip.

1.4   Consider a reaction $S \leftrightarrow P$. The standard free energy of the reaction was found to be as follows: $\Delta G^{\circ}_1 = -1840$ cal/mol. At the beginning of the reaction $[S] = 0.001$ M and $[P] = 0.1$ M. Which direction will the reaction proceed (from S to P or vice versa)? What is the free-energy change for S to P? *Ans:* $\Delta G = 887$ cal/mol, and the reaction proceeds from P to S.

1.5   Consider the reaction $A + B \leftrightarrow C$. The free-energy change can be written as

$$\Delta G = \Delta G^{\circ} + R T \ln\{[C]/([A][B])\},$$

where $R = 1.98 \times 10^{-3}$ kcal/(mol deg) and $T = 298$ K at 25°C. Determine the free-energy change (ΔG) for $[C] = 10^{-3}$ mol, $[B] = 0.7 \times 10^{-4}$ mol, and $[A] = 3 \times 10^{-3}$ mol when $\Delta G^{\circ} = -4.09$ kcal/mol.

1.6   Why do chemical reactions require enzymes in living systems? How do enzymes function? Give an example of an enzyme-coupled chemical reaction in a living organism.

1.7   Can energy be destroyed or created? What are the two components of free-energy change in a chemical reaction?

1.8   Why do phosphate bonds contain high free energy? Why is light the only electromagnetic wave form that can be used by living systems to produce chemical energy? Why are plant leaves green?

1.9   A list of small organic molecules and their most important chemical properties can be found in the database presented by the www.rrz.uni-hamburg.de/biologie/bonline/e16/16b.htm. Using this Web site, determine the structural formulas of acetone, acetic acid, and formic acid.

1.10   Photon-induced transitions of visible light involve individual atomic electrons. Magnetic radiation with a shorter wavelength can affect chemical bonds by giving the bonding electron too much energy to remain in the molecular orbital. Using the Web site http://www.physics.uq.edu.au/ph128/, determine the range of wavelengths that influence the covalent bonds of organic compounds.

1.11   Derive the Michealis–Menten equation starting with the reaction $E + S \rightarrow ES \rightarrow EP \rightarrow E + P$. Assume that the release of P is very rapid and therefore the reaction can be modified in the following form: $E + S \rightarrow ES \rightarrow E + P$. The time rate of change of P can then be expressed as follows:

$$d[P]/dt = k_{cat}[ES]; \qquad (1.1)$$
$$d[ES]/dt = k_1[E][S] - k_{-1}[ES] - k_{cat}[ES]. \qquad (1.2)$$

In these equations, the terms inside brackets denote molar concentration, the rates k_1 and $-k_{-1}$ represent the forward and reverse rates for the formation of the enzyme–substrate complex (ES), and k_{cat} denotes the rate constant for the conversion of ES to the enzyme-product complex (EP). After a short induction time, ES is usually present at a constant concentration (until substrate runs out); therefore, the rate of formation of ES becomes equal to the rate of disappearance of ES; that is, $d[ES]/dt = 0$. This steady state is governed by the equation

$$k_1[E][S] = (k_{-1} + k_{cat})[ES]. \qquad (1.3)$$

The enzyme concentration at any time t is equal to the total enzyme concentration at the beginning of the reaction $[E_t]$ minus the concentration of enzyme–substrate complex $[ES]$:

$$[E] = [E_t] - [ES]. \qquad (1.4)$$

Note that we assumed that the transformation from $[EP]$ to $[E] + [P]$ is so fast that $[EP]$ is practically zero for all purposes. Show that the dimensionless constant K_m in Michealis–Menten equation is given by the equation $K_m = (k_{-1} + k_{cat})/k_1$.

REFERENCES

Arnett EM, Amarnath K, Harvey NG, and Cheng JP. "Chemical bond-making, bond breaking, and electron transfer in solution." *Science* 1990, 247: 423–430.

Fruton JS. *Proteins, Enzymes, Genes: The Interplay of Chemistry and Biology*. Yale University Press, 2000.

Lazcano A and Miller SL. "The origin and early evolution of life: prebiotic chemistry, the pre-RNA world, and time." *Cell* 1996, 85: 793–798.

Ohnesorge F and Binnig G. "True atomic resolution by atomic force microscopy through repulsive and attractive forces." *Science* 1993, 260: 1451–1456.

Pace NR. "The universal nature of biochemistry." *Proc Natl Acad Sci* (2001), 98: 805–808.

2

Macromolecules of Life

2.1 | Introduction

Water constitutes about 70 percent of the weight of a living cell; the rest is mostly composed of *macromolecules* that each contain thousands of atoms. These very large molecules are made of chains of smaller units called *monomers*. The monomers that make up a biological macromolecule need not be identical. For example, the information macromolecule deoxyribonucleic acid (DNA) is made of four distinct monomers (nucleotides), whereas proteins can have up to twenty different monomers (amino acids). Even in the simplest living systems, hundreds of thousands of such macromolecules interact with each other at any instant of time and undergo or catalyze chemical changes.

The macromolecules of living organisms are classified into three groups: *proteins*, *nucleic acids*, and *carbohydrates*. Proteins make up most of the molecular machinery of all organisms. The word "protein" is derived from the Greek word "proteios," which means "of the first rank." Proteins are linear chains of at most twenty different amino acids (Fig. 2.1). Proteins constitute the building blocks of our tissues, facilitate complex chemical reactions, and act as sensors, transducers, and energy transformers. As enzymes, proteins bring substrates to appropriate configurations for chemical reactions to proceed. Figure 2.2 shows the number of proteins synthesized by various multicellular organisms grouped into major functional categories. All proteins contain the elements carbon, hydrogen, oxygen, nitrogen, and sulfur. When temporarily associated with a phosphate group, many proteins rapidly undergo shape changes and gain or lose enzymatic activity.

Nucleic acids are the information macromolecules of living systems. Like proteins, all nucleic acids contain carbon, hydrogen, oxygen, and nitrogen. Nucleic acids do not contain sulfur, but do contain phosphorus. Nucleic acids can be grouped into two sets: DNA (deoxyribonucleic acid) and RNA (ribonucleic acid). The chemical recipes of proteins are coded on long and fragile DNA molecules. Simple unicellular organisms such as bacteria store all their hereditary information

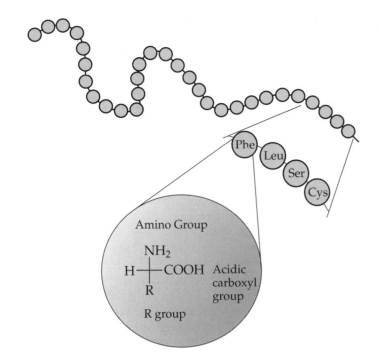

FIGURE 2.1 Polypeptides are long strings of molecular beads called amino acids. A protein is composed of one or more polypeptides. (Modified from http://www.nhgri.nih.gov/DIR/VIP/Glossary/Illustration/aminoacid.html.)

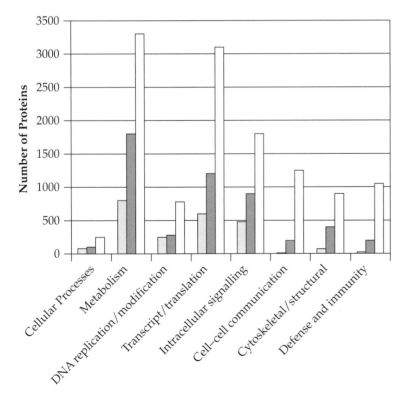

FIGURE 2.2 Number of different types of proteins generated by yeast, fruit fly, and human in major functional categories. This figure is based on that published by Eric S. Lander et al., *Nature* (2001, 409: 860–921) in an article entitled "Initial Sequencing and Analysis of the Human Genome."

in a single circular DNA, whereas animal and plant cells need multiple DNA molecules as their information database. In higher forms of life, from yeast cells to human, DNA is stored in a special cellular compartment called the nucleus. RNA molecules have chemical compositions that complement DNA and are involved in the synthesis of proteins. In a process called *transcription*, protein recipes encoded on DNA are copied onto messenger RNA (mRNA). The mRNA is then used as a template to build proteins from amino acids (Fig. 2.3). Transfer RNA molecules add amino acids to the growing chain of the protein according to the recipe encoded by mRNA. Therefore, in most modern life forms, information flows from DNA to RNA to protein. This complex process of information flow probably evolved gradually from simpler processes. Biochemical evidence points to short strands of RNA as the first information-carrying molecules. RNA molecules have the capacity to replicate without help from protein enzymes. If the first cells used RNA as the hereditary molecule, then RNA must have functioned as a template for the synthesis of DNA. Because DNA is double stranded and therefore stores two copies of its information, it probably evolved rapidly as the primary carrier of hereditary information in living cells. Meanwhile, RNA assumed its current role as a crucial intermediary in the translation of genetic recipes into proteins.

The third group of macromolecules found in biological systems is *carbohydrates*, which contain roughly equal amounts of carbon atoms and water molecules.

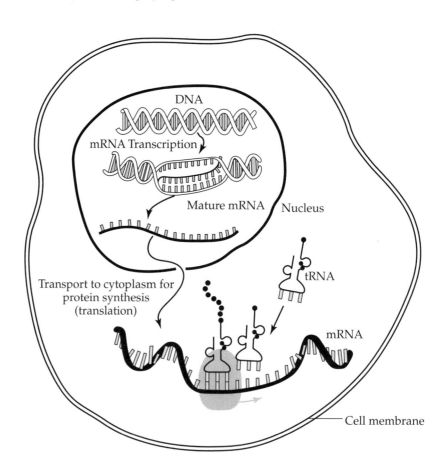

FIGURE 2.3 Information flow in a cell with a nucleus (eukaryotic cell). Hereditary information flows from DNA to RNA to protein. The protein recipes encoded by DNA are transcribed into mRNA and then these transcripts are used as templates in the synthesis of proteins. (Modified from http://www.nhgri.nih.gov/DIR/VIP/Glossary/Illustration/transcription.html.)

Carbohydrates such as glucose are an important source of energy for driving cellular processes. Carbohydrates are often present on proteins where they form covalent bonds with the free amino or hydroxyl groups on the amino acids asparagine and serine, respectively. There is considerable heterogeneity of these N- and O-linked carbohydrates among different proteins, and they can often form branched chains. Some carbohydrates such as sialic acid are negatively charged and when present on proteins exposed on the outside of cells form a charge barrier known as a glycocalyx. In some cases, the addition of a particular carbohydrate residue to a protein will determine its destination in the cell. For example, the presence of mannose-6-phosphate on a protein targets it to a subcellular compartment called the lysosome. Other carbohydrate-containing proteins are the blood group antigens A, B, A/B, and O. Differences in their carbohydrate makeup are responsible for the difference in the ability of a recipient to recognize the blood as "self." Carbohydrates on the surface of proteins are often the source of immune or allergic reactions.

Simple organisms such as bacteria and higher forms of life such as plants also use carbohydrates to build structures. The exoskeleton of all insects is made of a special carbohydrate called chitin (Fig. 2.4). Other polymeric carbohydrates include the plant carbohydrate, cellulose. The matrix surrounding most cellular structures is made up of carbohydrates and protein–carbohydrate complexes and is discussed in Chapter 6. Carbohydrates also play important roles in cell signaling and communication.

Another important group of molecules that form large structures is *lipids*. Lipids are mostly made up of carbon and hydrogen. Lipids are not true macromolecules, because they form large structures through associations other than covalent bonding. Lipid structures form membranes that separate cells from each other, create cellular compartments, and perform other complex tasks. Practically all biological membranes are made of lipid bilayers and associated proteins (Fig. 2.5).

Whereas each organism has its own unique set of deoxyribonucleic acids and proteins, all organisms use the same processes to build carbohydrates and lipids. A 65-kg adult male human is made of approximately 11 kg of protein, 9 kg of fat, 1 kg of carbohydrate, 4 kg of minerals and 40 kg of water. The weight of nucleic acids present in an organism is much less than the corresponding weights of other macromolecules.

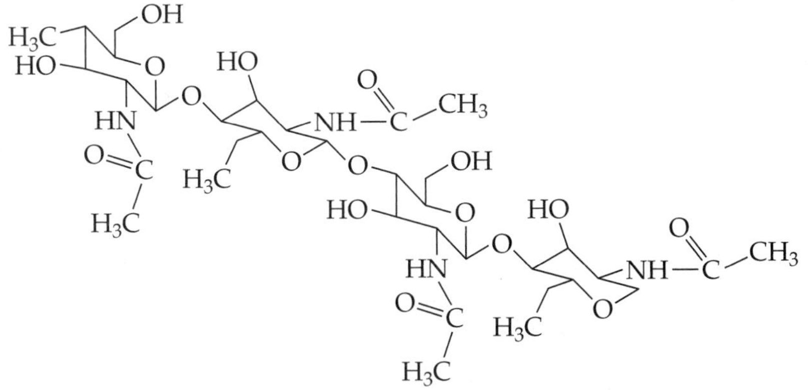

FIGURE 2.4 The exoskeleton of a scorpion is made of the carbohydrate chitin.

FIGURE 2.5 Schematic of a region of a cell membrane. Protein molecules embedded in a lipid bilayer carry out important functions in adhesion and signaling. Lipid bilayers and associated proteins control the entry of molecules into the cell.

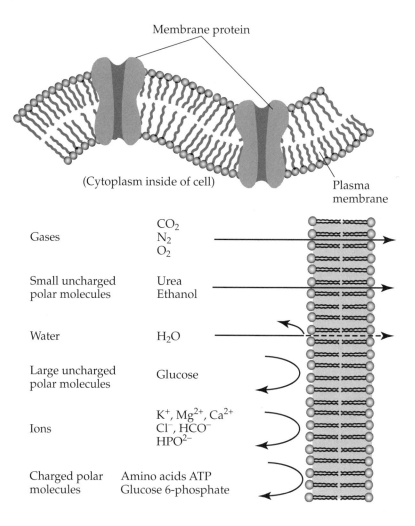

The four groups of macromolecules found in living systems are not mutually exclusive. Often macromolecules of different classes interact with each other through formation of covalent bonds and weaker bonds. Proteins bind to carbohydrates to form glycoproteins; carbohydrate chains bind to lipids, forming glycolipids; enzymes bind to their substrates using hydrogen bonding, electrostatic interactions, and van der Waal forces. These interactions are essential for the survival of living organisms.

2.2 | Fundamentals

2.2.1 Condensation Reactions

Macromolecules are constructed by a series of reactions identified as *condensation reactions* or *dehydration reactions*. Let *BABBA*—H be a polymer with reactive hydrogen at one end, and B—OH another monomer with a hydroxyl group (—OH) at one end. The condensation reaction between them can be written as

$$BABBA—H + B—OH \rightarrow BABBAB + H_2O.$$

The products of the reaction are the polymer *BABBAB* and the water molecule H_2O. A macromolecule is formed by adding one monomer at a time to a growing chain of monomers. The monomers *A* and *B* are the molecular beads that make up the macromolecule, and they are chemically different than the building blocks *A*—H and *B*—OH. Monomers are identified by the name of the building block followed by the word "residue." For example, nucleotides are the building blocks of information molecules, and nucleotide residues are the monomers in these molecules.

The synthesis of macromolecules occurs only if energy is added to the system. Therefore, the synthesis of macromolecules from their building blocks must be coupled to energy-releasing (exergonic) reactions. The reverse reaction of condensation, *hydrolysis*, involves the use of water in the disassembly of an organic molecule. Although hydrolysis releases energy to the environment, due to its high activation energy, it does not occur in the absence of mediating enzymes. In living systems, both hydrolysis and condensation reactions require the catalytic action of enzymes.

2.2.2 Structural Representation

The four types of macromolecules that are essential for living systems exhibit distinctly different three-dimensional structures. As in many other fields of science and engineering, biochemists have struggled to capture the best three-dimensional representations of a structure in the two-dimensional space of a page or on a computer screen. In the following treatment, we use two distinct forms of structural representation for macromolecules. The representation based on the so-called *Haworth* projections is the standard in many biochemistry books; therefore, we will call this representation the *standard* representation. The second representation is used most often by institutions that create digital databases for molecules. One such database is the *Kyoto Encyclopedia of Genes and Genomes* (KEGG). The KEGG database was formed by the Institute for the Chemical Research, Kyoto University of Japan. For brevity, we call this second representation of molecular structure, the KEGG representation. These two ways of visualizing three-dimensional structures of biomolecules are illustrated in Fig. 2.6.

1. *Standard Representation*:   (A) Simple molecules such as three-carbon sugars are presented as planar diagrams, with the understanding that atoms joining a carbon atom by horizontal bonds are in front of the page and those joined by vertical bonds are behind. This representation is called the *Fischer* representation and is illustrated in Fig. 2.6 for D-glyceraldehyde, a carbohydrate containing three carbon atoms. (B) In more complex organic molecules that form ring-like structures, the carbon atoms in the corners of the ring are not always explicitly shown. The approximate plane of the ring is considered to be perpendicular to the plane of the paper, with the thick line on the ring closest to the reader. This representation is called the Haworth representation and is illustrated in Fig. 2.6 for the biomolecule AMP.

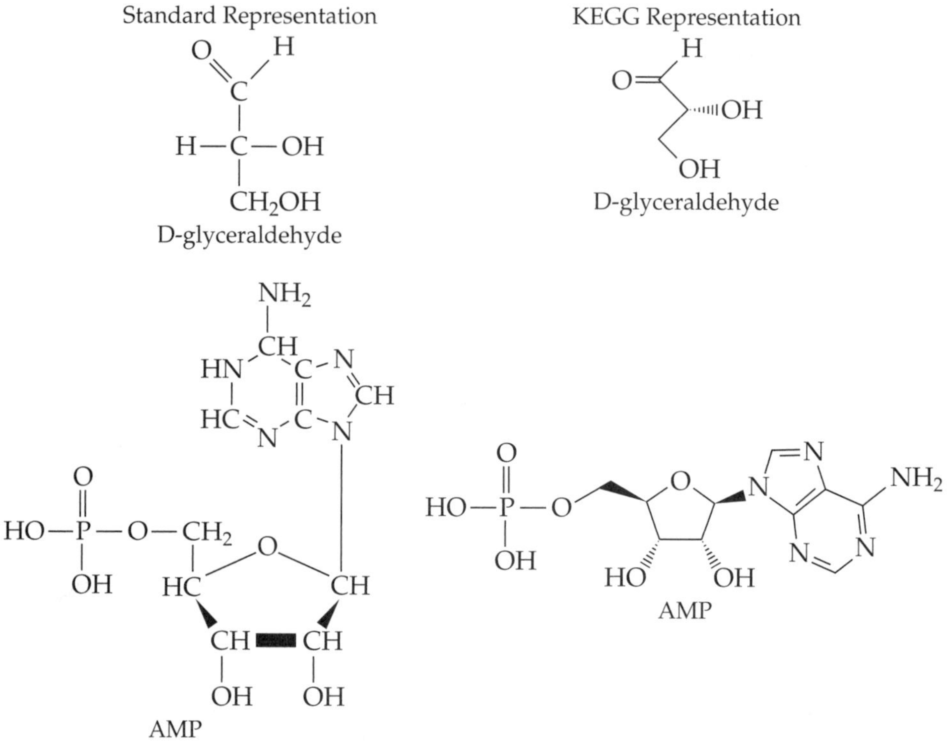

FIGURE 2.6 Standard (left column) and KEGG representation (right column) of the three-carbon sugar D-glyceraldehyde and AMP, a building block of DNA.

FIGURE 2.6 Standard (left column) and KEGG representation (right column) of the three-carbon sugar D-glyceraldehyde and AMP, a building block of DNA.

2. Compound formulas presented in the KEGG representation assume implicitly the presence of a carbon atom at each corner of a structural diagram if these corners are not explicitly marked for other atoms such as nitrogen or oxygen. Similarly, KEGG diagrams do not explicitly indicate the presence of hydrocarbon bonds. The presence of such bonds can be deduced from the fact that each carbon atom forms four bonds with neighboring atoms. The KEGG representations of the biomolecules D-glyceraldehyde and AMP are shown in Figure 2.6. The Web address for the KEGG compound database at the time of publication of this book was http://www.genome.ad.jp/kegg/catalog/compounds.html. The KEGG database is publicly accessible and contains the structural representations of more than 5000 biologically relevant molecules.

2.2.3 Stereoisomers and Biological Activity

A fundamental principle in chemistry is that molecules with identical chemical composition can have different spatial distributions of their atoms and therefore have significantly different physical and chemical properties. Molecules with the same chemical composition, but with different three-dimensional bond positioning are called *isomers*. Isomers that are mirror images of each other, like a left and a right hand, are called *stereoisomers*. Whenever a carbon atom binds to four different atoms or molecules, it gives rise to stereoisomers. Such a carbon atom is said to be

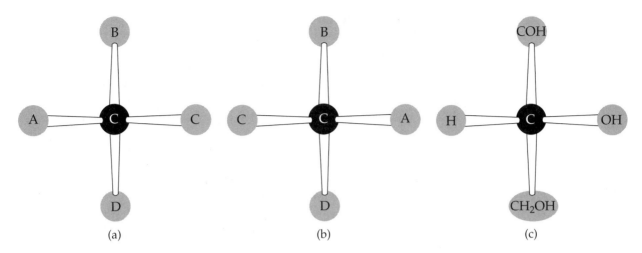

FIGURE 2.7   Molecules that are mirror images of each other (a, b). A specific example is that of D-glyceraldehyde, a three-carbon sugar (c).

an *asymmetric* or *chiral* carbon. The carbon atom shown in the black sphere in Fig. 2.7 is an asymmetric carbon. The molecular subunit with an asymmetric carbon may assume one of the two distinct structural configurations shown in Fig. 2.7a and Fig. 2.7b.

The macromolecule family of carbohydrates contains many examples of stereoisomers. For example, the three-carbon sugars D-glyceraldehyde and L-glyceraldehyde are stereoisomers. Both carbohydrates have the chemical formula $C_3H_6O_3$, but have three-dimensional shapes that are mirror images of each other, as shown by the following KEGG structural diagrams:

D-glyceraldehyde   L-glyceraldehyde

The bond indicated by the dashed arrow points into the plane of the diagram, whereas the bond shown by the black arrow protrudes away from the plane of the diagram, toward the reader. The central carbon atom in each molecule is the asymmetric carbon. These carbon atoms are linked to four different subunits: H, OH, CH_2OH, and CHO (Fig. 2.7c). The bonds of the central asymmetric carbon in a three-carbon sugar can be arranged in space in two different ways, D (*dextro*) and L (*levo*). According to the standard convention, when the carbohydrate under consideration is positioned on the paper as shown earlier, the hydroxyl group attached to the asymmetric carbon in the dextro molecule points inward from the plane of the paper whereas that of levo points outward. In general, a molecule with n asymmetric carbon atoms has 2^n stereoisomeric forms. As discussed in the next section, different stereoisomers of a molecule might have completely different biological activities.

2.3 | Carbohydrates

Carbohydrates are organic molecules constructed of carbon, hydrogen, and oxygen. Their chemical composition more or less follows the formula $C_n(H_2O)_m$, where n and m take the values of positive integers. As will be discussed in Chapter 3, carbohydrates provide chemical energy for powering biological processes. These macromolecules are metabolized in cells, and the energy contained in their covalent bonds is used in activities such as locomotion, cell division, and protein synthesis. Carbohydrates also have important functions as structural building blocks in organisms ranging from plants to bacteria. A subset of five-sugar carbohydrates forms part of the backbone of nucleic acids. Compounds that carbohydrates form with lipids and proteins play fundamental roles in protein trafficking and cell–cell recognition. Not all carbohydrates are sweet, but they are nevertheless called *saccharides* or sugars. Carbohydrates found in nature are mostly D-carbohydrates. Why the D-sugars are more abundant than L-sugars in living organisms is a mystery of molecular evolution.

2.3.1 Monosaccharides

The simplest carbohydrates are monosaccharides, the monomers from which all carbohydrates are built. The smallest monosaccharide molecules contain three and the largest contain seven carbon atoms. The three-carbon monosaccharides such as D-glyceraldehyde are found in living cells as intermediaries in the chemical reaction pathways that harvest energy from food. The metabolic process of reducing the hydrocarbon bond energy into readily usable free energy is discussed in Chapter 3.

Ribose and *deoxyribose* are five-carbon sugars with important biological functions. The molecular formulas of these sugars are as follows:

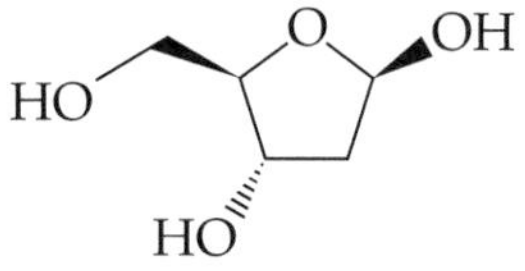

D-deoxyribose and D-ribose form parts of the backbones of the nucleic acids DNA and RNA, respectively. Both sugars are in the form of a five-membered ring. Because these sugars can undergo condensation reactions by linking at different molecular sites, a notation for identifying the carbon atoms in ring-like sugar molecules has been developed. The carbons in five-carbon sugars are identified using primed integer numbers 1′, 2′, 3′, 4′, and 5′. The values of these integers increase clockwise from the central oxygen atom as shown in Fig. 2.8.

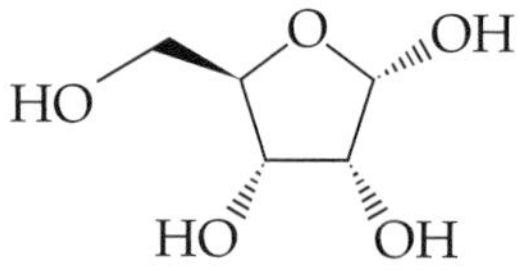

The five-carbon sugars, ribose and deoxyribose differ from each other by one atom: An oxygen atom is missing from carbon 2′ in deoxyribose in relation to ribose. In other words, the hydroxyl group (—OH) attached to carbon 2′ of the ring structure in ribose is replaced by a hydrogen atom in deoxyribose. As we shall see

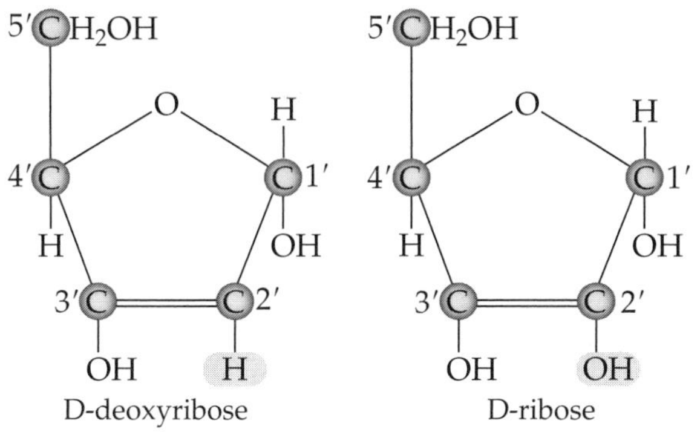

FIGURE 2.8 Notation used to identify carbon atoms in ring-like sugar molecules.

later in this chapter, this subtle difference in molecular formula is responsible for the significantly different functional roles RNA and DNA play in gene expression.

Let us next consider the six-carbon sugars. *D-glucose* is a six-carbon sugar with the molecular formula $C_6H_{12}O_6$. It is also called grape sugar. The common stereoisomers of glucose, α- and β-*glucose*, are given by the following structural formulas:

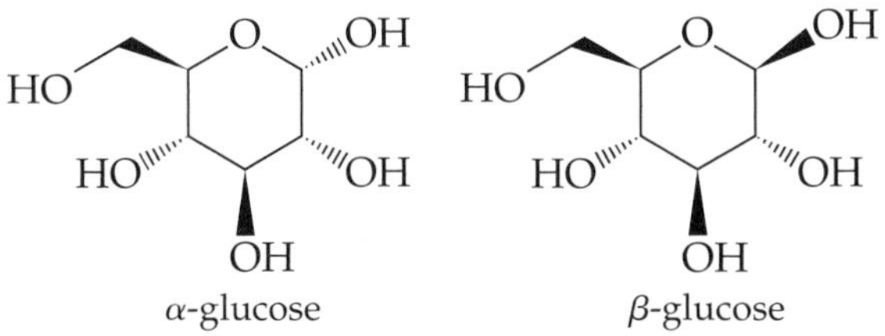

As indicated in the preceding diagram, these two forms differ from each other only in the spatial placement of the —H and —OH attached to carbon 1'. The ring structure forms α-glucose and β-glucose can be obtained from the chain form through the intermediate form shown in Fig. 2.9.

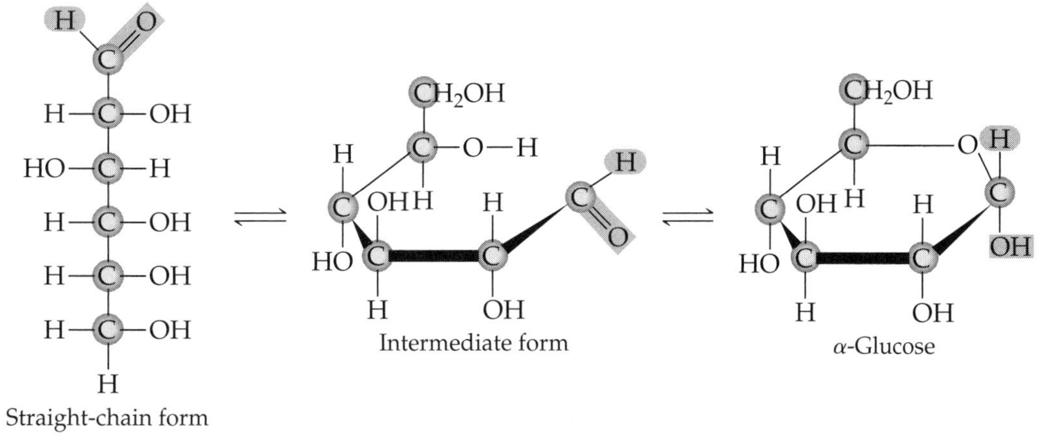

FIGURE 2.9 Various structural configurations of six-carbon sugar glucose. The ring form is the dominant form in an aqueous environment.

The predominant forms of sugars such as glucose in solution are the ring forms, not open chains. The six-carbon sugars have four asymmetric carbon atoms and therefore have 16 stereoisomers. These stereoisomers include the fruit sugar, *D-mannose*, a constituent of many glycoproteins, and *D-galactose*, the building block of the milk sugar, *lactose*:

D-mannose   D-galactose

D-glucose is contained in all living cells. Green plants produce this sugar by photosynthesis, and other organisms acquire it from plants. Cells use glucose as a primary energy source and harvest its chemical energy through a series of energy-releasing reactions.

2.3.2 Disaccharides

Disaccharides are sugars composed of two monosaccharide residues united through a condensation reaction. Figure 2.10 illustrates the formation of maltose from α-glucose and β-glucose through a condensation reaction.

The many versions of disaccharides and their structural formulas are presented in the KEGG database. Among these sugars is *sucrose*, the most abundant disaccharide throughout the plant kingdom (also known as table sugar), and lactose, the milk sugar:

Sucrose   Lactose

The chemical properties of disaccharides depend on the nature of the linked monosaccharides, the carbon atoms involved in bonding, and the form of the linkage. Disaccharides comprise an important food source for all organisms. Enzymes

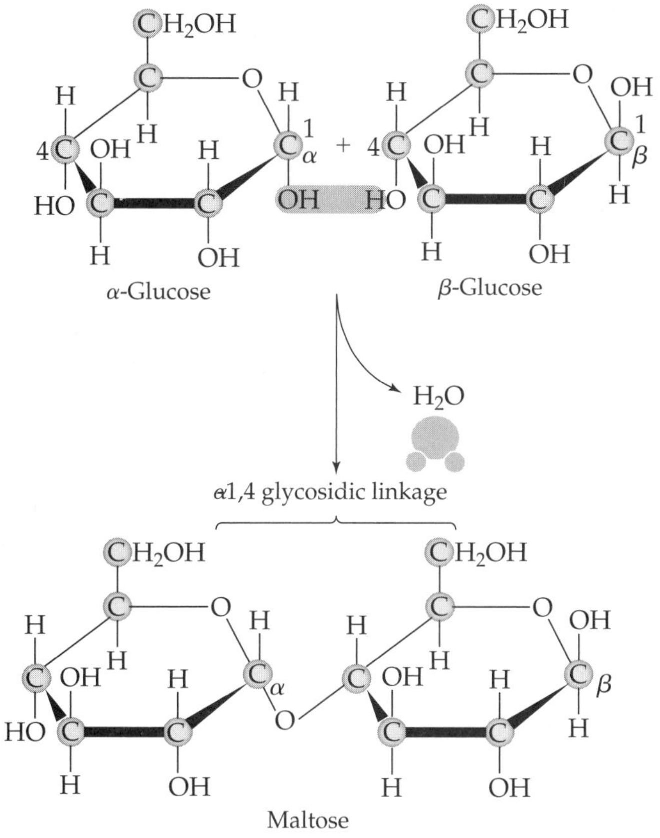

FIGURE 2.10
Condensation reaction for the
formation of a disaccharide
from two monosaccharides.

facilitate the degradation of these carbohydrates into smaller units to capture readily usable free energy.

2.3.3 Polysaccharides

Polysaccharides are giant chains of monosaccharide residues. Because monosaccharides can bind to each other in a number of ways, they can form polysaccharides of incredible diversity. These large carbohydrates may be linear polymers or have many branches. *Starch* and *cellulose* are among the most abundant polysaccharides found in nature. Starch is a long chain formed by the linkage of α-glucose, whereas cellulose is made up of long chains of β-glucose. Although cellulose and starch have the same chemical composition, they have strikingly different chemical structures and physical properties. For example, starch is soluble in water, but cellulose is not. Humans use starch as a primary food for providing energy. In contrast, humans lack the enzymes that digest cellulose, the structural material of choice for plants. Cellulose comprises more than half of all organic carbon and therefore is the most abundant organic compound in the biosphere. A third important polysaccharide is *chitin*, the material from which insect skeletons are built. Chitin is an amino sugar (a sugar containing nitrogen) and forms long straight chains that serve structural roles very similar to the bone tissue of mammals. In general, polysaccharides are not limited by a finite upper boundary. Although carbohydrates are not genetically prescribed, they are built and degraded in living organisms according to need through sequential actions of specific enzymes, which are regulated genetically.

2.4 | Lipids

Lipids have the common property of being insoluble in water. Strictly speaking, lipids are not macromolecules, but compounds of molecules whose structure depends on hydrophobic interactions and van der Waals forces. Another common feature of lipids is their high hydrocarbon content. *Simple lipids* such as fats, oils, and waxes are well known in everyday life.

Simple lipids are composed of two types of building blocks: *glycerol*, a small three-carbon molecule with three hydroxyl groups, and three hydrocarbon chains called *fatty acids*. The molecular formula of glycerol in the KEGG compound notation is

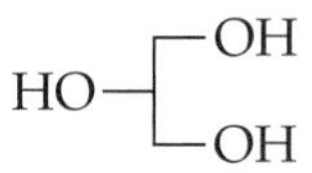

Glycerol

Structural formulas of selected fatty acids in the KEGG compound notation are as follows:

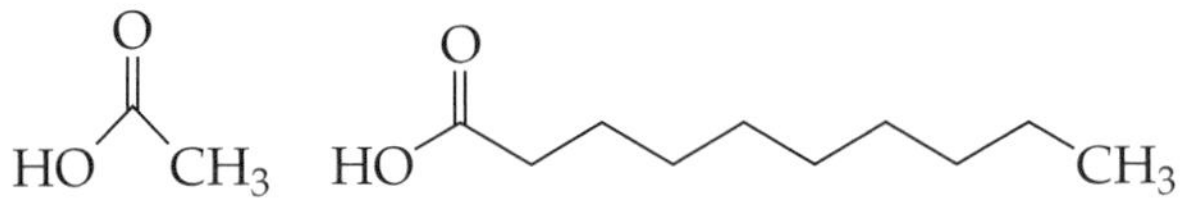

Acetic acid   Capric acid

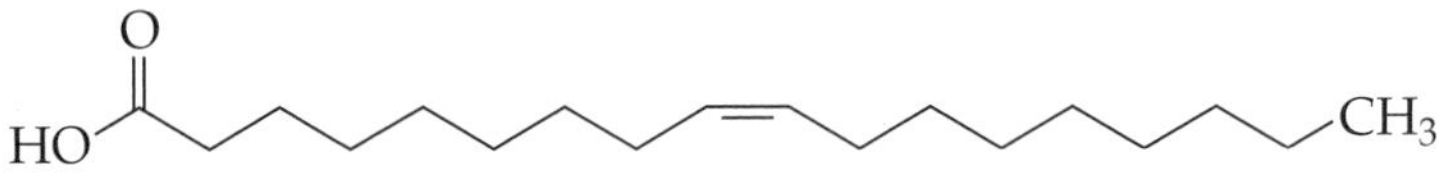

Oleic acid

A molecule of glycerol and three fatty acid molecules react to form a lipid and three molecules of water. The formation of a lipid from glycerol and three fatty acids is illustrated in Fig. 2.11.

The physical properties of lipids are highly dependent on the type of fatty acids used as building blocks. The tail of a fatty acid can be of different lengths and be either *saturated* or *unsaturated*, depending on the presence of double carbon bonds.

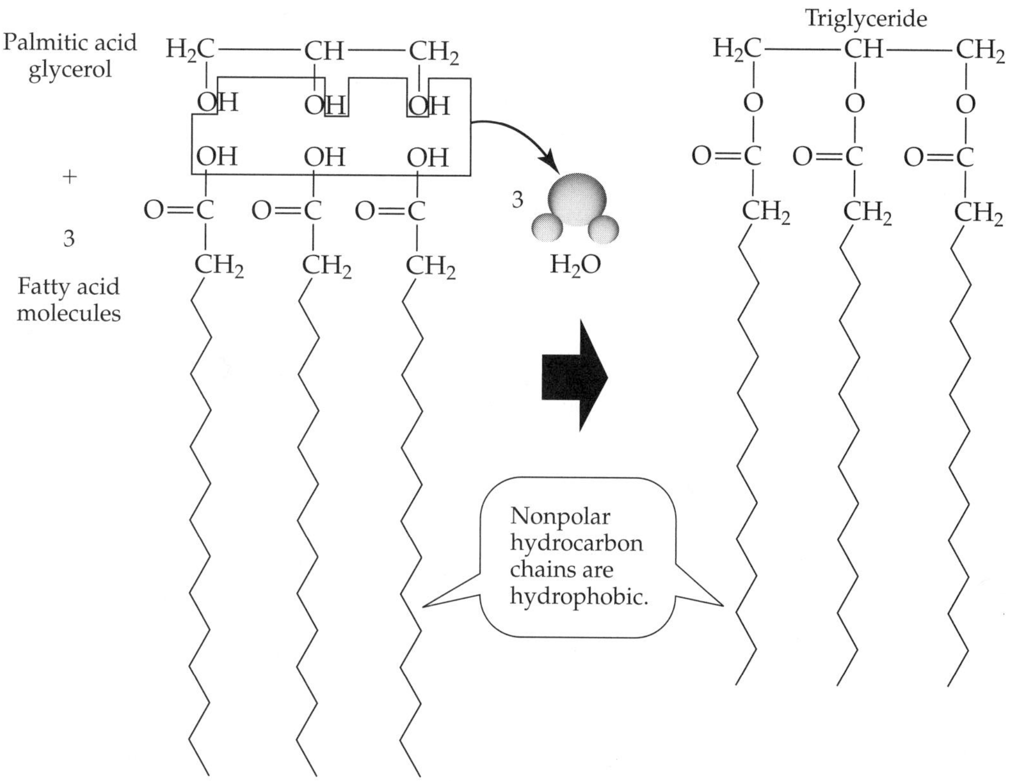

FIGURE 2.11   Condensation reaction leads to the formation of one lipid molecule from a glycerol molecule and three fatty acids.

The tail of a saturated fatty acid such as the capric acid shown earlier contains no double carbon bonds. The hydrocarbon chains of such saturated fatty acids are relatively rigid and straight, and they pack together tightly when forming fat. In *unsaturated fatty acids*, the hydrocarbon chain contains one or more double bonds. Each double bond causes a kink in the fatty acid, preventing the fatty acids from packing together tightly. The kinks associated with double bonds are important determinants of the fluidity and melting point of a lipid. Animal fats, which are solids at room temperature, tend to have long-chain, saturated fatty acids, whereas the liquid plant oils have short or unsaturated fatty acids. An example of unsaturated fatty acid is the oleic acid shown in the foregoing diagrams.

2.4.1 Phospholipids and Biomembranes

Phospholipids constitute the core elements of biological membranes. In the presence of low levels of detergents, phospholipids can also form other structures such as micelles and liposomes. Liposomes can be made in the presence of DNA and used as a vehicle for the delivery of genes in gene therapy. The lipid membrane can fuse with the plasma membrane of target cells and release the DNA into the cell. Like fats, phospholipids have fatty acids bound to glycerol; however, one of the three fatty acids in the molecule is replaced by a phosphate group. The two fatty acids bound to glycerol form hydrophobic (nonpolar) chains, whereas the phosphorus-containing compound linked to the third —OH bond of the glycerol forms a hydrophilic head. Two examples of phospholipids are as follows:

Phosphatidic acid   Phosphatidylserine

Note that in these diagrams the terms R_1 and R_2 represent the fatty acids attached to glycerol. Structural diagrams of other phopholipids can be found in the KEGG compound database. Phospholipids have the important property of having one hydrophilic and one hydrophobic region. The phosphate group attached to the glycerol has a negative charge and is therefore hydrophilic. These globular subunits are called phospholipid heads. The fatty acid tails are hydrocarbon chains, have no charge, and are hydrophobic. Each phospholipid molecule can be visualized as consisting of a hydrophilic globular head and a long hydrophobic tail.

In the presence of water, phospholipid molecules cluster together and form bilayer surfaces that shield the nonpolar fatty acid tails from water (Fig. 2.12). The hydrophilic heads face outward on both sides of the lipid bilayer, where they interact with the surrounding water molecules, and the hydrophobic tails remain in the interior of the bilayer. Phospholipids are needed to create a biochemical environment in the cell that is different from the external environment. Cellular structures such as nuclei and mitochondria are also enclosed in lipid bilayers.

FIGURE 2.12 Formation of lipid bilayers from phospholipid molecules. Hydrophobic interactions between lipids and water lead to the organization of lipid molecules into bilayers. The globular heads of lipids face the water, whereas the hydrophobic tails that remain in the interior of the bilayer create a barrier against the passive diffusion of polar (water loving) molecules across the membrane. (Modified from http://www.tulane.edu/~biochem/faculty/facfigs/bilayer.htm.)

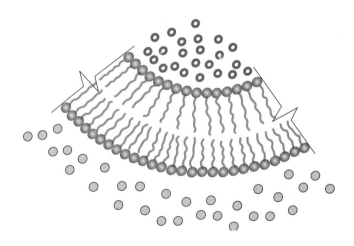

2.4.2 Other Subfamilies of Lipids

The lipid family of organic compounds has other biologically important subgroups: steroid hormones and steroids. These molecules are not composed of glycerol and fatty acids, but have ringlike structures similar to those observed in sugars. Unlike sugars, steroids consist mainly of hydrocarbons and are therefore hydrophobic. One important example of a steroid hormone is testosterone:

Testosterone   Cholesterol

This hormone is released into the blood stream from the testis and is involved in the development of male sexual characteristics. Testosterone is lipid soluble and therefore can diffuse across the plasma membranes of cells and enter the nucleus, where it can regulate gene expression.

Another biologically important lipid is cholesterol: Cholesterol can deposit on the inner surface of blood vessels, clogging the arteries and altering their mechanical properties. All steroids are synthesized from cholesterol and therefore have similar chemical structures.

Other types of lipids found in living systems include carotenoids (which trap light energy in plants), fat-soluble vitamins such as vitamins A and D, and glycolipids such as glucocerebroside. The structural formulas of these compounds as well as background information on them can be found in the KEGG database. Like carbohydrates, lipids are not directly specified genetically, but are formed from building blocks with the help of protein enzymes. However, the enzymes that catalyze the production of lipids are regulated genetically.

2.5 DNA and RNA

2.5.1 Deoxyribonucleic Acid (DNA)

DNA forms a double-stranded helix with a uniform radius and angle of twist (Fig. 2.13). Each strand is made of different combinations of the same four molecular beads, represented by the letters A, C, G, and T of the Latin alphabet. The beads in opposing strands complement each other according to base pairing combinations (A—T and C—G as shown in Fig. 2.13). The structure of DNA was discovered by

FIGURE 2.13 DNA forms a double-stranded helix with a uniform radius and angle of twist. The sugar–phosphate backbone forms the outer shell of the helix. The two strands of DNA run in opposite directions. Bases face toward each other and form hydrogen bonds. The types of base pairs found in DNA are restricted to those shown in the figure. (Modified from http://www.nhgri.nih.gov/DIR/VIP/Glossary/Illustration/basepair.html.)

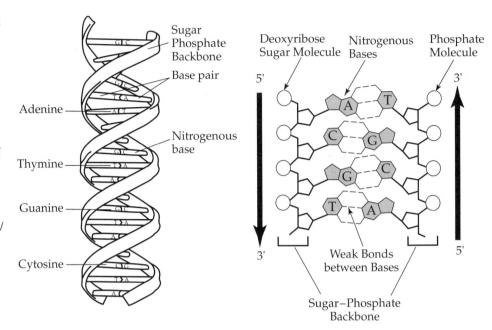

James Watson and Francis Crick and was published in *Nature* in 1953. Their article is widely considered to signal the beginning of the modern molecular era in biology. Chemical properties of DNA and the resulting structural and functional features are discussed next.

Building Blocks: Nucleotides of four different types constitute the building blocks of DNA. These nucleotides are called dAMP, dGMP, dTMP, and dCMP:

dAMP

dGMP

dCMP

dTMP

Each of these nucleotides is composed of three subunits: a phosphate group, a five-carbon sugar molecule (deoxyribose), and a nitrogenous base. The sugar and the phosphate group are identical in all four DNA nucleotides, but the nitrogen bases differ. The nitrogenous bases found in DNA are adenine (A) in dAMP, guanine (G) in dGMP, thymine (T) in dTMP, and cytosine (C) in dCMP:

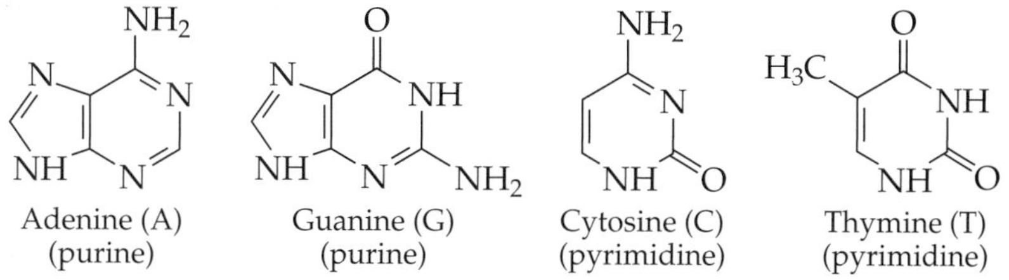

Adenine (A)   Guanine (G)   Cytosine (C)   Thymine (T)
(purine)   (purine)   (pyrimidine)   (pyrimidine)

The nitrogenous bases adenine and guanine are composed of two rings. The corresponding nucleotides are called *purines*. The bases of the other two nucleotides, dCMP and dTMP, have single rings and are called *pyrimidines*. The name for the nitrogenous base of each DNA monomer identifies the monomer itself. Thus, a DNA strand is composed of monomers identified as adenine (A), guanine (G), thymine (T), and cytosine (C).

Backbone: Alternating subunits of sugars and phosphates form the backbone of a DNA strand. The sugar subunit of one nucleotide binds to the phosphate group of the adjacent nucleotide. This backbone is illustrated as follows:

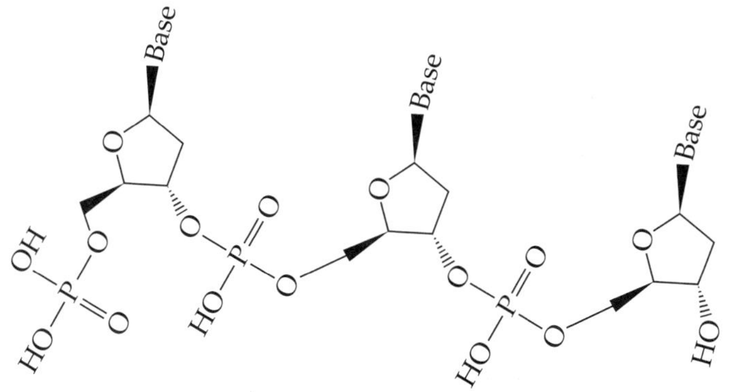

Sense of Direction: The resulting DNA strand has a sense of direction. The phosphate group attached to the 5′ carbon of the sugar of one nucleotide attaches to the sugar subunit of the adjacent nucleotide at the 3′ carbon site as shown in the preceding diagram. As discussed previously in the section about carbohydrates, the symbols 3′ and 5′ refer to the third and fifth carbon atoms in the sugar ring, counted clockwise from the corner of the ring occupied by the O atom (Fig. 2.13). The resulting strand has at one end a bound OH, but no phosphate. This end is called the 3′ end. The other end of the DNA strand has a bound phosphate group. This end is called the 5′ end and according to the convention used by biologists, the sequences of molecular beads that compose nucleic acids are listed from 5′ to 3′.

Complementary Base Pairs: DNA is a double-stranded helix with constant radius. The two strands of DNA twist like a screw, running in parallel, but opposite, directions. The nitrogen bases of DNA point toward the central axis of the helix. The two strands of the helical chain are held together by hydrogen bonding between monomers at the same positions in opposite strands. The same position refers to an identical number of monomers from one end of DNA. In particular, the monomer A on one strand always pairs with monomer T on the other strand at the same position. The monomers C and G are similarly paired. For this reason, the pairs (A, T) and (C, G) are called *complementary base pairs*. The constant diameter of the DNA helix is because these base pairs have identical physical dimensions in the direction normal to the axis of DNA. The complementary base pair rule assures that the information stored in DNA is in duplicate. The sequences of monomers on the opposing DNA strands are not identical, but complementary. For example, if one DNA strand has the sequence 5′ AACTTG 3′ at a certain location, the complementary strand will have the sequence 3′ TTGAAC 5′ at the same position. When a need arises to generate new copies of DNA (as occurs in cell division), each single-strand of DNA becomes a template to produce a complementary strand with a one-to-one correspondence between the two strands of DNA. The sequence of the DNA monomer is written either in capital or lowercase letters. For example, the sequences "ATTGCGGC" and "attgcggc" are identical. Capital letters are advantageous in distinguishing the DNA sequence from the rest of the text. Lowercase letters are used typically when large sequences from decoded DNA molecules are printed. For example, the following sequence is taken from one of the strands of a bacterial DNA molecule:

    catctgtcggccataccacttcgcgcaacagatcgcccagcagtggggccgccagtgcagaaatccactgttcg
    tcacgaaatccttcgcttaattgccgcactttgatggtcagtcgaaaactatcatcggaggggctacggcggaca
    tatccctcttcctgcagcgtctccagcagtcgccgcacagtggtgcgatgcaggccgctgagttccgccagcagc
    ccga

A survey of decoded DNA listed at the National Library of Medicine Gene Bank Web site shows that the content of the bases A, T, G, and C on DNA vary significantly from organism to organism. Because of base pairing, however, %A = %T and %G = %C when both strands of DNA are considered. In the human genome, G + C content is less than A + T content (Fig. 2.14). Although 5 percent of the genome has a G + C content of between 50 and 55 percent, this portion contains 15 percent of the genes.

2.5.2 Orientation of Genes along DNA

The hereditary information of an organism is distributed onto both strands of a DNA molecule. The molecular formulas of proteins synthesized by the organism constitute an important component of hereditary information. The term *gene* generally refers to those segments of one strand of DNA that are essential for synthesis of

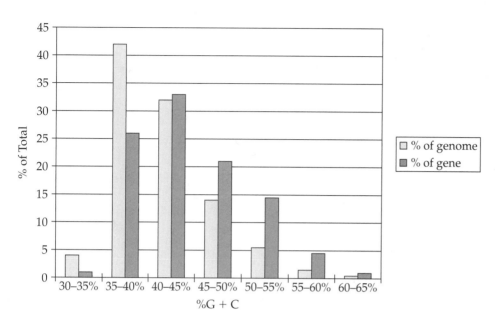

FIGURE 2.14 Distribution of G + C content in the human genome. (Modified from Venter et al., *Science* 2001, 291: 1304–1351.)

a functional protein or one of its major subunits (polypeptides). The term "gene" also refers to those segments of a DNA strand that encode a tRNA or rRNA molecule. As discussed later in this section, these two types of RNA molecules play pivotal roles in protein synthesis.

Figure 2.15 shows a schematic diagram of gene organization along both strands of DNA. All segments of a gene lie on the same strand of DNA. The strand on which the gene exists is sometimes called the sense strand and the noncoding strand the antisense strand.

The main features of gene position and orientation along DNA are discussed in the following list (a topic that is further expanded in Chapter 4, "Gene Circuits"):

1. In all organisms, the beginning of any gene faces the 3′ end of the DNA strand encoding it. Thus, when the DNA sequence of a strand is listed in the standard 5′ to 3′ direction, the sequence of a gene coded on the strand reads as in Arabic, from right to the left on the page of the text. In most instances, coding will be given of the complementary DNA strand (strand not containing the gene). Since the two strands have the opposite sense of direction, a regular 5′ to 3′ reading from the left to the right will give the complementary sequence of the actual gene.

2. The DNA strand segments that code for proteins are flanked by sequences that control the rate of their transcription. Depending on the type of cell, the production rates of gene products vary widely. The rate of *gene expression*, a term commonly used by biologists, refers to the rate at which a gene product is made. A gene is highly expressed when the rate at which it is transcribed and translated is high. Some genes are present in multiple copies in the genome further increasing the rate at which they can be made. Chapters 3–5 discuss the control mechanisms involved in the transcription of genes.

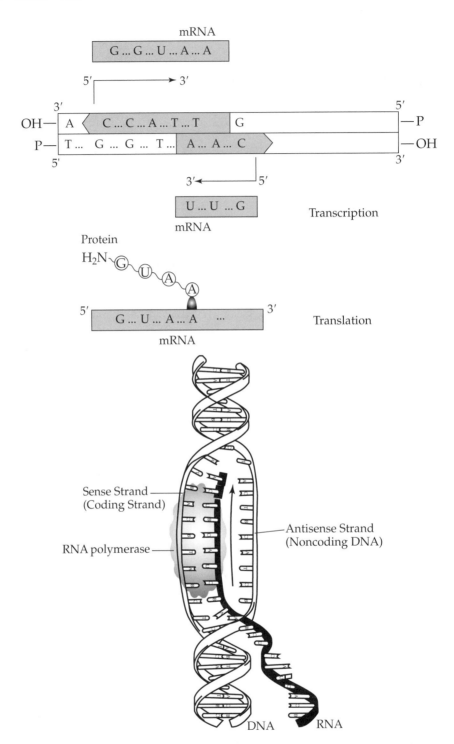

FIGURE 2.15 Schematic of the direction of genes along the two strands of a DNA molecule. The dark regions represent coding sections of DNA. The diagram also shows the two important steps in protein synthesis: transcription and translation. The figure on the bottom, depicting the event of transcription, is modified from http://www.nhgri.nih.gov/DIR/VIP/Glossary/Illustration/antisense.html.

DNA molecules from members of the same species vary slightly in their sequences of nucleotides. These variations often do not markedly affect the gene product and are known as polymorphisms. DNA differs more substantially across species. DNA harvested from members of hundreds of different organisms has been decoded. The sequence of DNA from a few individuals of the human race is also complete. Decoding the entire sequence of a human DNA is only the first step in genomic research. Among the billions of letters, scientists must identify those portions of the human DNA that actually encode protein-producing genes. Protein-coding regions make up about 1 percent of the human DNA; therefore, looking for the protein coding genes in the human genome is like looking for needles in a haystack. Scientists must not only identify the protein-coding sequences on DNA, but they must also sort out what these individual genes do. The genomic knowledge accumulated on simpler organisms can help identify the functions of many proteins found in higher organisms. Organisms such as yeast, fruit flies, and mice contain genes that are similar in sequence to those in humans. About 40 percent of all yeast genes have a human counterpart with interchangeable function. For example, many cancer-causing variants of human genes lead to uncontrolled growth of yeast colonies.

2.5.3 Ribonucleic Acid (RNA)

Like DNA, RNA molecules are chains of four molecular beads. These macromolecules play a fundamental role in protein synthesis. In the *transcription* phase of gene expression, the DNA sequence that ultimately encodes a protein is copied (transcribed) into RNA (Fig. 2.15). As noted previously, the resulting RNA molecule is called *messenger* RNA (mRNA). A single mRNA molecule may contain the instructions for a single protein, a protein subgroup, or a cluster of a few proteins. Transcription is tightly regulated to adjust the protein content of an organism to its needs.

The process of building proteins using mRNA molecules as templates is called *translation* (Fig. 2.15). The most critical step in translation is conducted by transfer RNA (tRNA). These molecules function as molecular adapters. The tRNA molecules match the sequences of beads on mRNA with the corresponding amino acids that make up the protein. Molecules in the third subgroup of RNA, ribosomal (r) RNA, associate with proteins to form ribosomes, which are factories for protein synthesis. Recent research has uncovered other types of RNA that play key roles in chromosome packing and gene expression. In the following, we present some of the important structural features of RNA and relate these features to their functions in living systems.

The structural and functional properties of RNA molecules are similar to those of DNA (Fig. 2.16). One chemical difference is that the deoxyribose sugar in DNA nucleotides is replaced by ribose in RNA, another five-carbon sugar. A second difference is that the DNA base thymine (T) is replaced by the base uracil (U) in RNA.

The building blocks of RNA are the nucleotides AMP, GMP, UMP, and CMP. Each of these nucleotides is composed of two nucleotide-invariant subunits (the phosphate group and ribose) and a nucleotide-dependent nitrogenous base. The bases found in RNA nucleotides are adenine (A), guanine (G), uracil (U), and cytosine (C).

Alternating sugars and phosphates form the backbone of RNA. This molecule is made of a single strand, but can bind in a complementary way to a DNA strand. In a DNA–RNA pair, the RNA strand runs in the opposite direction to the DNA

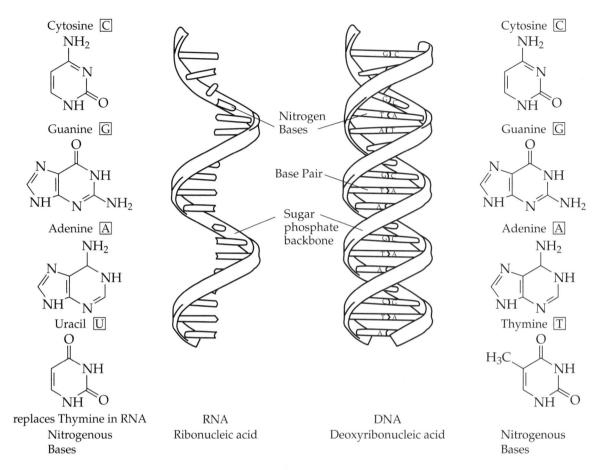

FIGURE 2.16 Comparison of RNA and DNA structures. (Modified from http://www.nhgri.nih.gov/DIR/VIP/Glossary/Illustration/RNA.html.)

strand such that the complementary base-pairing rule holds. The possible base pairs are (A, U), (T, A), (C, G), and (G, C). The complementary base pair rule between DNA and RNA assures that the information stored in DNA is accurately transcribed into RNA.

2.5.4 Types of RNA Molecules

Messenger RNA (mRNA) molecules are complementary copies of the instructions encoded on DNA (Fig. 2.15). As discussed in Chapter 3, mRNA is built by molecular machines with the use of a DNA strand as a template. The complementary-pair rule is illustrated by the following sequence:

3′ TACCGAAATGAGTGCGCTTA 5′ DNA

5′ AUGGCUUUACUCACGCGAAU 3′ mRNA

Whereas DNA contains the master copy of the genetic information, kept permanently on file, mRNA stores the working copy of this information for a brief period. After

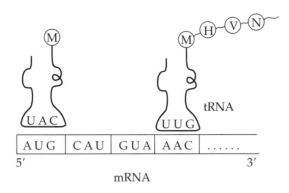

FIGURE 2.17 *Schematic diagram of tRNA as a molecular adapter. The sequence of three nucleotides of tRNA that pair with a codon on mRNA is called an anti-codon.*

the instructions are implemented by the cell, mRNA is degraded into its constituent nucleotides, which are then available for use in the synthesis of other molecules.

Transfer RNA (tRNA) functions as a structural adapter that matches each three-nucleotide amino acid codon sequence of the mRNA, with the amino acid specified by this codon (Fig. 2.17). A tRNA molecule consists of 75 to 80 nucleotides. The 3′ end of a tRNA molecule attaches to the specified amino acid, and in the middle of the tRNA are three bases (anticodon) that constitute the point of contact with mRNA. This binding site interacts with mRNA through complementary-base pairing. As illustrated in Fig. 2.17, each amino acid-specific tRNA binds to its amino acid in the cytoplasm and attaches it to the growing chain of amino acids (polypeptide) at the location prescribed by the mRNA.

Unlike mRNA, tRNA genes are directly encoded by DNA. The number of distinct tRNA molecules in many organisms is less than the $4^3 = 64$ possible combinations of three letters chosen from an alphabet of four letters. This is because some tRNA molecules attach the same amino acid to the growing chain of a polypeptide when they bind to different codons. The use of more than one codon to specify an amino acid means that there is some redundancy in the genetic code. However, the codons they bind to all share a common property, namely, that the first two letters in the codons must be the same. As discussed later in the chapter, it is possible that the very first tRNA molecules to emerge had specific affinity to nucleotide couplets, not triplets. However, adapters designed for nucleotide couplets, recognize and bind to at most $4^2 = 16$ different types of amino acids, apparently a subset of not enough diversity to sustain life.

The three-dimensional shapes of tRNA molecules enable them to bind not only amino acids and mRNA, but also enable them to interact with ribosomes, the mini-plants for protein synthesis. Information about the structures of macromolecules such as RNA comes from X-ray crystallography, which allows investigators to determine the positions of atoms in a crystalline substance by the diffraction patterns of X rays passed through a crystal. Figure 2.18 shows the backbone of the tRNA that acts as an adapter for the amino acid phenylalanine. Similar diagrams for many different tRNA molecules can be obtained by going to the National Library of Medicine Web site http://www.ncbi.nlm.nih.gov/ and pressing *Entrez*. Press *Structure*, type the keyword tRNA, and press *go*. The reader will then have access to the Web page for structure. The entry for the specific tRNA shown in Fig. 2.18 is represented by the PDB identification number "1EHZ." Pressing the PDB id number will take the reader to the structure summary page where the structure can be viewed by pressing *view*. The

FIGURE 2.18 Crystal structure of yeast phenylalanine tRNA at 1.93 A Resolution. The tube-like curve in the figure represents the sugar-phosphate backbone of tRNA. This molecule is associated with mineral ions. (From http://www.ncbi.nlm.nih.gov/Structure/.)

program offers different modes of viewing the structure and the reader can investigate these various modes by clicking *style* and going through the options. The National Library of Medicine Web page also has a "*help*" section that provides information for structural investigations.

Ribomosal RNA (rRNA) molecules are also encoded by DNA. These molecules bind to specific proteins and form *ribosomes*. As discussed in detail in Chapter 3, several different types of rRNA molecules (ones with different lengths and sequences) are needed in the synthesis of ribosomes. Ribosomal RNA molecules contain thousands of nucleotides. Figure 2.19 shows the three-dimensional configuration of a 2900-nucleotides rRNA molecule found in the bacteria *Escherichia coli* (*E. coli*).

Some RNA molecules not only have complex three-dimensional shapes, but also have enzymatic activity. The existence of these "ribozymes" and other experimental data suggest that RNA emerged before proteins and DNA in evolution. The earliest catalysts of biochemical reactions might have been RNA molecules with complex three-dimensional shapes. Such molecules may have acted as templates both to produce DNA and to synthesize proteins. However, proteins, with their incredibly diverse shapes and physicochemical properties, eventually replaced RNA as enzymes.

FIGURE 2.19 Structure of the RNA molecule 16s rRNA in the region around ribosomal protein S8 in *Escherichia coli*. A ribosomal RNA molecule (shown as a thick tube-like curve) interacts with protein subunits (helical cylinders, flat arrows, and connecting wire-like regions) in forming the ribosome. (Modified from http://www.ncbi.nlm.nih.gov/Structure/.)

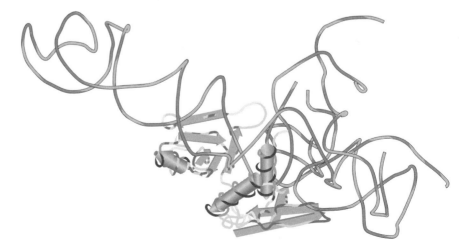

The capacity of DNA to form double helices may have made it a more stable and attractive molecule than RNA for storing hereditary information.

2.6 | Proteins

Protein networks guide the biochemistry of living cells. Metabolic and signaling pathways discussed later in the book are composed of networks of interacting proteins. As discussed in Chapter 9, elements of such networks have largely been identified in the yeast. The structure of protein networks, including connectivity of nodes as well as network logic, are currently the subject of intense research. In the following discussion, we focus on the common features of proteins, their composition, structure, and function.

All proteins in living systems follow a simple and universal blueprint: They are linear chains of amino acid residues (Fig. 2.20). Amino acids are a subset of organic molecules that serve as chemical messengers between cells or function as important intermediates in metabolic processes. Amino acids are composed of a central carbon atom with four subgroups attached to it: an amino group ($-NH_2$), a carboxyl group ($-COOH$), a hydrogen atom, and a distinctive side chain, represented by the letter R as shown in the following diagram:

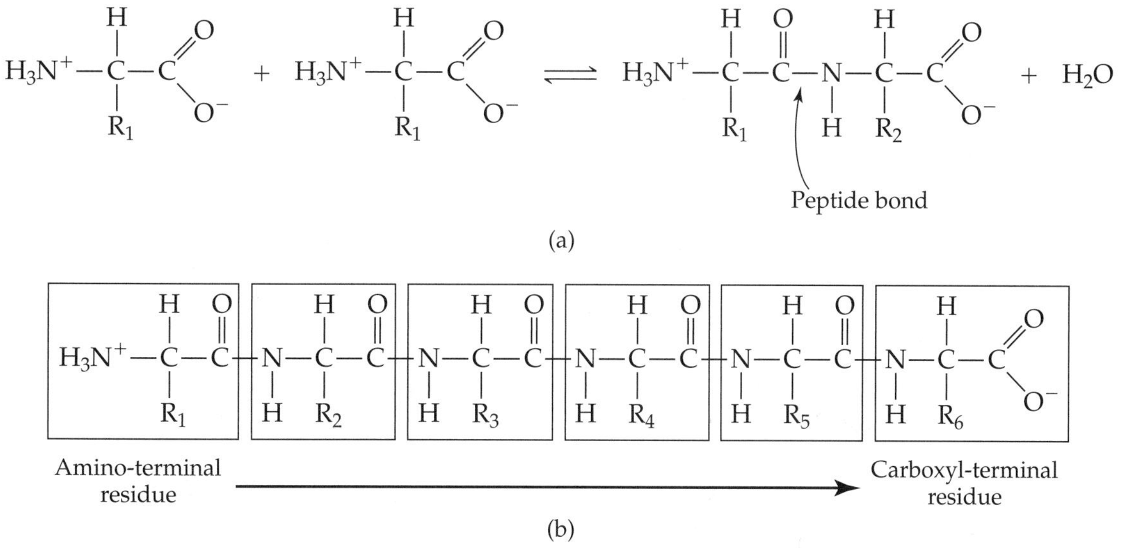

FIGURE 2.20 Synthesis of polypeptides from amino acids. Amino acids are composed of a central carbon atom with four subgroups attached to it: an amino group ($-NH_2$), a carboxyl group ($-COOH$), a hydrogen atom, and a distinctive side chain, represented by the symbol R.

The amino acids differ from each other only by their side chains (R). The tetrahedral array of four different groups about the central carbon atom yields two mirror-image forms: the L-isomer and the D-isomer. Only L-amino acids are found in proteins. D-amino acids are most widely found in bacterial cell walls.

About 300 amino acids are found in nature, but only 20 of these occur in proteins. Not every protein contains all of the 20 amino-acid types. Common elements in amino acids are carbon, hydrogen, oxygen, and nitrogen. In addition, all proteins have an amino acid that contains sulfur.

Serial linkage of amino acids in a condensation reaction results in proteins with intricate three-dimensional structures and a remarkable range of functions. The amino group of one amino acid is joined to the carboxyl group of the adjacent amino acid through the loss of one water molecule. This energy-requiring reaction is repeated over and over again in a growing polymer (polypeptide). An amino acid unit in a polypeptide is called a *residue*, and the covalent bond between two amino acid residues is called a peptide bond.

The peptide bonds formed by the amino and the carboxyl groups of amino-acid residues form the backbone of polypeptides and the distinctive *side chains* are exposed. Therefore, the side chains give the polypeptides their distinguishing properties. Polypeptides are linear polymers; that is, each amino acid is linked to its neighbors in a head-to-tail fashion rather than forming branched chains.

A polypeptide chain has a sense of direction such that one end has a terminal amino group and the other a terminal carboxyl group. Polypeptides grow from the terminal amino group toward the carboxyl terminal group. Thus, the sequence of amino acids in a polypeptide chain is written starting with the amino terminal residue. The first amino acid in most polypeptides is the sulfur-containing amino acid, methionine, which is identified with the letter M. Each of the 20 amino acids is represented by a letter in the Latin alphabet. A typical amino acid sequence for a protein composed of a single polypeptide reads as follows:

```
MLLVLVLIGLNMRPLLTSVGPLLPQLRQASGMSFSVAALLTALPVVTMGG
LALAGSWLHQHVSERRSVAISLLLIAVGALMRELYPQSALLLSSALLGGVG
IGIIQAVMPSVIKRRFQQRTPLVMGLWSAALMGGGGLGAAITPWLVQHS
ETWYQTLAWWALPAVVALFAWWWQSAREVASSHKTTTTPVRVVFTPRA
WTLGVYFGLINGGYASLIAWLPAFYIEIGASAQYSGSLLALMTLGQAAGA
LLMPAMARHQDRRKLLMLALVLQLVGFCGFIWLPMQLPVLWAMVCGL
GLGGAFPLCLLLALDHSVQPAIAGKLVAFMQGIGFIIAGLAPWFSGVLRSI
SGNYLMDWAFHALCVVGLMIITLRFAPVRFPQLWVKEA
```

Many proteins are single polypeptides. The shape and function of such proteins is then directly related to the primary sequence of amino acids present in the polypeptide. Other proteins are composed of multiple polypeptides that come together to form a complex. In some cases, each of these polypeptides is encoded by a single gene; in others, multiple genes may be involved.

2.6.1 Chemical Properties of the Twenty Amino Acids Found in Proteins

The 20 amino acids found in proteins vary in size, charge, and their capacity to form hydrogen bonds with other molecules. This variation is an important determinant

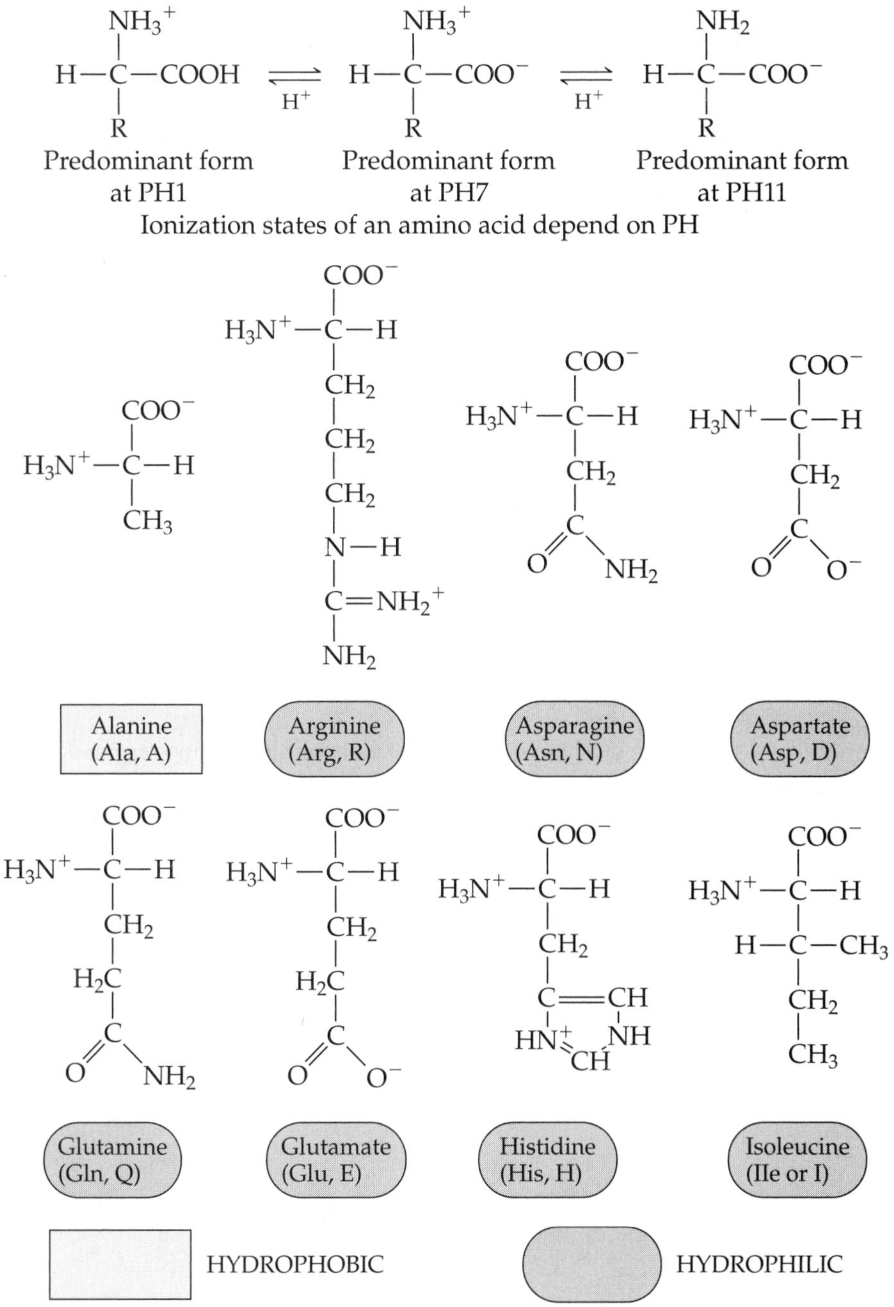

FIGURE 2.21 Structural diagrams and chemical classification of the 20 amino acids found in proteins.

Ionization states of an amino acid depend on PH

Predominant form at PH1 — Predominant form at PH7 — Predominant form at PH11

Alanine (Ala, A)

Arginine (Arg, R)

Asparagine (Asn, N)

Aspartate (Asp, D)

Glutamine (Gln, Q)

Glutamate (Glu, E)

Histidine (His, H)

Isoleucine (Ile or I)

HYDROPHOBIC

HYDROPHILIC

of the diversity found in proteins. Figure 2.21 shows the structural diagrams of the 20 amino acids found in proteins.

Amino Acids with Hydrocarbon Chains: The simplest amino acid of the 20 found in proteins is *glycine* in which the side chain is a single hydrogen atom. As with all amino acids, glycine has a three-letter and single-letter representation (gly, G).

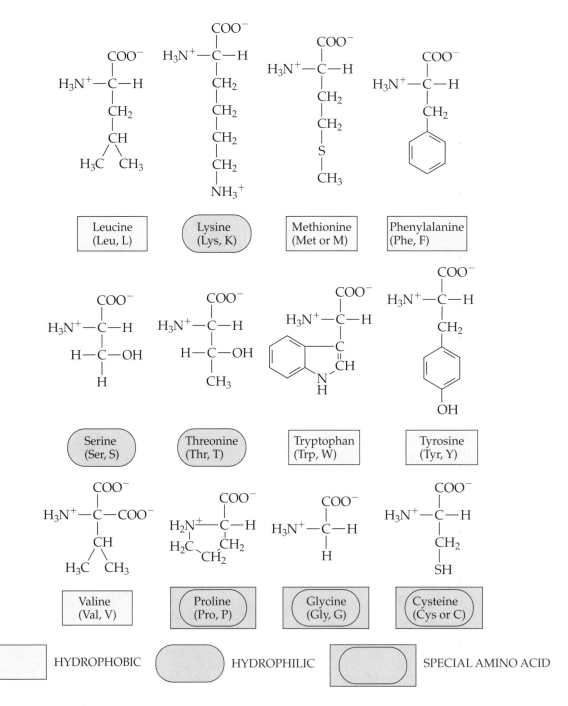

FIGURE 2.21   (*Continued*)

Glycine fits into tight corners in the interior of a protein molecule. Next in simplicity is *alanine* (ala, A), with a methyl group (CH_3) as its side chain. Larger hydrocarbon side chains, three or four carbons long, are found in *valine* (val, V), *leucine* (leu, L), and *isoleucine* (ile, I). These hydrocarbon chains are all hydrophobic. The different sizes and shapes of these hydrocarbon side chains enable them to pack together to form compact structures with few holes exposed to water. Regions of proteins rich in these hydrophobic residues often interact with lipid-containing membranes. *Proline* (pro, P) is often found in the bends of folded protein chains. Its hydrocarbon chain contains three carbon atoms, but unlike alanine, valine, leucine, and isoleucine, the side chain is bound to both the central carbon atom and the nitrogen atom. This configuration makes proline very rigid and its presence often creates a kink in a polypeptide chain.

Aromatic Amino Acids: Amino acids *phenylalanine* (phe, F), *tyrosine* (tyr, Y), and *tryptophan* (trp, W) have *aromatic side chains* in the form of rings. Tryptophan contains a nitrogen atom in its side chain. Phenylalanine and tryptophan are strongly hydrophobic. The side chain of tyrosine contains a hydroxyl group, which makes it less hydrophobic. This hydroxyl group is a potential site of addition of a phosphate group, a common post-translational modification of certain proteins.

Amino Acids Containing Sulfur: A sulfur atom is present in the side chains of *cysteine* (cys, C) and *methionine* (met, M). Both of these sulfur-containing side chains are hydrophobic. The side chain of cysteine is highly reactive, a property that enables cysteine to play a special role in shaping some proteins through disulfide links. Cysteine residues at different regions of a protein bind to each other and thus create folds and domains in the geometry of the protein.

Water-Loving Amino Acids: Amino acids *serine* (ser, S) and *threonine* (thr, T) are a hydroxylated version of alanine and valine, respectively. Replacement of a hydrogen atom in the side chain with a hydroxyl group leads to more reactive and more water-loving (hydrophilic) amino acids. Like tyrosine, these hydroxyl residues are potential sites of phosphate addition. *Lysine* (lys, K), *arginine* (arg, R), and *histidine* (his, H) possess polar side chains that contain nitrogen and are highly hydrophilic. The side chains of arginine and lysine are the longest of the 20 amino acids and are normally positively charged. Histidine can be uncharged or positively charged, depending on its environment. This amino acid is often found in the active sites of enzymes, where it can readily switch between these states to catalyze the making and breaking of bonds.

Aspartate (asp, D) and *glutamate* (glu, E) are polar and have negatively charged acidic side chains, carboxyl groups, at physiological pH. Uncharged derivatives of glutamate and aspartate are *glutamine* (gln, Q) and *asparagine* (asn, N), which contain a terminal amide group in place of a carboxylate. These amino acids are also polar molecules and the amide group of asn is a potential site of addition of sugar residues.

Bacteria can synthesize each of the 20 amino acids found in proteins using a carbon source and ammonium ions that exist in water. Plants use various simple nitrogen compounds and carbohydrates to make amino acids. In contrast, animals can synthesize only some of their amino acids using sugars and ammonia as starting materials. One could speculate that as higher organisms became more complex, they became more and more dependent on organic food. The enzymes used in synthesis of some amino acids were used more and more infrequently because of their availability in foodstuff. As a result, the genes for these enzymes became nonfunctional over the course of time. Amino acids that humans cannot synthesize are called *essential amino acids*. Out of the eight essential amino acids, six are hydrophobic. The amino acids with large hydrocarbon side chains (valine, leucine, and isoleucine), amino acids with aromatic side chains (phenylalanine and tryptophan), and the sulfur-containing methionine are in this group. The two essential hydrophilic amino acids are threonine and lysine. Essential amino acids must be obtained through the diet. Meat, fish, milk, and eggs contain all of the amino acids needed in making human proteins. Legumes, grains, and other plant sources typically contain only a partial set of essential amino acids. The needed amino acids can be obtained from plant sources by combining certain foods. Beans, for example, provide isoleucine and lysine, whereas rice contains adequate amounts of other essential amino acids.

2.7 | The Genetic Code

The nucleotide sequence of mRNA determines the sequence of amino acid residues of the protein it encodes. The set of relations that map mRNA coding onto amino-acid sequences of proteins is called the *genetic code*. The code provides one specific answer to the following question:

How do sequences of four letters (A,T,G,C) specify 20 distinct entities?

Clearly, single-letter and two-letter combinations do not provide enough choices to specify 20 amino acids. There are only $4^2 = 16$ distinct sequences of two letters in a four-letter language. Three-letter combinations, with each letter chosen from a pool of four, on the other hand, lead to $4^3 = 64$ distinct words, which is more than sufficient to code each of the twenty amino acids. Indeed, mRNA consists of a linear sequence of such three-letter words called *codons*. Table 2.1 shows the 64 mRNA codons and the entities they specify. Protein-coding genes all begin with a START codon and terminate with a STOP codon. The START codon specifies the sulfur-containing amino acid methionine (M).

Multiple nucleic acid codons correspond to the same amino acid. Amino acids represented by six codons are arginine (R), leucine (L), and serine (S). On the other hand, amino acids methionine (M) and tryptophan (W) are represented by single codons each. Codons specifying the same amino acid are said to be *synonymous*. Because the genetic code is not a one-to-one mapping between RNA and amino acids, it is called *degenerate*.

Table 2.1 indicates the following properties for the genetic code:

1. The first two letters in a codon are primary determinants of amino-acid identity. For example, all codons that begin with GU specify the amino acid valine (V) and are therefore synonymous. Similarly, codons that begin with GG specify glycine (G).

2. Codons that have U or C as the second nucleotide tend to specify the more hydrophobic amino acids. For example, codons beginning with GU and GC specify valine and alanine, respectively.

TABLE 2.1

Standard Genetic Code: Mapping between mRNA Codons and Amino Acids			
UUU F Phe	UCU S Ser	UAU Y Tyr	UGU C Cys
UUC F Phe	UCC S Ser	UAC Y Tyr	UGC C Cys
UUA L Leu	UCA S Ser	UAA STOP	UGA STOP
UUG L Leu	UCG S Ser	UAG STOP	UGG W Trp
CUU L Leu	CCU P Pro	CAU H His	CGU R Arg
CUC L Leu	CCC P Pro	CAC H His	CGC R Arg
CUA L Leu	CCA P Pro	CAA Q Gln	CGA R Arg
CUG L Leu	CCG P Pro	CAG Q Gln	CGG R Arg
AUU I Ile	ACU T Thr	AAU N Asn	AGU S Ser
AUC I Ile	ACC T Thr	AAC N Asn	AGC S Ser
AUA I Ile	ACA T Thr	AAA K Lys	AGA R Arg
AUG M, START	ACG T Thr	AAG K Lys	AGG R Arg
GUU V Val	GCU A Ala	GAU D Asp	GGU G Gly
GUC V Val	GCC A Ala	GAC D Asp	GGC G Gly
GUA V Val	GCA A Ala	GAA E Glu	GGA G Gly
GUG V Val	GCG A Ala	GAG E Glu	GGG G Gly

3. Codons that differ in the third nucleotide specify the same amino acid if the third nucleotide is either U or C. For example, histidine (H) is specified by codons CAU and CAC.

Slight differences from the standard genetic code just described occur in a few organisms. These differences tend toward even less dependence on the third letter of the codon to specify the identity of an amino acid. Let us next illustrate the flow of information from DNA to protein by considering the following DNA sequence representing the beginning segment of a protein-coding gene:

3′ TACTTGCAAATG 5′ DNA template

The complement mRNA sequence is as follows:

5′ AUGAACGUUUAC 3′ mRNA

Table 2.1 relates the mRNA codons AUG, AAC, GUU, and UAC to their corresponding amino acids, and the resulting peptide sequence is MNVY. The fact that both DNA and mRNA are specified in Gene Banks in the 5′ to 3′ direction might cause confusion in translating the codons on DNA to mRNA. DNA coding, when written according to the standard 5′ to 3′ direction, will be different from that in 3′ to 5′ direction:

3′ TACTTGCAAATG 5′   DNA template
5′ GTAAACGTTCAT 3′   DNA as read from 5′ to 3′
5′AUGAACGUUUAC 3′   mRNA read from 5′ to 3′
MNVY   encoded amino acids

2.8 | Protein Structure and Function

Protein folding is one of the most intriguing subjects of computational and experimental biology because of its significance in the design and discovery of new drugs. All proteins adopt a single unique confirmation (folded state). This folded state is encoded in the amino-acid sequence of the protein and is called the native state. The number of possible three-dimensional configurations of polypeptide chains increases geometrically with the length of the polypeptide. However, because protein folding occurs through a pathway that favors only a few intermediate steps, the native state is stable and the misfolding of proteins is rare. Furthermore, as discussed in Chapter 3, cellular systems prevent the formation of misfolded proteins.

The structure of the native state of a protein largely determines its function in a living system. Proteins are described and compared by depicting four levels of structure. The levels of protein structure that affect function are illustrated in Fig. 2.22.

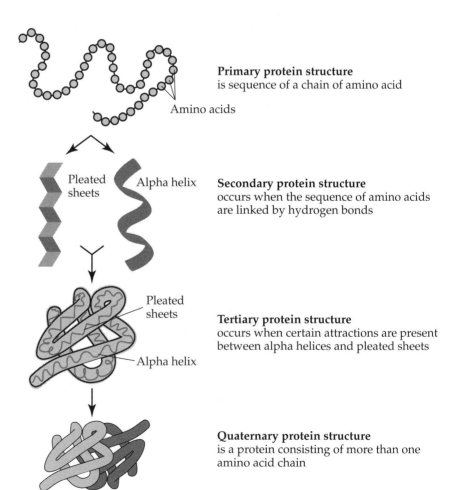

Primary protein structure
is sequence of a chain of amino acid

Amino acids

Pleated sheets

Alpha helix

Secondary protein structure
occurs when the sequence of amino acids are linked by hydrogen bonds

Pleated sheets

Alpha helix

Tertiary protein structure
occurs when certain attractions are present between alpha helices and pleated sheets

Quaternary protein structure
is a protein consisting of more than one amino acid chain

FIGURE 2.22 Structural levels of proteins. (Modified from http://www.nhgri.nih.gov/DIR/VIP/Glossary/Illustration/protein.html.)

The *primary structure* of a protein is its amino-acid sequence. It provides a complete description of the covalent connections of a protein. The primary structure is encoded on DNA, and the structural similarity of proteins can be assessed effectively by comparing amino-acid sequences. The *secondary structure* is a set of patterns concerning the spatial arrangement of amino acids that are near one another in the linear sequence. Secondary structures provide clues regarding the location of protein sites for interaction with other proteins and ligands. Figure 2.23 shows typical secondary structures such as *α-helix* and *β-pleated sheet* on the scallop myosin protein. The reader can obtain the structural diagram shown in the figure by going to the National Library of Medicine Web site http://www.ncbi.nlm.nih.gov/, pressing *Entrez*, pressing *Structure*, typing the keyword "myosin," and then pressing *go*. The particular myosin molecule shown in the figure has the PDB Id *1DFL*. Myosin is a very important contractile protein responsible for muscle contraction and is composed of six polypeptides: two identical polypeptides called heavy chains (A and B) and four shorter polypeptides called light chains (Y, Z, W, X). The *α-helical structure* is a corkscrew-like right-handed coil, with side chains extending outward from the peptide backbone of the helix. The α-helical structure is represented in protein graphics using a circular cylinder. An α-helical structure can be stretched, as the stretching requires only the breaking and rearrangement of hydrogen bonds; no covalent bonds are affected. When the tension in the helix is released, both the helix and the hydrogen bonds reform. Such behavior is called elastic in the field of mechanics. Fibrous structural proteins such as keratins are organized in α-helices. Keratin is found in fingernails, skin, and hair in humans. *β-pleated sheets* form when the protein chains extend and lie next to one another, forming flat sheets. This formation is also due to hydrogen bonding between the elements of the peptide linkages. A β-pleated sheet is shown with a flat arrow in protein graphics. The sense of direction of the arrow is from the beginning of the protein (the N-terminal) to its carboxyl terminal. Many proteins contain regions of α-helix and β-pleated sheet in the same polypeptide chain.

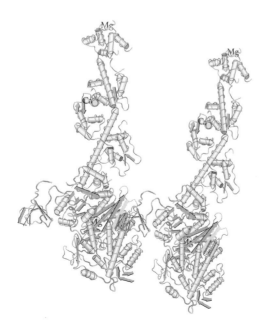

FIGURE 2.23 The three-dimensional molecular graphics of scallop myosin I. Myosin is the molecular machine supporting muscle contraction. Each head is composed of two different light chains as well as part of a heavy chain. The diagram indicates the presence of secondary structures such as α-helix and β-pleated sheets. The α-helix is shown as a circular cylinder decorated with helix and β-pleated sheet as a flat arrow pointing toward the carboxyl end of the polypeptide. The structure of the protein can be further studied by rotating the figure provided by the National Library of Medicine's Web page http://www.ncbi.nlm.nih.gov/Structure/.

Tertiary structure refers to the spatial arrangement of amino acids that are far apart in the linear sequence. The dividing line between secondary and tertiary structure is not precise. What is close and what is distant have not been defined in mathematical terms. However, in most proteins, a polypeptide chain is bent at specific sites and folded back and forth, enabling the interaction of amino acid residues that are far apart in the linear sequence. When cysteine residues from distant regions interact, they form disulfide bonds. A complete description of the tertiary structure of a protein requires the spatial location of every atom in the molecule in three-dimensional space. The identification of secondary and tertiary structures in a protein provides clues regarding its active (binding) sites and enable comparison with the reactive sites of other proteins. The MMDB structure summary for the scallop myosin shown in Fig. 2.23 identifies the regions of tertiary structures with symbols such as A.1, A.2, Y.6, and X.18. Pressing one of these symbols leads to the amino acid sequence of the tertiary structure.

Proteins containing more than one polypeptide chain exhibit an additional structure. *Quaternary structure* refers to the spatial arrangement of such subunits and the nature of their contacts. Hydrophobic interactions, hydrogen bonds, and ionic bonds all help hold the multiple polypeptide chains together to form a functional protein complex. In the scallop myosin example given earlier, the quaternary structure refers to the six polypeptide chains that compose the protein: A, Y, Z, B, W, and X. Pressing any of these symbols, one can obtain the corresponding amino-acid sequence. The National Library of Medicine has developed a database called VAST for assessing the structural similarities of various proteins and it is quite an effective tool in biotechnology because structural similarity often implies functional similarity. Manipulation of protein structure through changes in amino-acid sequence is a tool in modern drug design.

To accomplish all the fundamental functions that proteins carry out, many different types of protein are needed. A typical human cell contains about 100 million proteins of about 10,000 types. These cells all possess the same set of protein-coding genes (about 30,000), but different cell types (muscle cell, liver cell, neuron, and so on) express different subsets of these genes. The number of different types of proteins manufactured by an organism roughly correlates with the complexity of the organism itself. A typical gene in a vertebrate is composed of short sequences (exons) separated by long noncoding sequences (introns). Various spatial combinations of these genes correspond to different proteins. For example, a protein could be coded by exons 1 to 7, and another protein by exons 1 to 6 and 8 to 10 of the same gene (Fig. 2.24). Thus, a gene can code for multiple proteins in higher forms of life.

Many of the proteins made by humans and other vertebrates are similar in composition to those in simpler forms of life. Some of these proteins have additional domains that enable them to fulfill additional tasks (Fig. 2.23b). The enormous diversity in the types of protein arises primarily from the varying sequence of amino acids in the polypeptide. Permutations in the primary sequence allow for 20^r different protein types with r number of amino acid residues in the polypeptide. Protein lengths in most organisms range from 50 amino acids to tens of thousands. A typical protein molecule consists of a single polypeptide chain of about 100 residues. As discussed, some proteins have one or more polypeptide chains. These polypeptides can contain as many as 14,000 amino acids. The largest protein complex known has more than 40 separate polypeptide chains. Complicating the portrait of proteins is

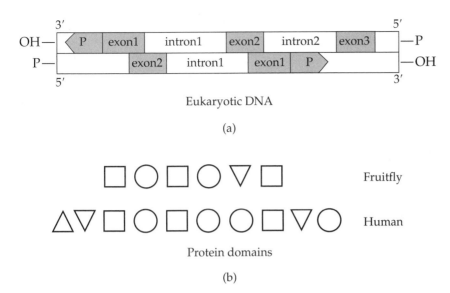

FIGURE 2.24 Schematic drawing of a human gene (a). The noncoding sections (introns) are much longer than the coding sections (exons). The existence of multiple exons for a gene leads to multidomain proteins in human and other mammals (b).

the presence of some proteins with carbohydrate, lipid, phosphate, and other types of attachments. These attachments and modifications occur not casually, but in controlled pathways after the formation of polypeptide chains. Such modifications lead to changes in shape that are necessary for specific functions. The diversity found in proteins is an example of how living organisms are able to create complex systems using the simple rules of permutation.

2.9 ASSIGNMENTS

2.1 Using the KEGG database, determine the structural formulas of two of the most common five-sugar carbons. Using the same database, explain the differences among ATP, ADP, AMP, and dAMP.

2.2 The primary carbohydrate found in milk is the disaccharide lactose. What are the monosaccharides that make up lactose? Conduct a literature search to determine the name of the enzyme vital for the degradation of lactose.

2.3 Use the KEGG database to elucidate the structural formulas of starch and cellulose?

2.4 Due to complementary base-pair formation, the number of purine nucleotides matches the number of pyrimidines in DNA. Pyrimidine nitrogen bases in DNA are thymine (T) and cytosine (C), and the purine bases are guanine (G) and adenine (A). These nucleotides are contained within the backbone of DNA.

The length of A-T and G-C bonds are equal so that the cylindrical helix formed by double-stranded DNA has a uniform radius. Using the genetic code of *Escherichia coli* presented at the National Library of Medicine Web site www.ncbi.nlm.nih.gov/, estimate the number of AT and CG base pairs in these bacterial cells.

Ans: E. coli genome is 48 percent adenine and thymine. Guanine and cytosine base pairing makes up 52 percent of the genome. The weight percentages are as follows: 26.8 percent guanine, 26.3 percent cytosine, 23.8 percent adenine, and 23.1 percent thymine.

2.5 Consider the hypothetical case in which a nucleotide can appear only once in a DNA codon. Determine the number of nucleotides such codons must have in order to express the 20 amino acids found in proteins. If codons were made of four nucleotides, how many different codons would represent the 20 amino acids?

2.6 Determine the codons corresponding to each of the essential amino acids and for hydrophobic amino acids. What are the chemical properties of amino acids that have six codons each?

2.7 Eukaryota is the name of one of the three domains of life. It contains a wide range of organisms including human and yeast. The cells of such organisms all have a nucleus, the organelle that contains DNA. Recent genomic studies focused on three of the simpler members of this group: yeast, fruit fly, and nematode worm. These organisms are a lot easier to study than people: They breed faster, their genes are easy to manipulate, and compared with animals like mice, they are cheaper to grow in large quantities. The genes of these organisms are close enough to those of the human that they can help reveal new cures for genetic diseases. Sixty percent of the 289 known human disease genes have counterparts in the fruit fly. Yeast, the simplest of the three organisms, has taught scientists a lot about cell division and DNA repair, processes that go wrong in cancer. Scientists grow yeast cells with mutated genes and expose them to a whole range of chemical compounds used in cancer therapy to find which ones will kill them. The results give clues as to how these drugs work and how they can be improved. Using the National Library of Medicine's search engine Entrez, determine the scientific names of these organisms. How many DNA molecules do yeast cells have? Determine how many base pairs each fruit fly chromosome contains?

Ans: Baker's yeast = Saccharomyces cerevisiae, nematode worm = Caenorhabditis elegans, and fruit fly = Drosophila melanogaster.

2.8 The DNA sequences of two other eukaryotes is of considerable interest to biologists. The mouse is, from a genetic point of view, almost human. Multicellular organisms like the mouse and similar higher forms of life have been used as model systems to study diseases such as Alzheimer's or diabetes. Drugs that help these and other conditions often interfere with so-called signaling pathways within cells. In these pathways, many different messages, usually proteins, are used, and each protein is a potential drug target. Another vertebrate that has been extensively studied is the zebrafish. One reason zebrafish are used in genomic research is that they are transparent when young, which allows researchers to watch the organs grow and readily detect any changes. Conduct a literature search on the genomes of these two eukaryotes. How many chromosomes does each of these species have? Go to the KEGG Web site, and find an example of a human and a mouse chemical pathway.

2.9 Human chromosome 21 contains only 225 genes. Using the National Library of Medicine's Entrez search tool, find a research article that presents the sequence of this chromosome. How many base pairs does the chromosome contain?

2.10 Find an article that presents the sequence of chromosome 21 and that also discusses what is known about the genes of this chromosome. Discuss the relation between Alzheimer's disease and mutations of some of the known genes on chromosome 21. What is the biological cause of Down's syndrome, a disease associated with more than 80 physical and mental problems, including congenital heart disease, an increased risk for certain leukemias, and immunological deficiencies? Speculate why having a third copy of chromosome 21, or trisomy 21, is not fatal before or immediately after birth?

2.11 Conduct a literature search for regions of chromosome 21 that share a lot in common with several regions of mouse chromosomes. Discuss why a high degree of conservation between mouse and human can be of significant importance to medical researchers.

Ans: Various regions of the human chromosome 21 share common sequences with chromosomes 10, 16 and 17 of mouse genome.

2.12 Arrange the 20 amino acids found in proteins according to their degrees of degeneracy in the genetic code. The degree of degeneracy associated with an amino acid is equal to the number of codons that code for the amino acid.

2.13 Conduct a literature search on the DNA sequence comparison of red wolves in the Southern United States with the gray wolves in Canada. Restrict your search to the *Journal of Zoology*. Further restrict the search by listing Bradley White as author. What criteria do the authors of the study use to show that red wolves are more closely related to coyotes than to gray wolves?

2.14 RNA is widely hypothesized to be the first molecule to emerge in the primeval soup of chemicals that eventually lead to the formation of self-replicating chemical systems. Discuss the experimental observations that provide credibility to this hypothesis.

2.15 Determine the DNA coding sequences of all the different tRNA molecules that exist in *E. coli*. Use the Web site www.ncbi.nlm.gov/ for this purpose.

2.16 Collagen is the most abundant protein of vertebrates. It exists in all multicellular animals. Collagen is an extracellular protein that is organized into insoluble fibers of great tensile strength. These fibers constitute the major stress-bearing component of connective tissues such as bone, teeth, cartilage, tendon, ligament, and the fibrous matrices of skin and blood vessels. A single molecule of collagen is composed of three polypeptide chains. Mammals have at least 17 genetically distinct polypeptide chains comprising 10 collagen variants that occur in different tissues of the same individual. Among these variants, type I collagen is found in skin, bone, tendon, blood vessels, and cornea. It is composed of two α1 and one α2 polypeptide subunits. Using the Web site www.ncbi.nlm.gov/, determine the sequences of these subunits for the human as well as for the mouse. Specifically, conduct a search using the Entrez search engine, restricted to nucleotides. Compare the first 20 nucleotides of the collagen polypeptides of these species, and determine the fractions of identical nucleotides in these partial sequences. What are the two most abundant amino acids in collagen? Determine the secondary structures of the two polypeptides found in collagen.

2.17 Collagen is insoluble even in solvents that disrupt hydrogen bonding and ionic interactions. Collagen is both intramolecularly and intermolecularly covalently linked. Which amino acid residues of collagen participate in covalent bonding? Using the Web site www.ncbi.nlm.gov/, conduct a literature survey to determine those molecular properties of collagen that make the protein fibrous. Go to Entrez, then to PubMed, type the keywords "collagen" and "fibrous." Restrict the search using the "limits" option to those articles whose titles contain these keywords. Then begin the search. Hit the first article, then hit books; you will then have access to several molecular cell biology books.

2.18 Keratin is a fibrous structural protein found in fingernails, claws, skin, hair, wool, and feathers. This protein has predominantly α-helical secondary structure. As a result of this protein feature, hair strands can be stretched because this stretching requires that only hydrogen bonds in the α-helix be broken. After release of tension in the hair the α-helix reforms. Keratin molecules in fingernails constitute a strong protective shell. The strength of this shell is due to the disulfide bonds formed between various cysteine residues. Using the Web site www.ncbi.nlm.gov/, determine the amino-acid sequence of keratin and the possible sites for disulfide bonds.

2.19 Silk is a protein composed primarily of β-pleated sheets. In silk, the polypeptide chains are extended and lie next to one another, stabilized by hydrogen bonds between the elements of the peptide linkages. The β-pleated sheet may be found between separate polypeptide chains as in silk or between different regions of the same polypeptide that is bent back on itself. Conduct a literature search for the polypeptides that compose silk.

2.20 Elastin is a protein with rubberlike elastic properties. It consists predominantly of small and nonpolar amino acid residues. Elastin fibers can stretch to several times their normal length and return to original configuration upon release of tension. Elastin forms a three-dimensional network of fibers with no recognizable periodicity. The fibers appear to be devoid of regular secondary structure. Determine the amino-acid residue sequences of elastin. Cite examples of organisms that contain this structural protein.

2.21 Until recently, it was believed that if two proteins were to have as little as 30 percent of their amino-acid sequences in common, their structures would be very similar. The validity of this common belief was put to test by Lynne Regan and colleagues at Yale using two proteins, one consisting of a series of α-helices and the other in the form of β sheet. These authors were able to transform α-helices into β sheets by replacing 39 percent of the original amino acids with new ones. Present a summary of the results of their study published in the July 1997 issue of the journal *Nature Structural Biology*.

2.22 Phosphorylation is the chemical reaction in which a phosphate group is added to one of the three types of amino acids in a protein. These amino acids are serine, threonine, and tyrosine. The enzyme that facilitates the phosphorylation of tyrosine is called tyrosine kinase. The other two amino acids are phosphorylated with the help of serine/threonine kinases. It is generally thought that phosphorylation leads to either a three-dimensional change in protein structure or the creation of a new binding site. Thus, phosphorylation is a method of altering protein function after the protein is made. Conduct a literature search to determine whether the physical effect of tyrosine phosphorylation is different from that of serine or threonine.

2.23 Protein structure is an important indicator of its function, thus helping scientists to decipher the biochemical interactions that add up to life. Biological molecules such as proteins and DNA contain both hydrophilic and hydrophobic regions arranged in long chains. The three-dimensional structures of these molecules are dictated by the way these chains fold into more compact arrangements, so that hydrophilic groups are on the surface where they can interact with water, and hydrophobic groups are buried in the interior, away from water. As a result of the presence of hydrogen bonds, the protein surface is much less tightly packed than the protein interior. In enzymes, the hydrogen-bonded water molecules have difficulty fitting into the grooves on the protein surface and are easily displaced by ligands. Computer simulations indicate that the arrangement of water molecules in an empty active site mimics the geometry and structure of the actual ligand. Search for Web addresses that describe three-dimensional modeling of proteins.

2.24 *Comparative genomics* refers to large-scale surveys of the rapidly expanding number of genome sequences. *Proteomics*, on the other hand, is an emerging field that studies changes in the protein composition of cells. Most disease processes manifest themselves at the level of protein activity. It is estimated that only a small fraction of human disease is caused by a defect in a single gene. Most of the remainder involves significant changes in the overall protein composition of cells. A systematic evaluation of these changes has the potential to be of great use in pharmaceutical industry. Using the information available in the Internet, discuss these two emerging fields of biology.

2.25 RNA molecules also have primary as well as secondary structures. Conduct a literature search for the primary and secondary structure of rDNA of Steptococcus mutans.

REFERENCES

Agalarov SC, Prasad GS, Funke PM, Stout PM, and Williamson JR. "Structure of the S15, S6, S18-RNA complex." *Science* 2000, 288; 107–113.

Alberts, B et al., *Molecular Biology of the Cell*. New York: Garland Science-Taylor and Francis, 2002.

Bassett DE, Eisen MB, and Boguski MS. "Gene expression informatics—it's all in your mine." *Nature Genetics Supplement* 1999, 21: 51–55.

Doerfler W. "In search of more complex genetic codes—can linquistics be a guide?" *Medical Hypothesis* 1982, 9: 563–579.

Houdusse A, Kalabokis VN, Himmel D, Szent–Gyorgyi AG, and Cohen C. "Atomic structure of scallop myosin subfragment S1 complexed with MgADP: a novel conformation of the myosin head." *Cell* 1999, 97: 459–470.

Jang H, Hall CK, and Zhou Y. "Protein folding pathways and kinetics: molecular dynamics simulations of β-Strand motifs." *Biophys. J.* 2002, 83: 819–835.

KEGG Web site for compounds and ligands: http://www.genome.ad.jp/dbget/.

Lander ES et al., "Initial sequencing and analysis of the human genome." *Nature* 2001, 409: 860–928.

Maslov S and Sneppen K. "Specificity and stability in topology of protein networks." *Science* 2002, 296: 910–913.

National Library of Medicine Web site: http://www.ncbi.nlm.nih.gov/structure/.

Purves WK, Orians GH, Heller HC, and Sadava D. *Life: The Science of Biology*, 5th ed. New York: W. H. Freeman and Company, 1998.

Shi H and Moore PB. "The crystal structure of yeast phenylalanine tRNA at 1.93 A resolution." *RNA* 2000, 6: 1051–1105.

Stryer and Lubert. *Biochemistry*, 3d ed. New York: W. H. Freeman and Company, 1988.

Teichmann SA, Chothia C, and Gerstein M. "Advances in structural genomics." *Current Opinion in Structural Biology* 1999, 9: 390–399.

Venter CJ. "The sequence of the human genome." *Science* 2001, 291: 1304–1351.

Voet D and Voet JG. *Biochemistry*. New York: John Wiley and Sons Co, 1990.

Cells and Their Housekeeping Functions

3.1 | Introduction

A cell is the basic unit of life in all organisms and is typically so small that it cannot be seen with the naked eye. Although most of the 30 million species that exist on Earth are unicellular, the visible life forms, animals and plants, contain billions of cells. In order to survive, several processes must be carried out by all cells, including the acquisition and assimilation of nutrients, the elimination of waste, the synthesis of new cellular material, movement, and replication. All cells have two common properties: (1) Each cell possesses the entire genetic information of the parent organism. This information, stored in DNA, is passed on to the daughter cells during cell division. (2) Each cell has a membrane or cell wall that separates its interior from the surrounding environment, enabling it to maintain a distinct chemical identity. The membrane is permeable to carbon dioxide, oxygen, and water and provides controlled access to sugars, amino acids, and ions.

In multicellular organisms, specialized cells carry out additional functions that are specific for each type of differentiated cell. For example, in addition to the "housekeeping" processes common to all cells, retinal cells are able to perform the chemical process that converts light into a signal that can be integrated by the nervous system. Other examples of "differentiated" function include the mechanical properties of muscle, bone, and skin cells; the ability of glands to secrete specific molecules; and the oxygen-carrying capability of red blood cells. In a human body, there are billions of cells, but nevertheless they can be classified into about 300 distinct cell types (muscle cells, retinal cells, and so on). All these cell types have identical genomes, but they differ from each other in the sets of genes they express. Cell differentiation (formation of cell types in a multicellular organism) is addressed in Chapter 8. In this chapter, we discuss the general (housekeeping) aspects of cell structure and function.

3.2 | Prokaryotes and Eukaryotes

Cells are grouped into two categories depending on the organization of the genetic material within them. The first category is composed of *prokaryotes*, unicellular organisms that account for much of the living biomass on Earth (Fig. 3.1). A prokaryotic cell has a single compartment and, by definition, no nucleus. On the other hand, in eukaryotes, cells are compartmentalized, and the DNA containing the genetic heritage of an organism is stored in a nucleus. Prokaryotes belong to one of two domains of life: bacteria and archaea. The third domain, eukaryota, consists of organisms with nuclei. The classification of life forms into these three distinct domains is not based on the details of the cell structure or the structure and content of the genome of an organism, but rather is derived from the assumption that all cells (organisms) descended from a "common ancestor cell." A tree of life with three distinct branches (domains) is constructed based on the similarity (evolution) of an rRNA molecule found in all organisms.

The hereditary information in prokaryotic cells is stored in a large circular DNA molecule that is compacted in an area of cytoplasm called the *nucleoid* region, but it is not enclosed within a separate nuclear envelope. The DNA is attached at one point to the cell membrane. During cell division, the DNA is replicated and one copy allocated to each daughter cell. Because reproduction in prokaryotes does not usually involve fusion with another individual and mixing of the genomes (sexual reproduction), prokaryotes are termed asexual. As a consequence, daughter cells are essentially clones and nearly always bear the exact genetic

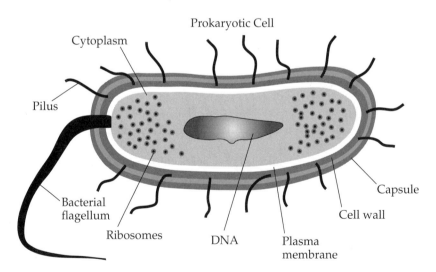

FIGURE 3.1   Schematic of a prokaryotic cell. Prokaryotes are unicellular organisms found in all environments including hot sulfur springs, on the ocean floor, and living within larger organisms. A prokaryotic cell is encased by a cell wall made up of polysaccharides. The DNA is dispersed in the cytoplasm and the cell contains no membrane-bound organelles. Many bacteria are capable of movement in a fluidlike environment using flagella (long, flexible, spiral-shaped structures). Flagella act also as sensors, aiding in the detection of concentration gradients and other signals.

characteristics of the parent cell. However, genetic variations do occur in prokaryotes. In a process known as recombination, prokaryotic cells conjugate and exchange genetic material. Another cause for genetic variation is mutation due to exposure to radiation or chemicals.

Many prokaryotes have considerable mobility in aqueous environments because of the action of a flagellum, a threadlike protrusion attached to the cell membrane, which either undulates like a whip or rotates like a propeller. The average diameter of a prokaryotic cell is about 2 μm, and its volume probably represents the lower limit that can accommodate the complex machinery of life. This constraint on the size of a bacterium illustrates the concept that physical laws can place limitations on the biological characteristics of organisms. In this case, the universal need to emit wastes, import nutrients, as well as the physical characteristics of diffusion give rise to the need to maintain a sufficient surface area/volume ratio. These physical constraints limit the size of bacteria, which must utilize simple diffusion as an adequate means for distributing nutrients throughout the cell.

Eukaryotes constitute the second category of living systems on Earth and can be defined as cells with nuclei. Eukaryotes range from unicellular yeast to plants and animals, which contain billions of cells. In contrast to prokaryotic cells, eukaryotes exhibit a large degree of compartmentalization. The genome and replication machinery is contained in a nucleus, protein synthesis is carried out in association with a system of internal membranes, and energy production is performed by membrane-bound structures called mitochondria.

The structural organization of a typical animal cell is shown in Fig. 3.2. The cells of *eukaryotic* organisms have compartments called *organelles*. Each organelle type has a distinct function and contains a collection of specific enzymes that catalyze the requisite chemical reactions. Because of the presence of these and other membrane-bound structures, the volume of a eukaryotic cell is several hundred times that of a bacterial cell. In bacterial cells, simple diffusion of molecules from one part of the cell to another can take place. However, as cell size increases, the ratio of cell surface area to cell volume decreases. This presents a disadvantage for bigger cells because exchange of raw materials and products across the plasma membrane as well as across the cell is less efficient. The presence of internal membranes increases the surface area/volume ratio many fold over that of prokaryotes and allows eukaryotic cells to be much larger than bacteria. In addition, eukaryotic cells have acquired many more active transport systems in the plasma membrane and within the cell.

Animal and plant cells are quite similar in structure, when they are studied with an electron microscope, but there are several significant differences between them. Plant cells have chloroplasts that convert the energy of light into chemical energy. Like bacteria, plant cells have a cell wall outside the plasma membrane that restricts shape change and mobility. Moreover, plant cells have a vacuole for collecting and storing nutrient molecules and waste products.

Because plants release oxygen during photosynthesis, they increase the oxygen content of the atmosphere, enabling the evolution of oxygen-breathing animals. Therefore, plants may have represented the earliest form of multicellular life. In the following, we will focus on common structures of animal and plant cells and continue the discussion on the organization of the DNA in the nucleus of a eukaryotic cell.

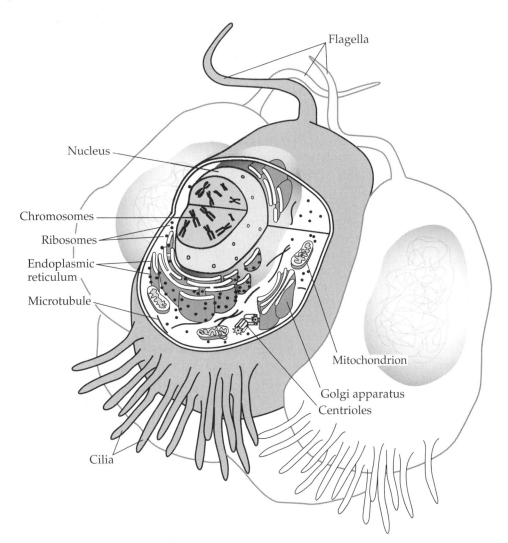

FIGURE 3.2   Organelles of an animal cell. The nucleus is the largest organelle in the cell and contains chromosomes, the storage sites of the genetic information database. Mitochondria are the powerplants of the cell where chemical energy is transformed into the high-energy bonds of ATP. Centrioles are involved in nuclear division during cell division and are sites of microtubule attachment. Ribosomes, the endoplasmic reticulum, and the Golgi apparatus work in synchrony in the synthesis, processing, and packaging of proteins. The chemical environment in the cell is distinct from the extracellular environment. (Modified from http://www.nhgri.nih.gov/DIR/VIP/Glossary/Illustration/cell.html.)

3.3 | Nucleus

The largest organelle within the cell is the *nucleus*, which contains DNA, the machinery for DNA replication and transcription and serves as the library of genetic information (Fig. 3.3). DNA replication is necessary for cell division so that both daughter cells have identical copies of DNA. Transcription, the formation of an RNA template of a gene, is essential for protein synthesis. The nucleus of a human cell is typically spherical, about 10 μm in diameter and is bounded by a *nuclear membrane*

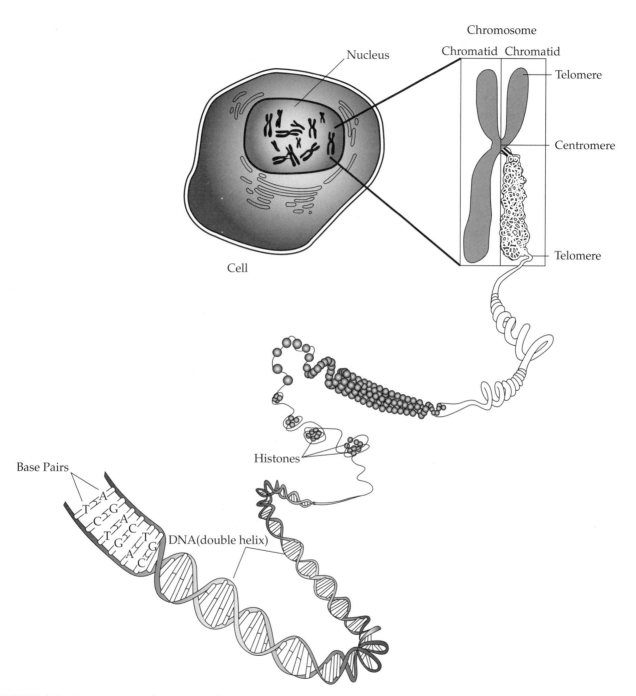

FIGURE 3.3   Organization of DNA into chromosomes in eukaryotic cells. Chromosomes are threadlike "packages" of separate DNA molecules in the nucleus of a cell. Different organisms have different numbers of chromosomes. Humans have 23 pairs of chromosomes, 46 in all: 44 autosomes and two sex chromosomes. Each parent contributes one chromosome to each pair, so a child gets half of its chromosomes from its mother and half from its father. In chromosomes, each DNA molecule is associated with protein molecules including a family of proteins called histones. Before cell division, DNA is replicated and then tightly bound in identical pairs called chromatids. Chromosomes are more loosely packaged in nondividing cells. (Modified from http://www.nhgri.nih.gov/DIR/VIP/Glossary/Illustration/chromosome.html.)

composed of two layers with many openings or pores for the traffic of small molecules and proteins. The two lipid bilayers of the double membrane are separated by a gap of 20 to 40 nm. Each cell of an animal or plant (as well as fungi) contains identical sets of chromosomes, long and thin DNA molecules that exist in association with a large number of proteins. Chromosomes store the genetic information of the organism. DNA molecules are packed around and attached to beadlike protein structures (nucleosomes) formed by the *histone* family of proteins. Just before a cell divides, its DNA becomes so tightly coiled that the resulting chromosomes can be seen using light microscopy. In that state, chromosomes are composed of two identical copies of DNA and associated proteins (*chromatins*). Organization of the nucleus with regard to DNA replication is discussed in Chapter 7. In the following, we focus on the DNA transcription machinery.

The first steps in protein synthesis begin in the nucleus with the binding of transcription factors to the regulatory sequence associated with the protein-coding gene. DNA protein interactions initiate a complex set of events that culminate in the production of mRNA. The key machinery involved in transcription is the protein–enzyme complex RNA polymerase II (Fig. 3.4).

Transcription requires the separation of the two DNA strands and the subsequent synthesis of RNA with the gene-coding region of a DNA strand used as the

Structure and Function of RNA Polymerase II

FIGURE 3.4 Gene transcription by RNA polymerase II (pol II). Transcription refers to the formation of a messenger RNA under the control of a specific segment of DNA. Transcription requires the enzyme RNA polymerase II, which separates the two strands of the double helix of DNA. One of the strands serves as a template for RNA. Transcription requires the appropriate ribonucleoside triphosphates (ATP, GTP, CTP, and UTP) and is powered by the detachment of a phosphate group from a triphosphate as a nucleotide becomes incorporated into the backbone of RNA. The enzyme binds to the promoter sequence on the DNA and begins the synthesis at a start site. It completes the synthesis at the termination site. See text for details.

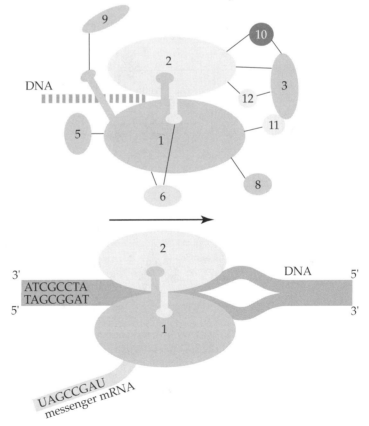

template. Molecular imaging using X-ray diffraction has revealed structural information about the components of the transcription machinery. RNA polymerase II (pol II) is the central machine for synthesis of all messenger RNA in eukaryotes (Fig. 3.4).

RNA polymerase transcribes RNA from a DNA template at an average rate of 30 nucleotides per second. The enzyme is composed of 12 different polypeptides with a total mass of about 500 kDa. These polypeptides are arranged in 10 subunits. The enzyme is highly conserved in evolution; at least 10 of the 12 mammalian pol II genes can be substituted for their counterpart in yeast. The multiunit enzymes that mediate the transcription of ribosomal and transfer RNA are called polymerase I and III, respectively. Polymerases I, II, and III are quite similar in structure; nine out of 10 subunits are identical in the three enzyme machines. Prokaryotes use the same enzyme to generate transcripts of protein-coding and RNA-coding genes. Transcription is an energy-consuming process and is driven by the energy released by ATP hydrolysis. When activated, pol II can unwind the DNA double helix, polymerize RNA, and proofread the resulting transcript. For promoter recognition and response to regulatory signals, pol II forms large assemblies containing as many as 60 proteins. In fact, eukaryotic pol II can only recognize the promoter region of genes, if the DNA is interacting with certain other proteins, called transcription factors. A pore in the central region of the enzyme may allow entry of substrates (RNA nucleotides) for polymerization and exit of the primary transcript. The pol II machinery includes a pair of jaws formed by subunits 1, 5, and 9 to grip DNA downstream of the active center. A clamp on DNA near the active center is formed by subunits 1, 2, and 6. This clamp may be locked into the tight position by the growing strand of mRNA, thus stabilizing the union of transcribing complexes. A similar enzyme serves the same purpose in prokaryotes. The regulation of gene expression and the DNA-binding proteins (transcription factors) involved in regulation are considered in more depth in Chapter 4.

The length of the complete RNA molecule produced by pol II from a single transcription site (primary RNA) can be as long as 20,000 nucleotides. This is much longer than the 1200 nucleotides of RNA needed to code for an average protein of 400 amino acids. The longer transcript is due to the presence of noncoding intron sequences in eukaryotic genes. The primary RNA transcript (pre mRNA) is transformed into mRNA with the excision of introns, an event called RNA splicing. The multimolecular machinery that recognizes exon/intron interface sequences of nucleotides and mediates excision is called a *spliceosome*. The excised intron transcripts are rapidly degraded in the nucleus, thus providing raw material for new transcripts. The spliceosome is also responsible for annealing the ends of exons. The resulting mRNA is transported out of the nucleus and interacts with ribosomes in the cytoplasm to begin the process of translation.

3.4 Organelles for Protein Synthesis and Transport

3.4.1 Ribosomes

The molecular machines that translate the sequence information on mRNA into actual polypeptides are called ribosomes. Ribosomes are small, darkly staining spherical structures, about 20 nm in diameter that are made up of 50 proteins and several long RNAs intricately bound together (Fig. 3.5). Ribosomes have no membrane and disassemble into two subunits when not actively synthesizing protein. Protein synthesis (translation) begins with the attachment of mRNA to a ribosome. The ribosome than begins knitting together amino acids, according to the template encoded

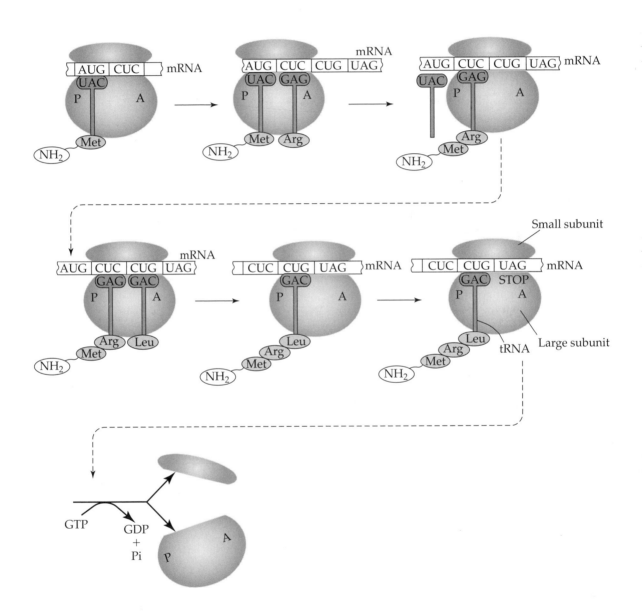

FIGURE 3.5   The translation of mRNA by the ribosome. Translation is inititated by tRNA molecules, which act as adapters. These are small RNA molecules that match each codon on the mRNA with the amino acid the codon prescribes. There are 40 different types of tRNA in a prokaryotic cell. There is at least one specific tRNA for each amino acid. Ribosomes, small beadlike structures, keep mRNA and tRNA in appropriate positions for the translation to occur.

in mRNA. Amino-acid-specific adapter molecules (tRNAs) match the code embedded in the mRNA with the corresponding amino acid. A ribosome has binding sites for two different tRNA molecules so that two amino acids can be joined to the growing polypeptide chain at one time. Because protein synthesis is extremely important, eukaryotic cells contain millions of ribosomes. A ribosome takes about 30 sec to synthesize a protein containing 400 amino acids, and a human cell might synthesize as many as 10^{10} proteins in 24 hours.

3.4.2 Endoplasmic Reticulum and Golgi Apparatus

Ribosomes are synthesized in the nucleolus, a compartment within the nucleus. Once constructed, ribosomes leave the nucleus through nuclear pores. Ribosomes that float freely in the cytoplasm to synthesize cytoplasmic proteins, which require no further modification. Ribosomes that manufacture membrane proteins or proteins that will be secreted or transported to other membranous organelles associate with a network of membranes called the *endoplasmic reticulum (ER)*. The ER is involved in several important cellular functions. When associated with ribosomes, the ER is responsible for protein synthesis, folding, and some posttranslational modifications such as addition of carbohydrates (glycosylation). The ER is also the site of lipid synthesis in the cell and is responsible for the production of most of the cell's membranes. In addition, the ER houses detoxifying enzymes (particularly important in the liver) and is responsible for the early stages of synthesis of steroid hormones such as testosterone. Cells that make large amounts of steroids for export to the rest of the body, such as those in the testes, ovaries, and adrenal glands, contain large amounts of SER (smooth endoplasmic reticulum).

The membranous system of the ER is continuous with the outer nuclear membrane (Fig. 3.6a), and its large surface area provides an appropriate template for enzyme-mediated chemical reactions. ER can be classified as smooth (SER) or rough (RER) according to its morphology. The membrane structures of both are identical, but rough ER (RER) has ribosomes bound to it; therefore, its surface looks rough when viewed with the electron microscope. SER is not involved in protein synthesis, but is the site of steroid and lipid synthesis and the formation of new membranes. It contains enzymes, which detoxify harmful organic molecules, and it acts as a storage site for calcium in skeletal cells. Unsaturated fatty acids and cholesterol are synthesized in SER. Membrane-bound polypeptides assume their three-dimensional shapes after being threaded through the membrane of the rough endoplasmic reticulum.

Following synthesis, small vesicles transport the proteins from the ER to another membranous compartment, the Golgi apparatus (Fig. 3.6b), which processes them further before transporting them to their ultimate destination in small vesicles. The information that dictates whether a polypeptide is synthesized on free or ER-bound ribosomes (whether it is secreted or remains in the membrane) is contained in the primary sequence of the gene and does not depend on the protein-synthesizing apparatus. Proteins are targeted to their specific cellular destinations by address labels. In the case of targeting to RER, the signal sequence consists of about 20 hydrophobic amino acids at the N terminal end of the protein. In the case of nuclear proteins synthesized on free ribosomes, the tag that directs them to the nucleus consists of a stretch or patch of basic amino acids. Transmembrane proteins that span

FIGURE 3.6 Endoplasmic reticulum (a) (modified from http://step.sdsc.edu/personal/vanderschaegen/cellorganelles/golgi.html) and the Golgi apparatus (b) (modified from http://gened.emc.maricopa.edu/bio/bio181/BIOBK/BioBookCELL2.html). The endoplasmic reticulum is a network of interconnected tubules, vesicles, and sacs. Ribosomes that are actively synthesizing proteins in need of further modification bind to the outer surface of the rough endoplasmic reticulum (RER). The newly synthesized proteins then pass through the ER membrane where they fold into their three-dimensional shape. The Golgi complex modifies the folded proteins by tagging them with polysaccharide chains and transports them to their destination in small vesicles. Glycolipids are processed in the smooth ER and subsequently in the Golgi complex.

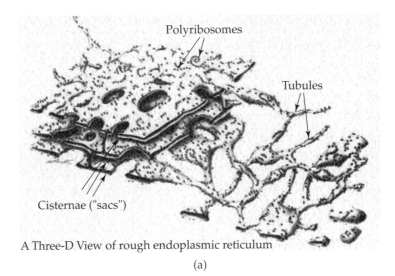

A Three-D View of rough endoplasmic reticulum

(a)

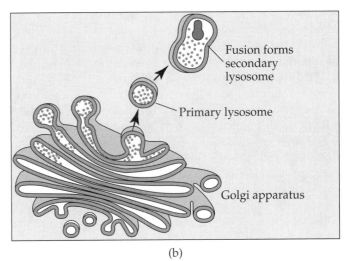

(b)

the membrane several times often have multiple signal sequences. The amino-acid sequence motifs for protein-sorting signals are well established. For example, the KEGG database http://www.genome.ad.jp/dbget/ offers PSORT search engine for prediction of protein-sorting signals, TSEG for prediction of membrane proteins, and SOSUI for prediction of transmembrane segments.

Ribosomes are generic machines in the sense that a ribosome may be involved in the translation of any mRNA. A ribosome that attaches to an mRNA with a signal-sequence coding sequence is directed to the outer surface of RER. As a protein is synthesized on an ER-bound ribosome, it passes across the ER membrane, where the protein is processed by enzymes and folded into its correct three-dimensional conformation. The mechanism whereby newly synthesized proteins pass through the ER membrane has been the subject of great interest and is quite well understood. Dr. Gunther Blobel of Rockefeller University was awarded the Nobel Prize in 1999 for his contribution to this field.

For the most part, proteins are inserted into the lumen of the ER as they are being synthesized rather than as fully made polypeptides. The simplest analogy for this situation is one of threading a needle. Trying to force a whole ball of thread through the eye of the needle is impossible, whereas it becomes a simple matter if the end of the thread is identified and the rest of the ball passed through the eye a little bit at a time. In order for a protein to be passed into the lumen of the ER, the end of the polypeptide must be identified by a *signal-recognition particle* (a receptor) and machinery must exist to pull the thread into the lumen (ATP and GTP-driven motors). In the specific instance of protein translocation across the ER membrane, a pore in the hydrophobic lipid membrane presents a hydrophilic interior to accommodate the newly synthesized protein. Proteins on the outside of the pore recognize the polypeptide and the ribosome. A ratchet-like mechanism is utilized by the ER to simultaneously bind and pull the polypeptide into the lumen. This ratchet machinery utilizes energy that derives from the hydrolysis of GTP and ATP.

As the translocated protein enters the ER lumen, the signal peptide is cleaved by an enzyme called signal peptidase. The emerging protein is then sequestered by *heat shock protein hsp70* and assisted with proper folding. This chaperone function is important to prevent aggregation in the ER lumen, which might occur because of the high concentration of unfolded proteins. An important rate-limiting step in protein folding in ER is the formation of correct disulfide bonds. The most extensive post-translation modification that occurs in secretory proteins is glycosylation, the addition of certain carbohydrate motifs. Glycosylation starts in the ER with further processing in the Golgi apparatus.

Once synthesized, both lumenal and transmembrane proteins are concentrated in lipid vesicles that bud from the endoplasmic reticulum and pass to the *Golgi apparatus*. The Golgi complex modifies proteins by adding to and modifying their carbohydrate chains to form mature glycoproteins. This organelle is composed of a stack of flattened membrane sacks. The Golgi vesicles pinch off from the edges of the flattened sacks to transport the modified proteins to their destinations in the cell and to its exterior. The mechanism of concentration, vesicle formation and subsequent fusion with the Golgi is an area of intense focus for cell biologists. The mechanical forces required to pinch off a vesicle are provided by special proteins that coat the vesicle as it is being formed (coatomer proteins or COPs) and the binding of GTP to a small G-protein. Once released from the ER, the vesicle must be protected from inappropriate fusion with other random membranes and allowed to fuse with the Golgi.

This is accomplished by covering the naked lipid membrane with coat proteins and by the selective recruitment of some proteins that target the vesicle to the correct Golgi stack and others that provide a protein scaffold that catalyzes the fusion of vesicle and Golgi membranes. This latter process requires the simultaneous uncoating and fusion of the vesicle and requires hydrolysis of GTP and ATP. A similar process mediates vesicular transport between the different Golgi stacks and also between the Golgi and the plasma membrane. Proteins that normally reside in the ER or Golgi must either be selectively retained during vesicle formation or retrieved from distal compartments.

Evidence for both mechanisms is quite strong. For example, ER proteins possess a short amino-acid sequence that marks them for retention in the ER. On the other hand, a class of membrane vesicle that is specifically involved in retrieval of ER proteins from the Golgi apparatus has been isolated. These vesicles have a different coat protein on their surface. An alternative mechanism to explain intra-Golgi transport is the cisternal maturation model.

In this model, rather than vesicular transport being the major mode of transport of cargo from cis to trans Golgi cisternae, the membranous stacks themselves continuously move upward from cis to trans faces of the Golgi, changing their molecular makeup and carrying cargo with them. Proponents of this model hypothesize that vesicles are used only in retrograde transport, to retrieve ER components that pass into the Golgi.

As is the case with many biological theories, recent studies reveal that both the vesicular transport theory and the cisternal maturation models are correct (Pelham and Rothman, *Cell* 2000, 102: 713). There is no doubt that cisternal maturation occurs, but the rate at which this takes place is too slow to account for the speed at which many proteins pass through the Golgi stack. However, not all proteins pass rapidly through the Golgi. For example, large aggregates of extracellular matrix proteins such as pro-collagen require a lot of posttranslational modification. These aggregates pass through the Golgi without entering vesicles and their rate of passage coincides with the rate of cisternal maturation. Other proteins are concentrated in vesicles and pass rapidly from one stack to another. However, rather than being specifically targeted to particular cisternae, these vesicles "percolate" both up and down the maturing stack, the final cis–trans directionality of cargo transport being dependent on a sort of iterative progression.

At the same time as this is occurring, small vesicles that only contain ER proteins with the ER-retrieval label are passing from the Golgi to the ER. Just as slow moving freight trains and trucks are used to ferry large unwieldy cargos from one destination to another, rapidly moving cars transport selected cargo (passengers). A picture now emerges in which a rapid vesicular transport system is superimposed on a more slowly maturing cisternal system.

3.5 | Cytoskeleton

Organelles of eukaryotic cells are not freely suspended in the cell cytoplasm, but are anchored to the *cytoskeleton*. The cytoplasm is an aqueous medium containing many enzymes and other compounds needed by the cell. The cytoplasm also contains a dynamic *cytoskeletal* network consisting of microfilaments, intermediate filaments, and microtubules. These cytoskeletal fibers give the cell physical strength and rigidity and hold intracellular structures in place. The cytoskeleton also facilitates and controls movement within the cell and locomotion of the cell. In this section, we will briefly review the cytoskeleton of an animal cell. Cytoskeletal networks composed of microtubules play fundamental roles in cell division and are also discussed in Chapter 7. See assignments for further information on intermediate filaments.

3.5.1 Actin Microfilaments

The actin cytoskeleton is required for cell motility as well as phagocytosis (eating of microbes by specialized cells called macrophages). The interaction of actin with the molecular motor myosin results in contractile alterations of the actin cytoskeleton and enables cells to move and change shape. (See Section 3.7.) The actin cytoskeleton is also associated with adhesion of cells to other cells and to the extracellular matrix. Signaling receptors on cell surfaces are anchored to the actin cytoskeleton either directly or through adapter proteins. This aspect is discussed in more detail in Chapter 7. Actin microfilaments are 5 nm in diameter and are made of polymers of the protein actin. Actin is a highly conserved ancient protein of eukaryotes and is the most abundant protein in the cytoplasm of mammalian cells, accounting for up to 20 percent of the total cytoplasmic protein. Actin is found in the cytoplasm either as a globular monomer (called G actin) or as a filament (designated F-actin). F-actin is formed by the head-to-tail polymerization of the asymmetric G-actin. The resulting actin fibers create an intricate cytoskeletal network that is dynamic (continuously remodeling) and also contractile due to the association of F-actin with myosin, a protein that acts as a motility engine (Figs. 3.7–3.8).

Actin-filament formation begins when three actin monomers combine simultaneously to form an unstable trimeric nucleus. This nucleus may dissociate back into monomers rapidly or may survive long enough to permit subsequent binding of additional actin molecules. Nucleation of actin filaments is regulated in the cells by other proteins such as the Arp2/3 complex. Monomeric actin molecules (each containing ATP) add serially to both ends of the elongating filament. Monomer addition occurs more rapidly at the (+) barbed than at the (−) pointed end. After monomer incorporation into growing filaments, ATP-actin undergoes nucleotide hydrolysis to form ADP-actin subunits within the flament's helical lattice. ADP-actin is much more likely to dissociate than ATP-actin, and once active filaments reach a steady-state length, ADP-actin monomers are released at the (−) end at about the same rate as new ATP-actin monomers are added to the (+) end. This process is called treadmilling. Actin filaments can assemble and disassemble very

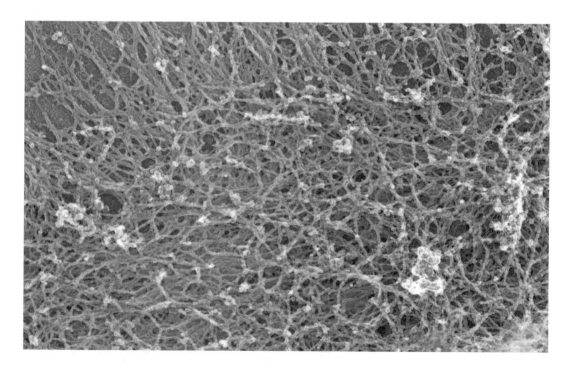

FIGURE 3.7 This transmission electron micrograph in the figure (by Dr. John Hartwig) shows the crosslinked actin cytoskeleton of a spreading macrophage lamellipodium.

FIGURE 3.8 Actin can polymerize from both ends in vitro, but the rate of polymerization is much higher at the so-called (+) barbed end than the (−) pointed end (a). Monomers add mainly at the (+) end and dissociate from actin filaments at the (−) end. Thus, actin filaments are polarized. Actin filament polarity is defined by the angle at which myosin proteins bind to actin filaments. When myosin head subunits are allowed to bind to F-actin they decorate the filament, which appears like the flight of an arrow, with a barbed (+) end and a pointed (−) end.

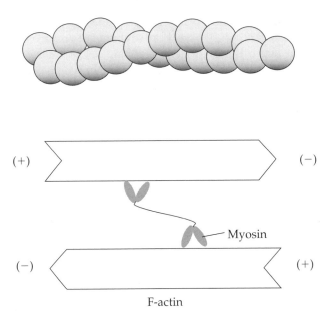

rapidly, processes that are highly regulated in cells by a number of actin-binding proteins. One such actin-binding proteins (called α-actinin) binds to the sides of actin filaments, linking them into bundles. Such crosslinking creates rigid structures needed for the cell to resist physical forces exerted by the environment (such as the fluid force exerted on endothelial cells covering the lumen of blood vessels). Some actin-binding proteins "cap" the (+) ends to prevent depolymerization, and others act to sequester G-actin to prevent it from polymerizing.

Fibrous actin is found in muscle cells in association with the regulatory proteins tropomyosin and troponin (Fig. 3.9). The so-called thin filaments of muscle cells consist of two actin strands twisted into a double helix together with troponin and tropomyosin. The thin filaments hardly change length during muscle contraction (or muscle stretching), but are arranged in such a way that sliding past each other alters the length of muscle cells. This aspect is discussed in more detail in Section 3.7.

Cells suspended in a fluid are usually spherical in shape, suggesting the existence of a uniform surface tension on the periphery of the suspended cells. The cells

FIGURE 3.9 Schematic drawing of an actin filament (thin filament) found in muscle cells. The filament is in association with regulatory proteins troponin and tropomyosin.

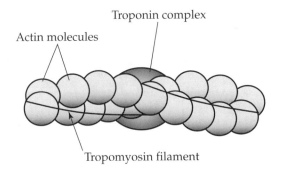

that compose tissues deviate from a spherical shape, because the forces of cell–cell and cell substratum adhesion pull these cells in different directions. When actin filaments supporting the cell membrane are disrupted by agents such as cytochalasin B, the cultured cells adherent to a substrate lose their shape and resemble a plastic bag full of organelles and fibers.

3.5.2 Microtubules and Tubulin

Microtubules comprise another important component of the cell cytoskeleton. Important functions of microtubules include the transport of vesicles and other organelles such as those that pass through the Golgi apparatus during protein sorting (Section 3.4). Microtubules are also components of cell appendages such as flagella that are used to propel spermatozoa. In this case, the microtubules are in a relatively stable configuration, the beating motion of the flagellum depending upon the sliding of microtubules along one another. In other situations, such as the mitotic spindle so important during cell division, microtubules are remarkably dynamic and grow and shrink at the same time as they move chromosomes and segregate daughter cells. (See Chapter 7.) In all of these situations, microtubules associate with other proteins such as the ATP-hydrolyzing motor proteins kinesin and dynein discussed later in this chapter.

Microtubules are hollow tubes 20 nm in diameter composed of polymers of α- and β-tubulin (Fig. 3.10). Each microtubule consists of 13 strands of polymerized tubulin dimers called protofilaments in which α- and β-tubulins interact head to tail. The protofilaments first organize as a sheet, which then folds to form the hollow microtubule. Microtubules are nucleated and anchored in a region of the cell called the microtubule-organizing center (MTOC), alternatively known as the centrosome. Like actin filaments, microtubules are inherently polarized with a + end that grows from the MTOC into the cytoplasm. Also like actin filaments, microtubules grow by the reversible addition of subunits catalyzed by nucleotide hydrolysis. In the case of microtubules, the nucleotide is GTP. Remarkably, the + end of microtubules can switch from a rapidly growing state to a rapidly shrinking state, a phenomenon known as dynamic instability. During polymerization, a tubulin dimer in which β-tubulin is bound to GTP assembles with other dimers. GTP hydrolysis only occurs *after* the subunits have polymerized and a "cap" of GTP-bound subunits is found at the + end of growing microtubules; the size of the cap then depends upon the rate of new subunit addition. Consequently, if the rate at which GTP-bound tubulin is formed or added to the growing microtubule, is reduced, the size of the GTP cap diminishes and the GDP-bound tubulin is exposed. Because GDP-bound subunits dissociate much more readily than GTP-bound subunits, the microtubule depolymerizes rapidly, a process called a microtubule catastrophe. In the presence of the appropriate enzymes, the now free, GDP-bound subunits are converted to GTP-bound subunits, which can then be added to the same microtubule or to neighboring ones. Because of this dynamic instability, it is very important that a cell is able to "capture" the + ends of the microtubules in order to generate more stable structures. The identification of molecules that act to capture microtubule ends is a very active area in modern cell biology, with implications in cell division, cancer, neurobiology, and cell motility.

FIGURE 3.10
Microtubules are hollow
cylindrical fibers that are
composed of 13 parallel, but
staggered, protofilaments
containing alternating α- and
β-tubulin subunits (a).
Microtubules are the building blocks of spindle structure during cell division (b).
The newly divided chromo-
somes move toward oppos-
ing poles along microtubular
tracks prior to cell division.
(Modified from Rieder and
Salmon, *Trends in Cell
Biology* 1998, 8: 310.)

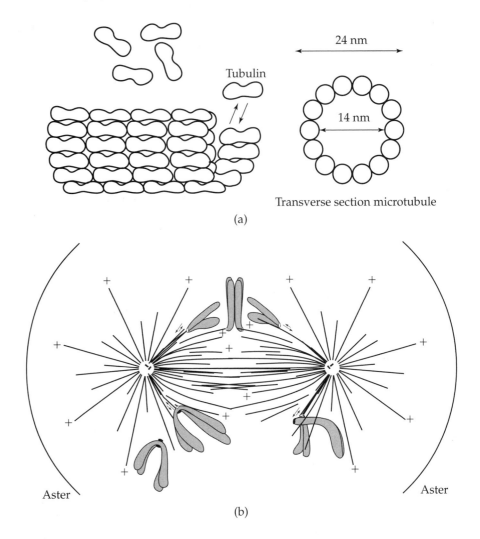

3.6 | Metabolic Processes and Mitochondria

Cell metabolism is defined as the sum of all the chemical reactions in the living cell that are used for the production of useful energy and the subsequent synthesis of cell constituents. Almost all cellular reactions are catalyzed by enzymes, which are capable of speeding reaction rates by a factor of 10^2 to 10^6. Living cells steadily remodel and replace their structures. The process of building new molecules as building blocks is called anabolism. Structures that are worn out or no longer needed are broken down into smaller molecules and either reused or excreted; this process is called catabolism. Great quantities of energy are required not only to produce the work needed for the pumping of the heart, for muscular contraction, and for nerve conduction, but also to provide the chemical work needed to make the large molecules characteristic of living cells. Anabolism and catabolism are aspects of overall metabolism, and they occur interdependently and continuously. For these processes to occur, cells extract energy from their environment. Animal cells derive energy from the fats, carbohydrates, and proteins they consume as food. Part of the energy obtained from chemical degradation of the foodstuff is used to synthesize the building blocks of macromolecules. Extraction of energy and synthesis of building blocks are mediated by a highly integrated network of enzymatic reactions. Thousands of metabolic reactions take place even in the simplest bacterial cells.

The chemical energy used in order to derive the processes vital for life is derived originally from the sun. The photosynthetic activities of plants and some bacteria transform the energy stored in the photons into the chemical energy of carbon-containing organic compounds such as starches and sugars. In contrast, most other organisms (animals, fungi, and most prokaryotes) depend on a ready supply of carbon-containing compounds as their food source.

The principal stages in the generation of useful energy from food are shown in Fig. 3.11. Briefly, the degradation of food begins in the digestive tract where large molecules are split into smaller units. In this *first stage*, proteins are hydrolyzed to their 20 constituent amino acids, carbohydrates are hydrolyzed to simple sugars like glucose, and fats are hydrolyzed to glycerol and fatty acids. No useful energy is produced in this step. Cells internalize these smaller units and degrade them further to extract useful energy. Glucose is the preferred choice for cells to utilize as raw energy-containing material.

The intake of food by cells from the bloodstream is tightly controlled. Messenger substances (hormones) released from endocrine glands into the blood stream are carried throughout the body and affect the metabolism of those cells that have receptors for that hormone.

Hormones can alter permeability of the cell membrane to extracellular substances such as glucose or alter the activity of key intracellular enzymes (also called

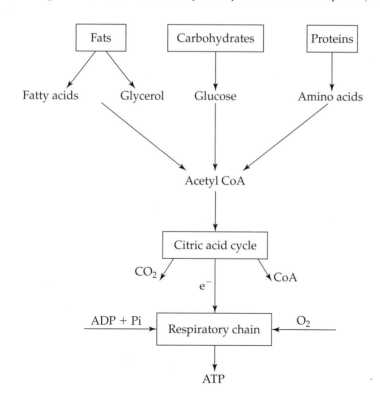

FIGURE 3.11 Stages in the extraction of energy from food. In the first stage, proteins are degraded to amino acids, carbohydrates to glucose and other sugars, and fats to fatty acids and glycerol. In the second stage, these smaller units are converted into the acetyl unit of acetyl CoA. The third stage consists of the citric-acid cycle and the respiration cycle.

pacemaker enzymes) that control major chemical pathways. Insulin, for example, increases glucose uptake by muscle cells and increases the storage of glycogen. Insulin deficiency (the basic disturbance in type 1 diabetes) depresses glucose uptake by certain cells and increases glycogen breakdown, thereby causing abnormally high levels of glucose in the blood. High sugar levels in blood increase its osmotic pressure, remove tissue water, and cause cellular dehydration and electrolyte loss. In the face of glucose starvation, cells break down structural lipids and proteins. This is why the onset of type 1 diabetes is typically associated with protein deficiency and weight loss. Animals respond to low blood glucose levels by secreting epinephrine (adrenaline) from the adrenal gland and glucagon from the pancreas. The release of these hormones leads to an increase in the conversion of glycogen to glucose in the liver. By working in opposite directions, glucagon and insulin establish the levels of various fuels in the circulation and prevent violent departures from these levels. Such metabolic control mechanisms assure that excess nutrients not immediately used to meet energy needs are stored as glycogen (a carbohydrate stored mostly in the liver and skeletal muscle) and in a fat reserve called triglyceride. Carbohydrate, stored as glycogen, is sufficient for energy needs for only a few hours, whereas a typical adult has sufficient fat stored for several weeks of starvation.

The second stage of metabolism occurs in the cytoplasm where small organic units obtained in the first stage are converted into a few simple units that play a central role in metabolism. In this *second stage*, sugars, fatty acids, glycerol, and several amino acids are converted into the acetyl unit of the acetyl CoA. The process does not require oxygen, but yields only small amounts of ATP. When the supply of oxygen to actively contracting muscle cells is insufficient, pyruvate is converted to lactate releasing useful energy for a limited duration. However, the resulting accumulation of lactate in muscle tissues is responsible for muscle cramps. In anaerobic organisms such as yeast, pyruvate is transformed into ethanol, like lactate (a product of fermentation). In animal cells, most of the useful food energy is extracted in the *third stage*, which consists of the citric acid cycle and oxidative phosphorylation and is carried out under aerobic conditions in mitochondria.

3.6.1 Oxidation–Reduction Reactions

One important motif in food degradation is what is known as *oxidation–reduction* or *redox* reactions. In these reactions, one or more electrons from one substance are transferred to another. The gain of electrons by a ligand (atom, ion, or molecule) is called *reduction*. The loss of electrons is called *oxidation*. Gain or loss of an electron is equivalent to the gain or loss of a hydrogen atom because transfers of hydrogen atoms involve the transfer of electrons. In a redox reaction, the reactant that accepts electrons (hydrogen atoms) is called an oxidizing agent and the one that gives up electrons is called reducing agent. Oxygen is the most effective oxidizing agent receiving two electrons for every oxygen atom. The oxidization/reduction of a number of phosphate-containing molecules plays important roles in glucose metabolism. These substances can exist in an oxidized form and a reduced form. Such a pair is then called a *redox* couple. Consider, for example, the oxidization reduction of *nicotinamide adenine dinucleotide* (NAD):

$$NAD^+ + 2\ H\lambda\ NADH + H^+. \tag{3.1}$$

In this equation, NAD^+ is the oxidized form and NADH is the reduced form. The structural formulas for NAD^+ and NADH are as follows:

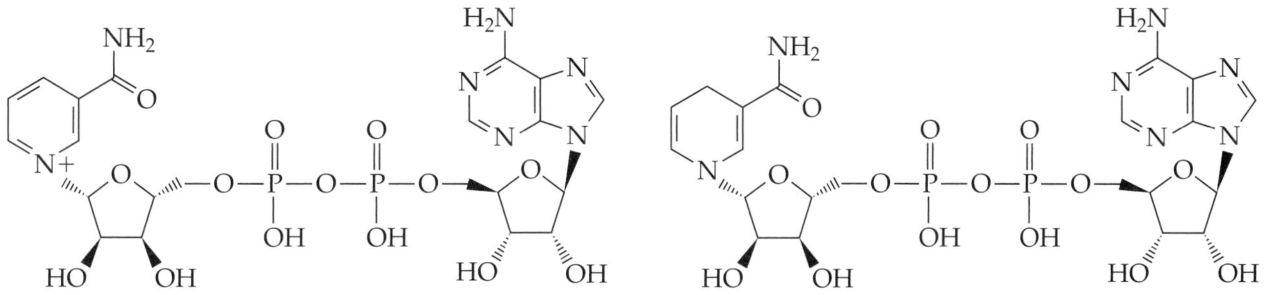

Note that NAD is a ribonucleotide, much like ATP. It has an adenine unit, two ribose sugars, and two phosphate units and a reactive nicotinamide ring, shown in the upper left corner of the molecule in the figure. This ring accepts a hydrogen ion and two electrons for NAD^+ to be reduced to NADH. Part of the energy originally present in the electron-donating reactant becomes associated with the compound that is receiving the electron. In glucose metabolism, the electron-transfer potential of NADH is converted into the phosphate-transfer potential of ATP. The electron-transfer potential of a substance, which can exist in an oxidized form X and a reduced form X^-, is called the redox potential. This parameter is a measure of the readiness with which an atom or molecule takes up an electron. The change in standard free energy $\Delta G^{o\prime}$ is related to the change in the standard redox potential $\Delta E^{o\prime}$ by the equation

$$\Delta G^{o\prime} = -nF \, \Delta E^{o\prime}, \qquad (3.2)$$

where n is the number of electrons transferred, F is Faraday's constant, and $\Delta E^{o\prime}$ is measured in volts. The standard redox potential $\Delta E^{o\prime}$ of the redox couple X and X^- is defined as the voltage difference between a cell containing 1 mol of X and X^- each and the other cell containing 1 mol of H^+ at 1 atm H_2 gas. A negative reduction potential means that the substance X^- has lower affinity for electrons than H_2. Oxygen has the highest value of redox potential. An oxidation–reduction reaction with a positive $\Delta E^{o\prime}$ value, will have a negative $\Delta G^{o\prime}$ value. Chemical reactions that take place during degradation of foodstuff are exergonic redox reactions: The overall change in free energy (ΔG) is negative, and part of the chemical energy in the reactant is stored in the product that receives the electron.

3.6.2 Degradation of Glucose

The overall metabolic equation concerning the degradation of glucose can be written as

$$C_6H_{12}O_6 + 6\,H_2O + 6\,O_2 \rightarrow 6\,CO_2 + 12\,H_2O. \qquad (3.3)$$

The cellular pathways involved in the degradation of glucose are shown in Fig. 3.12. The four pathways of the glucose metabolism are *glycolysis*, *pyruvate oxidation*, the

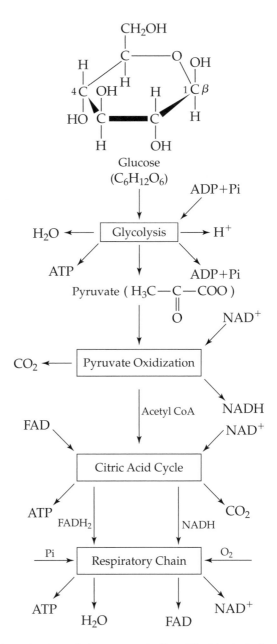

FIGURE 3.12 Schematic view of glucose metabolism. The blocks indicate the four separate pathways involved in the processing of cellular energy. Each pathway is composed of multiple consecutive reactions catalyzed by enzymes. The reactions in the glycolysis pathway occur in the cell cytoplasm. The other three pathways describe reactions and processes that occur in the mitochondria.

citric acid cycle, and the *respiratory chain*. The metabolic reactions in the first pathway occur in the cell cytoplasm. The latter three take place in the mitochondria and are cumulatively called *cellular respiration*. The products of one pathway feed into the subsequent pathway as shown in Fig. 3.12.

Glycolysis: The first pathway in the complex network of reactions consists of 10 reactions that convert the six-carbon glucose molecule ($C_2H_{12}O_6$) into two molecules of the three-carbon compound pyruvic acid or pyruvate ($C_2H_2O_3^-$). Degradation of glucose to pyruvate results in a reduction in free energy of about 140 kcal/mol. About a third of this energy is captured in the formation of ATP and the reduction of NAD. The first five reactions in the pathway consume energy. Separate reactions use two phosphate groups from two different ATP molecules to convert a six-carbon glucose molecule into two molecules of three-carbon sugar phosphate. The second set of reactions produces four ATP molecules and also reduces two molecules of NAD^+ to form NADH. The reduction of NAD^+ to NADH is equivalent to the transfer of two electrons and a proton. Chemical reactions of the glycolytic pathway do not involve oxygen. The net reaction in the transformation of glucose to pyruvate is

$$\text{Glucose} + 2\,\text{Pi} + 2\,\text{ADP} + 2\,NAD^+ \rightarrow$$

$$2\,\text{pyruvate} + 2\,\text{ATP} + 2\,\text{NADH} + 2\,H^+ + 2\,H_2O. \quad (3.4)$$

Glycolysis is highly regulated. Cells acquire just enough glucose from blood to meet the need for ATP. All the metabolic intermediates between glucose and pyruvate are phosphorylated compounds. Detailed information about the enzymes and ligands of this pathway can be found at the KEGG database (http://www.genome.ad.jp/kegg/). Enzymes that catalyze the reactions in the subsequent oxidation of pyruvate are associated with the inner membrane of the mitochondria.

3.6.3 Mitochondria: Power Plants of Eukaryotic Cells

Mitochondria (singular: mitochondrion) are the power stations of the cell in which food molecules are burned (oxidized) to release energy in a process called aerobic respiration (Fig. 3.13). These are the organelles in which products of glucose and other small molecules, such as pyruvate, are oxidized to produce ATP. Mitochondria are different from other cell organelles in that they have their own DNA and the machinery to synthesize some of their own proteins. A mitochondrion is roughly of the size of a bacterial cell, and its ribosomes are similar in composition to prokaryotic ribosomes. Although mitochondria depend on the nucleus and cytoplasm for viability and maintenance, they divide asexually to produce additional mitochondria much like bacterial cells. These findings suggest that mitochondria are the descendents of prokaryotes engulfed by larger prokaryotes.

Highly metabolic cells such as liver cells contain up to 2000 mitochondria. These organelles change shape easily and move efficiently within the cell in order to provide ATP directly to sites of high ATP consumption. As shown in Fig. 3.13, the mitochondrion is composed of a smooth outer membrane and an inner membrane with many folds. These folds separate the interior into two compartments: the internal matrix space and a much narrower intermembrane space. The outer membrane has a structural and protective function, but offers little resistance to the movement of substances into and out of the mitochondrion. Membrane channels composed of the transport protein *porin* allow the flux of small molecules into the interspace between the two membranes. The inner membrane folds inward in many places, forming regular shelf-like structures and providing a large surface area where a significant portion of the energy transfer occurs. The inner membrane exerts strict controls on the flux of

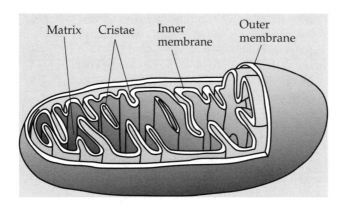

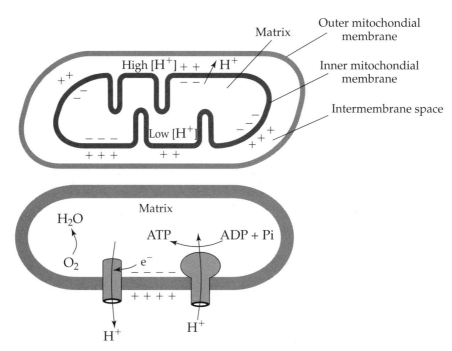

FIGURE 3.13   A mitochondrion is a sausage-shaped organelle. It is composed of two membranes. The many folds of the inner membrane divide the interior space into two components: the matrix and the intermembrane space (modified from http:/gened.emc.maricopa.edu/bio/bio181/BIOBK/BioBookCELL2.html). The large aqueous channels of the outer membrane are made of the transport protein *porin* and allow small proteins and other small molecules to enter the intermembrane space. The inner membrane is not permeable to most of these molecules and is especially impermeable to ions. A variety of inner-membrane proteins selectively transport some small molecules into the matrix space.

molecules between the matrix and the intermembrane space. The inner membrane has much less lipid and much more protein than the cell membrane, presumably because these membrane-bound proteins act as enzymes in the breakdown of food and production of ATP. The inner membrane is especially impermeable to ions. Proteins associated with the inner membrane selectively facilitate the entry of pyruvate and fatty acids into the matrix. The matrix and the inner membrane possess large numbers of enzymes that mediate the oxidization of high-energy chemicals such as pyruvate.

3.6.4 Pyruvate Oxidization

Enzymes that catalyze the reactions in the pyruvate-oxidation pathway are associated with the inner mitochondrial membrane. These enzymes oxidize the three-carbon pyruvate $(CH_3CO_2O^-)$ to the two-carbon acetate (CH_3COO^-) and yield free energy and CO_2. Part of the energy from the oxidation is saved by the reduction of NAD^+ to $NADH + H^+$. Some of the remaining energy is stored temporarily by linking acetate to an enzyme called coenzyme A (CoA) to produce the energy-rich compound acetyl CoA $(C_{23}H_{38}N_7O_{17}P_3S)$:

$$\text{Pyruvate} + \text{CoA} + \text{NAD}^+ \rightarrow \text{acetyl CoA} + CO_2 + \text{NADH}. \qquad (3.5)$$

Coenzyme A is another ribonucleotide that plays a central role in metabolism. The structural formulas of CoA and Acetyl CoA are as follows:

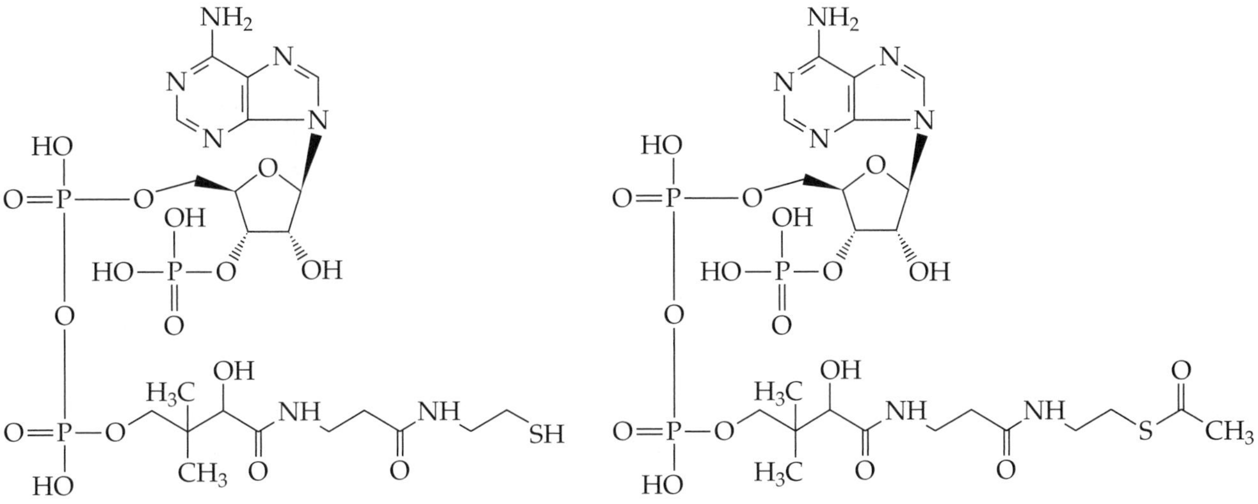

The terminal sulfhydryl group in CoA is the reactive site. Acetyl CoA can donate acetate to acceptors, much as ATP can donate phosphate to various acceptors:

$$\text{Acetyl CoA} + H_2O \rightarrow \text{acetate} + \text{CoA} + H^+ \quad \Delta G^{o\prime} = -7.5 \text{ kcal/mol.} \qquad (3.6)$$

The $\Delta G^{o\prime}$ for the hydrolysis of acetyl CoA has a large value comparable to that of ATP. Carriers such as ATP, NAD and CoA mediate the interchange of activated groups in many biochemical reactions common to all forms of life. The combined reactions that result in the conversion of pyruvate to acetyl CoA are catalyzed by a complex of enzymes consisting of 72 subunits and 24 different proteins. This enzyme complex cooperates with coenzymes, some of which are vitamins such as thiamin. Vitamins are organic molecules that are needed in small amounts in all forms of life, but higher animals have lost the capacity to synthesize them and must obtain them in their diet. Most water-soluble vitamins are components of coenzymes such as CoA. Detailed information about this pathway can be found at http://www.genome.ad.jp/kegg/.

3.6.5 Citric-Acid Cycle

The principal inputs to the citric-acid cycle are the two-carbon acetate in the form of acetyl CoA, water, oxidized electron carriers NAD (which were previously introduced), and flavin adenine dinucleotide (FAD). The structural formulas of the oxidized (FAD^+) and reduced ($FADH_2$) forms of FAD are as follows:

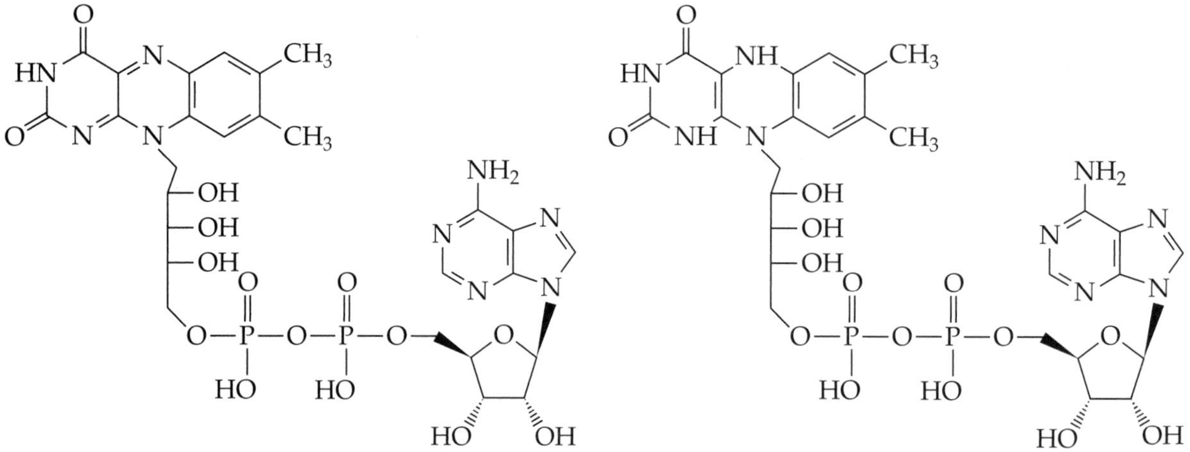

The reactive part of FAD is its isoalloxazine ring shown at the top of the structural formulas. FAD, like NAD, accepts two electrons. For each glucose molecule metabolized, the citric-acid cycle produces four CO_2, two ATP, six NADH + H^+, and two $FADH_2$. The net reaction of the citric-acid cycle is

$$\text{Acetyl CoA} + 3\,NAD^+ + FAD + GDP + Pi + 2\,H_2O$$

$$\rightarrow 2\,CO_2 + 3\,NADH + FADH_2 + H^+. \tag{3.7}$$

Most of the enzymes involved in citric-acid cycle are dissolved in the matrix of the mitochondria. Others are large protein complexes associated with the inner membrane.

3.6.6 Respiratory Chain

The reduced forms of FAD and NAD are energy carriers. The energy they store is used in the production of ATP. The most direct way to produce ATP from ADP and Pi would be to couple the oxidization of NAD (or FAD) to the production of ATP. Consider the reaction

$$NADH + H^+ + \tfrac{1}{2}\,O_2 \rightarrow NAD^+ + H_2O \qquad \Delta G = -52.4\ \text{kcal/mol} \tag{3.8}$$

The oxidization of NADH by oxygen is highly exergonic. If ATP is considered as a molecule that packs readily usable energy into bundles of 12 kcal/mol, NAD can be thought similarly as an energy-carrier molecule that can give up 52.4 kcal/mol to derive an energy-requiring process. If NADH were oxidized in a single reaction, the energy released would be too much to contain and put it into good use. Thus, the release of energy during oxidization must be allocated to a series of reactions, each releasing a small amount of energy. These reactions occur on the surface of the inner membrane and involve the sequential transfer of electrons through membrane-associated molecules. On the inner membrane of the mitochondrion are large enzymes capable of rapid electron-exchange (oxidation and reduction) reactions. The totality of these molecules makes up the electron-transport system (ETS). An electron-carrying enzyme has high energy and tends to deliver its electrons to a more electronegative enzyme. Enzymes in the ETS sequence have an increasingly greater affinity for electrons so that the flow of electrons from one ETS enzyme to the next runs downhill and energy is produced. As electrons pass down the ETS, enough energy is trapped to synthesize ATP. The ATP generated is used throughout the cell to drive most of the otherwise energetically unfavorable reactions. Each transfer results in a reduction of free energy, and part of the energy released is used to pump protons from the matrix into the intermembrane space through protein channels (Fig. 3.14). Since the nonpolar interior of the lipid bilayer is impermeable to hydrogen ions, proton accumulation in the intermembrane space causes both a concentration gradient and charge gradient across the membrane. This concentration gradient forces protons to pass through the channel protein ATP synthase. The relaxation of the proton imbalance is coupled to the formation of ATP in the complex (Fig. 3.14).

The free-energy change (ΔG) in transporting an uncharged solute molecule from side 1 where it is present at concentration 1 to side 2 where it is present at concentration 2 is given by the formula

$$\Delta G = RT \ln(c2/c1). \tag{3.9}$$

As usual, R denotes the gas constant and T is the absolute temperature in Kelvin. For charged particles, the membrane electric potential and the ion concentration gradient provide the driving force for movement of ions across a membrane. These forces may act in opposite directions. In the case where both forces act in the same direction down the concentration gradient, the free energy of change corresponding to transport of the ion might be large enough to drive other processes. For charged particles, the free-energy change is given by

$$\Delta G = RT \ln(c2/c1) + ZF\,\Delta V, \tag{3.10}$$

where Z is the electrical charge of the transported species, ΔV is the potential in volts (V) across the membrane (from side 1 to side 2) due to the charge difference across the membrane, and F is the Faraday constant ($F = 23.062$ kcal/(V mol)). The large difference in proton concentration between inside and outside the matrix results in a negative ΔG for flow of protons through the ATPase. In the presence of proton flow, ADP and Pi bind to ATP synthase and the enzyme catalyzes the formation of ATP. ATP formed in this way detaches from the ATP synthase only in the presence of proton flux. Thus, the proton flux does not directly affect the formation of ATP, but causes ATP to detach from the ATP synthase, enabling the continuation of ATP–ADP turnover.

Mitochondria are highly efficient as energy-processing plants. The conversion of glucose to two molecules of pyruvate in the cytoplasm produces two molecules of ATP for each glucose molecule. Pyruvate molecules are then imported into the mitochondrion and oxidized by molecular oxygen to yield 30 molecules of ATP per

FIGURE 3.14 ATP synthase catalyzes the formation of ATP. The protein-conducting unit F_o spans the lipid bilayer. ATP-synthesizing unit F_1 faces the matrix. Hydrogens flow from the intermembrane space to the matrix through F_o. According to the currently accepted model, the three catalytical β subunits of F_1 are structurally identical, but are in different configurations at any particular point. One catalytic site is in the O form (configuration), which has little affinity for substrates. A second subunit is in the L form, which binds to either ATP (or ADP and Pi) loosely and is catalytically inactive. The third β subunit is in the T configuration and binds ATP (ADP and Pi) tightly and is active. Energy input from the proton flux converts the T site into an O site, an O site into an L site, and an L site into a T site. The formation of ATP from ADP and Pi does not require proton flux. However, proton flux leads to the release of tightly bound ATP and the cycle continues.

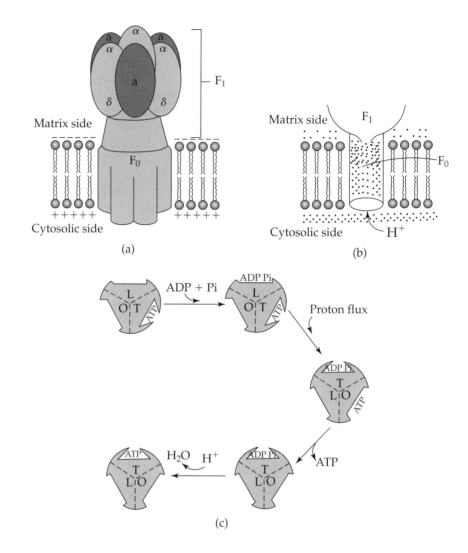

molecule of glucose used as fuel. Acetyl CoA brings two-carbon acetyl units into the citric-acid cycle where they are completely oxidized to CO_2. Four pairs of electrons are transferred to NAD+ and FAD for each acetyl group that is oxidized. ATP is then generated as these electrons flow from the reduced forms of these carriers to O_2 in a series of reactions.

3.6.7 Photosynthesis

Photosynthesis is the process that converts light energy to the chemical energy of organic compounds. Plants and some other simpler organisms use sunlight, water, and an inorganic carbon source (CO_2) to make carbon compounds from which their bodies are built and their food needs met. A side product of this chemical process is the oxygen gas (O_2) used by heterotrophs in converting organic compounds to CO_2 during the process of transforming chemical energy into work and heat. The overall chemical equation of photosynthesis is

$$6\ CO_2 + 12\ H_2O \rightarrow C_6H_{12}O_6 + 6\ H_2O + 6O_2.$$

In this chemical reaction, the term $(C_6H_{12}O_6)$ represents glucose. Water appears on both sides of the equation because it is used as reactant and released as a product. The overall reaction just described does not proceed in a single step, but in two sequential pathways, with each pathway containing a series of chemical reactions. The chemical reactions in the first pathway are driven by light energy. The subsequent pathway involves the entrapment of energy received from photons into ATP molecules. Light is a form of radiant energy that comes in pockets called *photons*. The amount of energy, E, contained in a photon is inversely proportional to its wavelength λ:

$$E = h\ c/\lambda.$$

In this equation, $c = 3 \times 10^{10}$ cm/s is the velocity of light in vacuum and h is the universal Planck's constant (1.584×10^{-34} cal s). The wavelength of visible light ranges from 400 nm to 700 nm. Humans perceive light as having distinct colors and these colors are directly related to the wavelength. Colors range from violet at 400 nm, the shortest wavelength for visible length, through blue, green, yellow, orange, to red at 700 nm, the longest wavelength for visible light. The lower the wavelength, the higher is the level of energy stored in a photon. When a photon meets a molecule, it may bounce off, pass through, or be absorbed by it. The energy of a molecule absorbing a photon rises from a ground state to an excited state by the amount of energy stored in the photon. The increase in energy drives one of the electrons of the molecule to an orbital further away from the nucleus. This loosely held electron then gets transferred to other molecules in subsequent reactions, ultimately resulting in the production of ATP from ADP. Many of the steps in photosynthesis are common to the degradation of glucose. For more on photosynthesis, refer to the assignments at the end of the chapter.

3.6.8 Common Themes in Metabolic Pathways

Although the number of chemical reactions in cells is large, they contain a number of common motifs. The central themes of metabolism are as follows:

1. ATP is the universal currency of energy. The hydrolysis of an ATP molecule may increase the equilibrium ratio of products to reactants in a coupled energy-requiring reaction by a factor of about 10^8.

2. ATP is generated by the oxidation of fuel molecules such as glucose. The chemical energy in carbon bonds is used through electron carriers to create a proton gradient across the inner membrane of mitochondria, which is then used to synthesize ATP.

3. The metabolic pathways that generate ATP and transfer high-potential electrons to electron carriers such as NADPH also provide building blocks for macromolecules. Biomolecules are constructed from a relatively small set of building blocks obtained through degradation of foodstuff.

4. Biosynthetic and degradative pathways are almost always separate. These pathways utilize different enzymes. A biosynthetic pathway is made exergonic

by coupling it to the hydrolysis of sufficient number of ATP. For example, four more ATP molecules are used in converting pyruvate into glucose in biosynthesis than converting glucose into pyruvate in glycolysis.

Recurring motifs in these reactions can be enumerated as follows:

1. The flow of molecules down a metabolic pathway is determined primarily by the amounts and activities of certain enzymes rather than by the substrate available. Examples of this will be discussed in Chapter 4. The first irreversible reaction in a metabolic pathway is called the *committed step* and constitutes an important point of control. Enzymes catalyzing committed steps are typically regulated by the end product. Bacteria use this mode of control in regulating the machinery that digests the milk sugar lactose. We will discuss lactose-mediated control of gene expression in Chapter 4.

2. Some regulatory enzymes in a metabolic pathway are controlled by phosphorylation. Examples of protein phosphorylation are discussed in the next three chapters. Metabolic pathways can be rapidly switched on or off through this type of enzyme modification. A similar mode of metabolic regulation is the degradation of some enzymes by protein-processing machines called proteosomes. Examples of control of protein concentration by degradation are presented in Chapter 6.

3. Metabolic pathways involve compartmentalization of chemical reactions. The fates of certain molecules depend on whether they are in the cytoplasm or in the mitochondria. Compartmentalization is a useful tool in separating degradative pathways from biosynthetic pathways.

The pathways central to the metabolism of foodstuff and biosynthesis have largely been identified and are illustrated for a variety of organisms at the KEGG database (http://www.genome.ad.jp/kegg/).

3.7 | Molecular Motors

Once synthesized, ATP hydrolysis can be used to drive any one of many reactions that require energy. Some of these reactions are metabolic or anabolic as already discussed, and others are actually used by the cell to carry out mechanical work. Perhaps the best understood cellular processes that involve the coupling of ATP hydrolysis to mechanical work are those utilized by motor proteins such as kinesin and myosin. An enzyme that converts chemical energy into mechanical energy is called a motor protein. An enzyme that can convert ATP into ADP and use the resulting energy to drive a chemical or mechanical change is known as an ATPase. Motor proteins such as myosin and kinesin have ATPase activity, which can be utilized to drive a mechanical process. Mechanical energy can be in the form of kinetic energy of motion or takes the form of elastic energy stored in the system. It is likely that part of the free energy released during transitions between chemical states is stored as elastic energy in the protein motor.

The driving force for forward movement arises during transformation of this internal energy into kinetic energy. The energy-transformation mechanism employed by protein motors is similar to that utilized by a pole vaulter trying to clear a crossbar. During the pre-jump running phase, the vaulter transmits part of its kinetic energy into the elastic energy of the pole. The pole vaulter then uses this elastic energy to gain upward velocity against gravity and clear the bar. Protein motors become strained during transitions between chemical states. Part of the free energy released is stored as internal elastic energy. The release of this internal energy is the driving force for forward, or in this case, upward movement.

The simplest cellular motors are proteins that use the chemical energy stored in ATP to cause a positional change relative to a substrate or a track. A major protein motor in cells is myosin, which causes movement in relation to actin-rich thin filaments in muscle. In this system, myosin is the motor, actin filaments are the tracks along which myosin moves, and ATP is the fuel that powers movement. Another motor protein is kinesin, which uses ATP energy to cause movement along microtubules. When kinesin motors bind to their substrate, they can move molecules, chromosomes, and even organelles. As discussed in a recent review article by Ronald D. Vale and Ronald A. Milligan (*Science*, 288: 88–95), myosin and kinesin have striking structural similarity, suggesting that they may be linked in evolution to a protein predecessor.

3.7.1 Myosin Motors and the Crossbridge Cycle

Myosin is the molecular motor of muscle (Fig. 3.15). Together with actin and regulatory proteins, it is responsible for muscle contraction, cell motility, and extension of cell projections. Actin and myosin have contractile roles in all eukaryotic cells. Actin is a highly conserved ancient protein of eukaryotes. It is usually the most abundant protein in a cell, comprising about 10 percent of the cell content. Actin from different species polymerizes readily to form filaments under appropriate conditions. Myosin

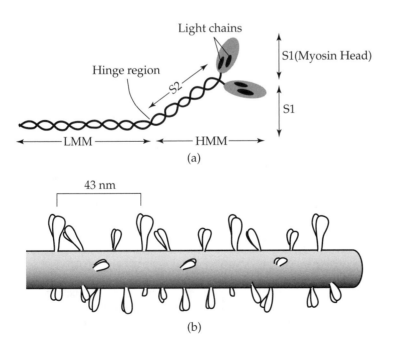

FIGURE 3.15 Schematic of a myosin molecule (a). Myosin molecules in muscle cells aggregate and form thick filaments. Their tails form the backbone of the filament, whereas the heads protrude and interact with active sites on thin actin filaments pulling them against the external load thereby causing muscle contraction (b). LMM refers to part of the myosin molecule that is embedded into the body of the thick filaments of muscle fibers whereas HMM is the portion of the myosin molecule that protrudes from the body of the thick filament.

is a member of a family of proteins that is less conserved than actin in evolution. Myosins in muscle readily form thick filaments (Fig. 3.15). Nonmuscle cells contain ten to a hundredfold less myosin than actin. The "unconventional" myosins also have motor domains that hydrolyze ATP, but have many variations in the tail regions, which allow them to interact with different proteins. These nonmuscle myosins are used in a variety of different processes including actin-based vesicular transport and hearing.

The arrangement of actin monomers in a microfilament is assymetrical such that the end to which actin monomers are added differs from the other end (Fig. 3.8). As discussed earlier, cell biologists refer to these poles as the plus end and the minus end. Consequently, actin microfilaments are directional: For example, when decorated with the head region of myosin molecules, all heads on the same filament point in the same direction and give the decorated filament the appearance of the flight of an arrow. In this situation, the decorated filament has a barbed (+) end and a pointed (−) end.

Skeletal Muscle Structure: The interaction of the protein motor myosin with actin-rich thin filaments illustrates the mechanism of force generation and movement in cells. The repeating contraction unit in muscle cells is called a sarcomere, which is composed of thick myosin and thin actin filaments arranged in parallel (Fig. 3.16).

Myosin II is the building block of thick filaments (Fig. 3.15). This protein is made of six polypeptide chains: two identical heavy chains (each weighing 230 kDa) and four light chains (20 kDa each). Electron microscopy shows that the molecule is composed of two oval-shaped heads of about 60 nm in length and a long tail of about 130 nm in length. The tails bind together and form the backbone of thick filaments. The heads contain binding sites for actin and ATP and its hydolysis products (ADP and Pi). An α-helical neck region between the head and tail regulates the activity of the head region.

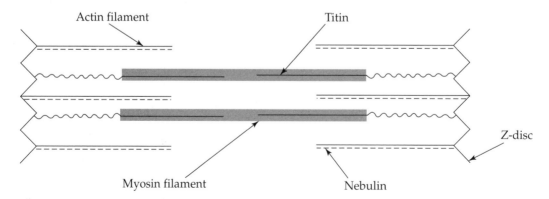

FIGURE 3.16   The elementary repeating unit of a muscle cell is a sarcomere. The protein nebulin strengthens the thin filaments and the springlike titin stretches when the muscle is stretched. In relaxed muscle (low Ca^{2+}), tropomyosin prevents actin from interacting with the myosin heads to generate force. During muscle contraction, myosin binds to the actin filament. As the muscle contracts, myosin heads pull the thin filaments toward the centers of sarcomeres, resulting in the shortening of muscle. Muscle is a unique tissue because, unlike nonliving materials, muscle length shortens while carrying axial tension.

Thick filaments are 16 nm in diameter and 1.5 μm long. Myosin heads (cross-bridges) protrude from the filament in a regular array at intervals of 14.3 nm along the filament axis. The central region of the thick filament spans a length of 150 nm and is devoid of projecting crossbridges. The myosin heads on each side of the bare zone point toward the center of the filament; hence, the thick filament is inherently bipolar.

The major component of thin filaments is actin, which has high affinity for myosin. A tropomyosin and troponin complex constitute about one-third of the mass of thin filaments. Each thick filament is surrounded by six thin filaments that overlap the thick filament. In relaxed muscle (low Ca^{2+}), tropomyosin prevents actin from interacting with the myosin heads to generate force. A signal for muscle contraction is provided by nervous input through neuromuscular junctions and results in the release of intracellular Ca^{2+} from the specialized endoplasmic reticulum found in muscle cells. In the presence of elevated Ca^{2+}, myosin binds to the actin filament and the muscle contracts. As the muscle contracts, myosin heads pull the thin filaments toward the centers of sarcomeres, resulting in the shortening of muscle. Muscle is a unique tissue because, unlike nonliving materials, muscle length shortens while carrying axial tension.

Work Stroke of a Crossbridge: Muscle shortening during contraction is due to the relative movement of actin filaments with respect to the myosin filaments. The relative movement is caused by the reversible interaction of myosin heads protruding from thick filaments with the actin binding sites on the thin filament. The myosin head has strong affinity to actin and to ATP, but not to both at the same instant. Given a choice, myosin prefers ATP to actin. According to the presently accepted model of muscle contraction, the steps involved in the action of myosin motor are shown in Fig. 3.17. The steps (a), (b), and (c) indicate states where myosin interacts with actin. These are called the attached states. Myosin binds to actin and then produces force and movement as it tilts into one or more subsequent conformations. The states (e) and (d) are states when myosin is free of actin. ATP and ADP are shown as chains of circles, with smaller circles representing phosphate groups.

As shown in Fig. 3.17, ATP hydrolysis occurs while the myosin head is free of actin and induces a change in configuration of the myosin head, allowing it to interact with actin. Detachment of the hydrolysis product Pi from myosin leads to the strongly attached state in which the myosin head pulls the thin filament toward the center of sarcomere. Detachment of ADP from myosin leads to binding of ATP to the same site and subsequent detachment of myosin from actin and continuation of the cycle. Recent studies of myosin motor kinetics using in vitro motility systems revealed information on the forces generated by a single motor (Fig. 3.18). These and other experiments indicate that a myosin head can generate a force of about 5 pN and that a myosin molecule can catalyze the hydrolysis about 20 ATP molecules per second during muscle contraction.

3.7.2 Kinesin Cycle

Kinesin is a motor protein that powers the transport of intracellular organelles along microtubules (Fig. 3.19). Kinesin also facilitates chromosome movement during cell division, a function that is addressed again in Chapter 6. Kinesins constitute a family

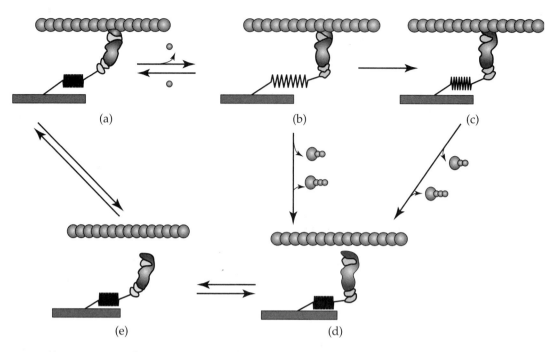

FIGURE 3.17 The interaction of a myosin motor with an actin fiber. Muscle is composed of thick and thin filaments (a), which are organized in contractile units called sarcomeres. The heads of myosin (crossbridges) interact with actin, converting the chemical energy of ATP to mechanical work and heat (b). The free energy of ATP hydrolysis is used in two different tasks: (1) to dissociate myosin from actin and (2) to release the ATP hydrolysis product Pi from myosin so that myosin can bind strongly to actin, generating force.

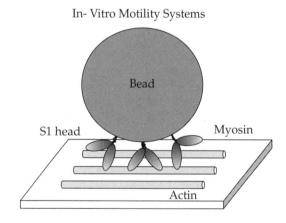

FIGURE 3.18 In vitro motility system. Myosin-coated beads move unidirectionally on oriented actin cables. Actin cables from algea *nitrella* are fixed on a substrate, with all fibers having the same sense of direction. Myosin-coated beads move unidirectionally along these actin cables in the presence of ATP. Beads coated with skeletal muscle myosin move with a speed of 5 µm/s and it is estimated that 25 myosin molecules on each bead are sufficient to actuate the motion. Alternatively, actin filaments can move on glass decorated with myosin in the presence of ATP.

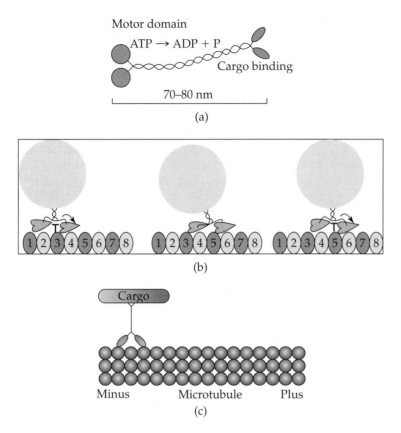

FIGURE 3.19 A kinesin motor moves along micro-tubules in the presence of ATP. The large globular re-gions of kinesin interact with microtubular tracks, whereas the two smaller domains bind to the organelles to be transported (a). Kinesin "walks" on the microtubule without losing contact for long distances compared to the step size of the walk (b). For most kinesins, the mo-tion is directed from the minus end of the molecule toward the plus end (c).

of about 100 eukaryotic motor proteins that interact directly with microtubules. Motile organelles such as mitochondria and vesicles use *kinesins* to propel them along microtubules.

Conventional kinesin is the prototypical member of the kinesin family of motor proteins and consists of two heavy (globular) and two light chains, resulting in a molecular weight of 400 kDa (Fig. 3.19a). The two large globular regions contain both the ATPase active site and the microtubule-binding site. The ATPase activity is strongly promoted by the microtubules. In the presence of microtubules, kinesin can hydrolyze up to 100 ATP molecules per second. The globular tail domain contains the binding site for the cargo (organelle, chromosome, etc.). The various kinesin genes all contain microtubule-binding sites and have ATPase activity, but differ in their cargo-binding sites. Most kinesins direct movement toward the plus end of microtubules, but some are minus-end-directed motors.

Kinesin molecules and their associated cargo can move along microtubules for several micrometers. High-resolution video microscopy has revealed that an or-ganelle that uses a kinesin for transport can fall off the microtubule and reattach if confronted with an obstacle. The same microtubule can simultaneously support plus and minus end-directed motion. If two organelles moving along the same micro-tubule in opposite directions bump into one another, one of them detaches, then reattaches after the other organelle has passed. Importantly, since directionality is conferred by the kinesin molecule, the displaced organelle resumes moving in the original direction. Kinesin motion is reminiscent of human walking. Both heads translocate in turn by 16-nm steps, and each translocation moves the center of mass 8 nm forward. During the translocation motion of one head, the other head remains bound to the microtubule.

Glass surfaces coated with kinesin support gliding of individual microtubules in vitro when viewed by high-resolution video microscopy. In this case, the cargo-binding region of the kinesin molecule is immobilized to the slide and the microtubule moves. Binding of a single kinesin molecule to each tubulin heterodimer can move a microtubule at a rate of about 1 μm/s. Increasing the density of kinesin molecules above this does not alter the speed of microtubule movement. The maximum force produced by a kinesin molecule was estimated to be about 5 pN, and the force produced decreases inversely with the forward speed of the kinesin. For more information on kinesin-induced microtubule motion, see the assignments in the back of the book and the kinesin home page (www.proweb.org/kinesin//index.html).

Another family of microtubule-based motor proteins is the dyneins. Ciliary dynein powers flagella motion in some motile single-cell eukaryotes and is responsible for the beating of cilia in some tissues of multicelluar organisms. For example, cilia on the surface of cells in the oviduct are used to move the egg, and cilia in the cells of the respiratory system are used to move mucus.

3.8 | Cell Membrane

The cell membrane separates the cell interior from the surrounding environment and is composed of a *phospholipid bilayer* and associated proteins and carbohydrates (Fig. 3.20). The cell membrane is also called the plasma membrane. It is composed of a double layer of phospholipid molecules (called the phospholipid or lipid bilayer), protein molecules associated with the lipid bilayer, and a carbohydrate-containing cell coat called the glycocalyx. The plasma membrane separates the cell interior from the surrounding environment and carries out several important functions. The plasma membrane allows nutrients to enter the cell, keeps out unwanted molecules and particles, transports waste out into the extracellular fluid, and prevents needed metabolites and ions from leaving the cell. Many of these reactions occur, not in solution, but on the surfaces of the membrane. The trapping of reactants and catalytic enzymes onto a membrane scaffold influences the rate of catalysis as well as the ability to control chemical reactions. The phospholipid bilayer is about 7–9 nM thick and can be visualized with the electron microscope as a continuous thin, double line. Because of their inherent amphipathic nature, the phospholipid molecules in the bilayer are arranged with the hydrophobic fatty-acid tails of one side of the membrane leaflet facing inward toward the tails of the other leaflet. The hydrophilic phosphate ends point outward from the central surface of the plasma membrane. Large numbers of cholesterol molecules are embedded in the hydrophobic interior of the bilayer.

Small nonpolar molecules such as carbon dioxide and oxygen diffuse freely across the bilayer (Fig. 3.20). Carbon dioxide and oxygen are both nonpolar and lipid soluble. The lipid bilayer is also permeable to small, uncharged polar molecules, such as urea and ethanol, that move easily in both directions across the lipid bilayer. In diffusion, the net flux of a solute is a function of the local concentration gradient

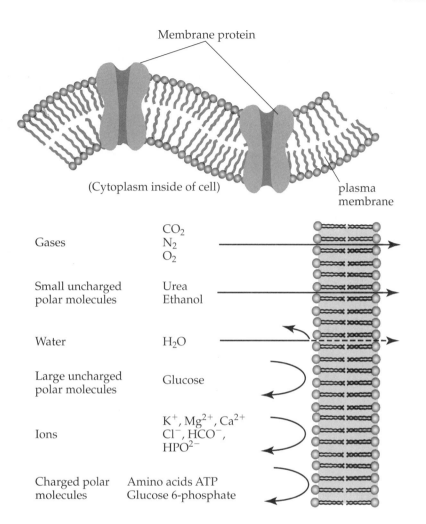

FIGURE 3.20 Plasma membrane of an animal cell. The plasma membrane is composed of a thin, continuous phospholipid and cholesterol-containing bilayer with proteins embedded in it. In many respects the bilayer behaves as a two-dimensional fluid. Some integral membrane proteins such as protein-transport molecules cross the entire bilayer, whereas other proteins only partially penetrate it. Cell-surface receptors interact with extracellular molecules and are components of the system of signal transduction across the membrane.

of the solvent. For example, carbon dioxide is continuously produced as the cell uses up fuel molecules. Thus, there is a higher concentration of carbon dioxide inside the cell than out, resulting in net flux of carbon dioxide out of the cell. Oxygen, on the other hand, is used up by the cell in carrying out its activities and has a lower concentration in cells than in blood. Therefore, there is typically a net flux of oxygen into the cell down the concentration gradient. When the concentration gradient is zero, the net flux across the membrane is also equal to zero. Even in this equilibrium case, there is a continuous flux of molecules across the bilayer in both directions. No metabolic energy is expended when the net movement is from a high to low concentration of the molecule.

In addition to oxygen and carbon dioxide, many other small hydrophobic molecules can diffuse across phospholipid bilayers. Because the hydrophobic core of a typical cell membrane is 100 to 1000 times more viscous than water, the diffusion rate of nonpolar particles across the bilayer is much smaller than their corresponding diffusion rate in water. Thus, movement across the hydrophobic portion of a membrane is said to be the rate-limiting step in passive diffusion.

3.8.1 Diffusion of Water

Water moves across the bilayer through momentary openings created by random movements of the membrane lipids. Water also moves through channel proteins that are integrated into the cell membrane. The cores of these proteins are hydrophilic, and multiple water molecules can pass through them at a rate of 10^8 molecules per second. The gradient of particles across the cell membrane is the principle driving force for the diffusion of water across the bilayer. Water tends to move across a membrane from a solution of low solute concentration to one of high solute concentration. From another perspective, water moves from high water concentration (low solute concentration) to low water concentration. This process is known as *osmotic flow* (Fig. 3.21). The particulate concentration gradient across a membrane becomes zero when the two regions separated by the membrane have an equal number of dissolved particles per unit volume. In the count of particles, a small particle (e.g., a sodium ion) is as important as a larger particle (e.g., glucose). Therefore, the

FIGURE 3.21 Osmosis and living organisms. When two compartments separated by a membrane have different particulate concentrations, water from the compartment with smaller particulate concentration would tend to flow into the compartment with the higher concentration (a). In living systems, such compartments are of finite size and cannot stretch indefinitely to accommodate the incoming water. The pressure in the compartment with lower particulate concentration increases and eventually halts the flow of water in the compartment. The pressure difference between the two compartments corresponding to zero flux is called the turgor pressure. Within the cytoplasm of the cell, the concentration of various species vary spatially, according to the law of diffusion. Diffusive forces work toward creating uniform concentration (b).

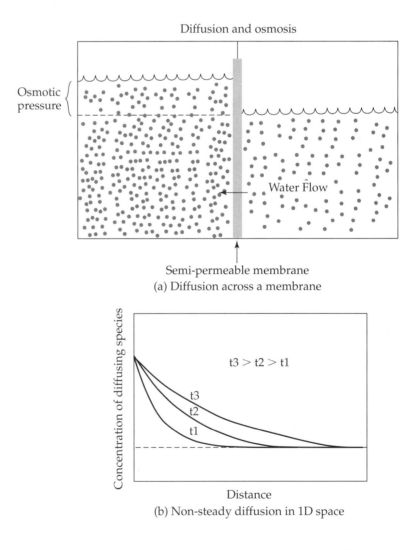

Diffusion and osmosis

Osmotic pressure

Water Flow

Semi-permeable membrane
(a) Diffusion across a membrane

Concentration of diffusing species

t3 > t2 > t1

t3
t2
t1

Distance
(b) Non-steady diffusion in 1D space

diffusion of water is affected not by what is dissolved in the water, but by how many molecules are dissolved. When the cell interior has the same number of dissolved particles per unit volume as the extracellular environment, these two solutions are *isotonic* and there is no net movement of water across the cell membrane. Single-celled organisms inhabiting the oceans are usually isotonic with seawater. Similarly, the cells of animals are isotonic with the blood, lymph, and extracellular fluid surrounding them.

When a solution has fewer solute molecules per unit volume with respect to another solution, it is said to be *hypotonic*. The solution with the higher concentration of particles is *hypertonic*. Water molecules diffuse from a hypotonic solution to the hypertonic solution across the cell membrane. The body fluids of many marine fish are hypotonic in relation to the surrounding seawater, resulting in a net movement of flow out of their bodies. Fish counter this consequence of *osmosis* by drinking seawater and excreting the excess solutes. In contrast, plant cells are usually hypertonic to their surrounding environment, and so water tends to diffuse into them. In plants, which have a cell wall, the plasma membrane can be extended only to a certain extent; the movement of water into the cell increases the pressure within the cell wall. In young plant cells, the pressure difference across the cell membrane causes the membrane to expand and the cell to grow. As the cell matures however, the cell wall stops growing and resists expansion. This cell wall rigidity leads to further increases of pressure in the cell interior. The pressure difference between the inside and outside of the cell prevents the net movement of additional water into the cell. This pressure difference is called *turgor* or *osmotic* pressure. In other words, osmotic pressure is the pressure required to stop the net flow of water across a membrane separating solutions of different particulate concentrations. This is the pressure that keeps the plant cell wall stiff and the plant body turgid. Note that at the turgor pressure, the concentration of molecules in the cell interior is greater than in the exterior. Turgor pressure decreases as a consequence of water loss, and the plant wilts. Because animal cells do not have rigid cell walls like plant cells, they begin to swell when placed in a hypotonic solution. Some cells like red blood cells actually burst as water enters them by osmotic flow. Immersion in a hypertonic medium causes the cells to shrink as water leaves them by osmosis. For these reasons, cultured animal cells are maintained in isotonic cell culture medium, where the concentration of solutes is close to that of the cell cytoplasm.

3.9 | Active Membrane Transport

Polar molecules, such as glucose, nucleosides, and amino acids and ions of hydrogen, sodium, calcium, and potassium, cannot pass through the hydrophobic interior of the cell membrane and diffuse to the other side. Proteins are required to transport such molecules and ions across all cellular membranes (Fig. 3.22). The transport of ions and polar molecules occurs through *membrane channels* made of such transport proteins. These channels are key to maintaining the right balance of ions in cells and are vital to transmit signals from nerves to tissues. Defects in ion channels and transporter proteins can cause serious diseases. In the heart, defects in potassium channels do not allow proper transmission of electrical pulses, resulting in the arrhythmia. Cystic fibrosis is caused by a defective gene encoding a sodium and

FIGURE 3.22 Examples of active transport of ions and polar molecules across cell membranes include ATP-powered pumps, transporter proteins, and ion channels. The arrows indicate the direction from high to low concentration of the ion or polar molecule across the lipid bilayer.

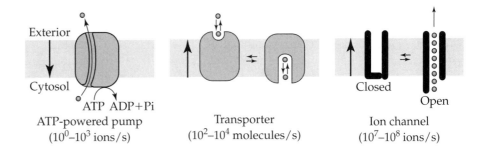

Exterior

Cytosol

ATP ADP+Pi

ATP-powered pump
(10^0–10^3 ions/s)

Transporter
(10^2–10^4 molecules/s)

Closed

Open

Ion channel
(10^7–10^8 ions/s)

chloride transporter found on the surface of epithelial cells that line the lungs and other organs. The defect causes the body to produce thick mucus that clogs the lungs, leading to infection. Another defective transporter protein blocks the pancreas from sending digestive enzymes into intestines.

These channels are made of cell-surface proteins that span the entire thickness of the cell membrane (*integral membrane proteins*). The tertiary structures of these proteins form hydrophilic channels through which the transported molecules pass. Transport proteins form highly selective channels; depending on its three-dimensional geometry, a transport molecule may accept one molecule, but exclude a nearly identical one.

There are two types of protein-mediated transport. In *facilitated transport*, transport proteins can move substances across a membrane only down a concentration gradient. Here the transport protein serves as a passive channel. Although the process of transport requires no energy outlay by the cell, the rate of *facilitated transport* is far higher than passive diffusion across the viscous and hydrophobic lipid bilayer of the cell membrane. It occurs via a limited number of transporter proteins, rather than throughout the phospholipid bilayer. Each type of transport protein facilitates the movement of a single species of molecule or a single group of closely related molecules. The plasma membranes of cells contain transport proteins that enable molecules such as amino acids, nucleosides, and sugars to enter or leave the cell down their chemical concentration gradients.

In *active transport*, proteins move molecules up a concentration gradient. This process requires expenditure of energy by the cell. Depending on the situation, the transport of a molecule can take either of these two paths. Consider, for example, the transport of glucose across the cell membrane. Because cells oxidize glucose to meet their energy needs, their glucose concentration tends to fall below that in blood; a steady supply of glucose is carried into these cells down the concentration gradient by facilitated diffusion. Liver cells, on the other hand, store glucose. Although these cells already have a high concentration of glucose (stored in the form of glycogen), additional glucose is transported into them by active transport. In starvation, liver cells can synthesize glucose from fatty acids, amino acids, and other small molecules and release it into blood down the concentration gradient via facilitated transport.

An important example of active transport is the sodium–potassium pump (Fig. 3.23). Animal cells maintain steep concentration gradients of sodium ions (Na^+) and potassium ions (K^+) across the plasma membrane. The concentration of Na^+ outside the cell is 14 times higher than the concentration inside the cell. The low

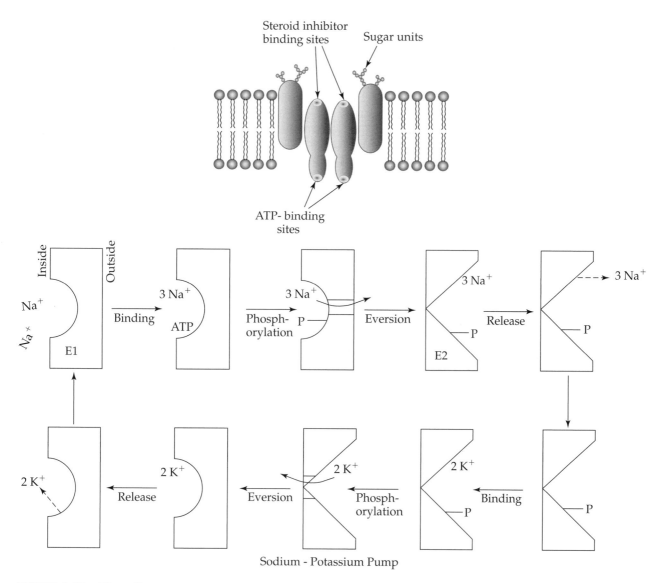

FIGURE 3.23   The sodium–potassium pump is an example of active transport. In this system, energy is used to move a solute against its concentration gradient. For each molecule of ATP used, two potassium ions are pumped into the cell and three sodium ions are pumped out of the cells.

concentration of Na^+ within the cell enables this ion to serve a signaling role. The flow of Na^+ into a nerve cell is a key event in the propagation of the nerve impulse. In contrast to the Na^+, the concentration of the K^+ is 10–30 times higher inside the cell than the concentration outside. This pairing of the sodium and potassium ion concentration gradients is extremely important in maintaining osmotic balance and thereby controlling cell volume.

The strict controls of sodium and potassium content of a cell are exerted by an active transport system known as the sodium–potassium ATPase (Fig. 3.23). The transport protein in this system is an ATP-powered pump, which uses the energy of ATP hydrolysis to move ions against a chemical concentration gradient. According to the present understanding of this pump, the transport protein exists in two alternative configurations. One configuration has the cavity opening to the inside of the cell, into which Na^+ can fit (configuration 1). Configuration 2 has the cavity opening to the exterior of the cell, with K^+ fitting to the opening. The transport protein cycles between these configurations, and this cycle is coupled to the hydrolysis of ATP. The cycle is as follows: A sodium ion binds to the transport protein in configuration 1. An ATP molecule associates with the transport protein. After ATP hydrolysis, the phosphate group is transferred to the transport protein, causing it to switch to configuration 2. The sodium ion is released to the outside of the cell, and a potassium ion is bound to the attachment site of the transport protein. Binding of the potassium ion results in the release of the phosphate group and the protein goes back to configuration 1. The potassium ion is released into the cell, and the cycle is completed with the attachment of a sodium ion into the cavity of the transport protein.

Active transport similar to that described for sodium and potassium exchange is also used to regulate pH concentration inside the lysosome. In a lysosomal membrane, a proton pump keeps the concentration of protons (H^+) about 1000 times higher than in the cell cytoplasm. This high proton concentration is important for the function of the lysosome in degrading proteins and other macromolecules, but would be detrimental to the cell if all of its proteins and other macromolecules were exposed to it. Other membrane compartments in the cell also maintain a somewhat acidic pH of about 6–6.5. These compartments are involved in the trafficking and digestion of proteins and particles such as bacteria and viruses. An important stage in the infectious cycle of malaria relies on this pH gradient in the cell. The antimalarial drugs quinine and chloroquine are thought to work by inhibiting the generation of this proton gradient and are often used by cell biologists to study the role of pH gradients in protein trafficking.

3.9.1 Membrane Electrical Potential

Specific ionic concentrations in the cell *cytosol* (the liquid component of cytoplasm) differ from that of the surrounding fluid. In addition to the widely different sodium and potassium ion concentrations, calcium and chloride ions also have different concentrations in and out of the cell. The average concentration of calcium ions in the cytosol (Ca^{2+}) is 1000 times lower than that in blood. In most cells, a rise in Ca^{2+} concentration is an important regulating signal. A transient rise in calcium concentration initiates important cell functions such as the contraction of muscle cells and secretion of digestive enzymes in pancreatic cells.

Ionic concentration gradients create a difference of net charge (electric potential) between the inside and outside of the cell. The cell interior is slightly more negative than the cell exterior in multicellular organisms. The electrical charge across the membrane is not zero because protein channels permit selective transport. Usually, cell membranes contain many open K^+ channels, but few open Na^+ and Ca^+ channels. As a result, the major ionic movement across the plasma membrane is that of K^+ from the inside out.

This leaves an excess of negative charge on the inside and an excess of positive charge on the outside of the membrane. The hydrophobic interior of the lipid bilayer of the plasma membrane serves as a thin sheet of nonconducting material, enabling the storage of electrical charge across it. The ability of the lipid bilayer to maintain a charge gradient across it is a general mechanism used by the cell to actually store energy. The resulting electrical potential is equal to 0.07 V per unit thickness of the bilayer. Because of the extreme thinness of the bilayer (3.5 nm), this potential is 10^5 times greater than the voltage gradient in high-voltage transmission lines that carry electricity. Changes in the potential difference across nerve membranes occur during the transmission of electrical impulses during the activity of the nervous system and are directly responsible for mechanical events such as muscle contraction.

3.10   ASSIGNMENTS

3.1   An adult human male of 80 kg contains about 50 billion cells. Assuming that the mass density of humans is about that of water, determine the average volume of a human cell.

3.2   Chloroplasts are members of a family of organelles called plastids. They are similar in structure and function to mitochondria, as they also have a double membrane, an intricately folded inner membrane, and their own genetic material. They are involved in a complex process called photosynthesis that involves energy conversion. Chloroplasts use the energy of light to fix carbon dioxide to synthesize sugars and other food molecules and release oxygen, whereas mitochondria break down food molecules using oxygen, to release energy and carbon dioxide. Conduct a literature search to describe important steps in photosynthesis and associate these steps with different compartments of chloroplasts.

3.3   The wall of a plant cell is a rigid structure that restricts swelling and relative movement. If it were not for the rigid casing of the cell wall, plant cells would continue to absorb water by osmosis, until they burst. What is the composition of plant cell walls?

3.4   The *vacuole* in a plant cell is a large fluid-filled space, separated from the rest of the cell by a single membrane. Depending upon the hydration conditions, the vacuole occupies 5 percent to 95 percent of the cell volume. Expansion of vacuole volume is associated with a corresponding expansion of the plant cell wall by growth. Vacuoles are used to store nutrients, waste products, and other molecules. Why do plant cells store poisonous alkaloids in vacuoles?

3.5   Search the literature to write an essay on the sliding filament theory of muscle contraction. Discuss the Huxley–Simmons model of crossbridge action, the simplest model of crossbridge kinetics, and the recent revisions of the model. What is the primary structural difference between insect flight muscle and skeletal muscle of animals? How does the sliding filament model apply to flight muscles of insects such as the fruit fly?

3.6   Many types of cellular motility are driven by cyclical interactions of myosin with actin filaments. Optical tweezer transducers are increasingly being used to investigate the mechanical transitions made by a single myosin head while it is attached to actin. How do optical tweezers work? Conduct a literature survey on the working stroke of myosin. Myosin-coated beads of 2 μm in diameter move unidirectionally on oriented actin cables at 1.2 μm/s in the presence of ATP. Determine the force exerted on the bead by the surrounding fluid, assuming slow viscous motion. How is this force related to the force generated by myosin?

3.7   DNA replication, the process in which the nucleotide sequence of a DNA strand is copied by complementary base-pairing into a complementary sequence is catalyzed by the enzyme DNA polymerase III. This is an enzyme composed of 10 subunits and functions in cooperation with other replication proteins to carry out the duplication of the entire 4.4 mB *Escherichia coli* chromosome in 30–40 minutes. As discussed in this chapter, each of the two DNA strands serves as a template for the formation of a new strand. Conduct a literature search to review the latest in the structure and function of DNA polymerase III. Polymerase I is one of three DNA polymerases found in *E. coli* and is directly involved in DNA repair. Conduct a literature search to determine the three distinct catalytical activities of this enzyme.

3.8   What is the definition of pigment? Why do the plant leaves appear green?

3.9 Discuss why having more than two copies of a chromosome usually leads to severe abnormalities.

3.10 Discuss how the complex DNA structure of eukaryotes could have evolved from prokaryotic DNA.

3.11 Biologists measure the changes in membrane potential using microelectrodes. The interior of a cell has a negative charge compared to the surrounding medium. In a plant cell, ΔV is at least -120 mV. In animal cells, ΔV is about half that of plant cells. A transport process is active when $(\Delta G) > 0$ and passive when $(\Delta G) < 0$. Determine the free-energy change for the transport of H+ from inside a cell where $c1 = 10^{-3}$ mm to outside where $c2 = 10^{-1}$ mm across a negative membrane potential of $\Delta V = -59$ mV at 25°C.

3.12 *ATP Hydrolysis as a Molecular Clock.* Living systems utilize ATP hydrolysis as a timing mechanism that ensures turnover of the actin filaments. Other examples of timing use of nucleotide hydrolysis can be found in signal transduction and cytoskeletal structures. Conduct a literature search on the effect of methylation of actin on its ATP hydrolysis-mediated polymerization.

3.13 *ATP Hydrolysis and Active Transport.* Animal cells have a high concentration of sodium and low concentration of potassium relative to the external medium. Active transport of Na^+ and K^+ is required for, among other things, controlling cell volume and driving the active transport of sugars and amino acids. More than a third of the ATP used in a resting animal is utilized to pump these ions. Referring to a biochemistry book, discuss the interplay between ATP hydrolysis and the Na^+ and K^+ transport in animal cells.

3.14 In most living systems, the Na^+ inside cells is kept low. The transport of Na^+ to cells is thermodynamically favored and is used by cells to power uphill movement of several ions and molecules across the cell membrane. Intestinal epithelial cells import glucose and amino acids from the lumen of the small intestine by coupling this uphill transport with the energetically favorable transport of sodium ions. Later, epithelial cells export these substances to the blood. Identify the protein channels involved in these processes.

3.15 When sodium ions enter cells during simulation of a nerve cell, the electric potential of the membrane changes from -0.06 V to $+5$ V, with the cell cytosol becoming positive compared to extracellular fluid. This depolarization is followed by a rapid repolarization or return to the resting potential. The result is an impulse of an electrical current, which propagates through the axon (the long segment of a nerve cell) at a speed of 100 meters per second. Thus, even when an axon is 1 m long, it takes only a few ms for signal to move along the length of the nerve. Discuss the signal processing through nerve cells in skeletal muscle cells.

3.16 Ion pumps and other protein channels are examples of feedback control systems. The basic idea of feedback control is to use sensors to make the performance of the molecular machinery dependent on the outcome, in this case the concentration of a certain ion. Conduct a literature search to discuss the feedback mechanisms involved in the regulation of the sodium channel.

3.17 Proteins, polysaccharides, and other larger particles such as microorganisms and cellular debris are transported across the cell membrane by means of vesicles that bud off from or that fuse with the plasma membrane. In *endocytosis*, material to be taken into the cell induces the plasma membrane to bulge inward, producing a vesicle enclosing the substance. This vesicle separates from the surface of the cell and migrates to the interior of the cell. The materials carried are then released into the cytoplasm. Conduct a literature search to identify the three different forms of endocytosis: phagocytosis (cell eating), pinocytosis (cell drinking), and receptor-mediated endocytosis.

3.18 Fusion between common intracellular membranes shares fundamental properties with fusion that occurs during viral infection such as that of HIV. Conduct a literature search to find out about the minimal machinery for vesicle fusion. In other words, what forces the two lipid bilayers to coalesce? What is the role of the so-called SNAP and SNARE proteins in catalyzing membrane fusion?

3.19 *Bio-Nano Technologies.* This chapter presented illustrations of cellular organelles and machinery resulting from self-assembly of molecular building blocks. Biomolecules are particularly promising components in self-assembly processes, because their binding capabilities have been tailored to perfection by billions of years of evolution. "Bottom-up" technologies based on the self-assembly of molecular building blocks to form larger functional elements are currently being explored as potential ways to fabricate nanometer-size devices. Conduct a literature search on bio-nano technologies, and discuss the potential of such technologies in creating nano-scale circuits, motors, and other devices.

REFERENCES

Blobel G and Dobberstein B. "Transfer to proteins across membranes. II. Reconstitution of functional rough microsomes from heterologous components." *J Cell Biol* 1975, 67(3): 852–862.

Chevet E, Cameron PH, Pelletier MF, Thomas DY, and Bergeron JJ. "The endoplasmic reticulum: integration of protein folding, quality control, signaling and degradation." *Curr Opin Struct Biol* 2001, 11: 120–124.

Cramer P, Bushnell DA, Fu J, Gnatt AL, Maier–Davis B, Thompson NE, Burgess RR, Edwards AM, David PR, and Kornberg RD. "Architecture of RNA Polymerase II and Implications for the Transcription Mechanism." *Science* 2000, 288: 640–648.

Custodio N and Carmo–Forseac M. "Quality control of gene expression in the nucleus." *J Cell Mol Med* 2001, 5: 267–275.

Hamm–Alvarez SF, Kim PY, and Sheetz MP. "Regulation of vesicle transport in CV-1 cells and extracts." *J Cell Science* 1993, 106: 955–966.

Herendeen DR and TJ Kelly. "DNA Polymerase III: Running rings around the fork." *Cell* 1996, 84: 5–8.

Keiser CA and Schekman R. "Distinct sets of SEC genes govern transport vesicle formation and fusion early in secretory pathway." *Cell* 1990, 61: 723–733.

Keren K, Krueger M, Gilad R, Ben–Yoseph G, Sivan U, and Braun E. "Sequence-specific molecular lithography on single DNA molecules." *Science* 2002, 297: 72–75.

Lazcano A and Miller SL. "On the origin of metabolic pathways." *J Mol Evol* 1999, 49: 424–431.

Morgan JA and Rhodes D. "Mathematical modeling of plant metabolic pathways." *Metab Eng* 2002, 4: 80–89.

Pelham HRB. The Croonian Lecture 1999. "Intracellular membrane traffic: getting proteins sorted." *Philos Trans R Soc Lond B Biol Sci* 1999 Aug 29, 354(1388): 1471–1478. Review. PMID: 10515003; UI: 99444592.

Pelham HRB and Rothman JE. "The debate about Golgi transport-two sides of the same coin." *Cell* 2000, 102: 713–719.

Vale RD and Milligan RA. "The way things move: looking under the hood of molecular motor proteins." *Science* 2000, 288: 88–95.

Weber T, Zemelman BV, McNew JA, Westermann B, Gmacl M, Parlati F, Sollner TH, and Rothman JE. "SNAREpins: minimal machinery for membrane fusion." *Cell* 92: 759–772.

CHAPTER

4

Gene Circuits

4.1 | Introduction

This chapter discusses the rules that living systems use to determine gene-expression characteristics and protein profiles. As discussed in Chapter 2, the flow of information in gene expression is from genes encoded by DNA to mRNA by transcription and from mRNA to protein by translation. However, a gene is not simply a recipe for the amino-acid sequence of a protein because a protein-coding gene also contains crucial information about the conditions for which the gene is expressed. Genes that encode proteins are composed of coding sequences and regulatory sequences; the latter controls the rate of transcription of the coding sequence. Regulatory sequences are similar in function to electrical circuits in that they process multiple input signals to determine output in a brief time interval. Binding of special proteins to the DNA-regulator sites comprises the signals. Such interactions between genes and proteins are central to the functioning of living organisms.

Our exposition begins with protein-coding genes and their anatomy. Prokaryotic genes are clustered on DNA in units called operons. A common regulatory element describes the conditions under which the proteins encoded by the genes in an operon are synthesized (expressed). In contrast, protein-coding genes in eukaryotes each have their own regulatory circuits. The logic controlling transcription of genes involved in the development of an organism is especially complex and can be analyzed best using the tools of circuit engineering. The input signals to these genetic circuits are the products of other genes; thus, the genome of an organism may be viewed as a network that controls gene activity and phenotype. The chapter concludes with a discussion of a bioinformatics approach in the investigation of biochemical pathways. Recent advances in experimental large-scale biology have created an exciting new role for engineers in deciphering the structure and function of large gene networks.

4.2 | Gene to Protein

DNA structure was discussed in Chapter 2. Briefly, genomic DNA is composed of double-stranded DNA. Each strand is a polymer of four different types of nucleotide bases: adenine (A), guanine (G), cytosine (C), and thymine (T). The backbone of a DNA molecule consists of the linked phosphate and sugar groups that connects the 3' carbon of one sugar to the 5' carbon of another, establishing a sense of direction for the DNA strand.

The sugar–phosphate backbones of two strands of DNA form a uniform helix, with strands placed in opposite directions. The bases of each strand are placed inside the helix. The strands are held together by hydrogen bonds between opposing bases according to the base pair rule: A is always paired with T and G is always paired with C. Consequently, the sequence of bases along one strand completely determines the sequence in the other strand (Fig. 4.1).

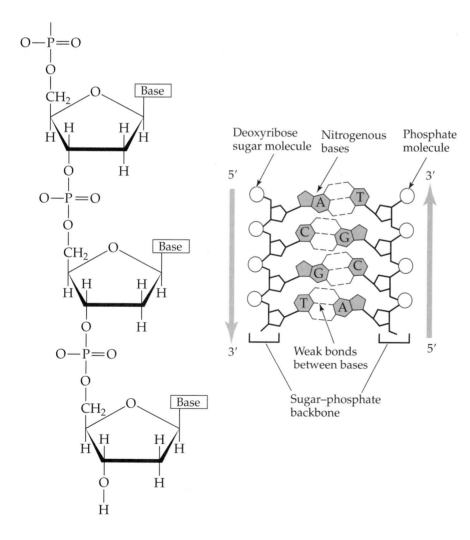

FIGURE 4.1 A DNA double helix is composed of two strands (chains) with opposite orientation. The backbone of a DNA strand consists of sequential links of sugar and phosphate units. The carbons of this sugar are identified as 1', 2', 3', 4', and 5' clockwise from the oxygen atom. The phosphate bonds connect the 3' carbon of one sugar to the 5' carbon of the adjacent sugar. The nucleotide bases are positioned on the inside of the DNA helix and are held together by hydrogen bonds. The base sequence is read by convention from the 5' to 3' end.

Double-stranded DNA is circular in prokaryotes, whereas the genome of eukaryotic cells is composed of linear DNA molecules, each forming a chromosome with associated proteins. Identical copies of DNA are present in the nuclei of cells of a multicellular organism. Any slight differences that may exist are due to random mutations or exposure to radiation or chemical mutagens.

Even for the simplest organisms, the genomic text encoded by DNA is partially understood and includes the recipes for proteins and the conditions under which these recipes are produced. DNA also codes for RNA molecules involved in protein synthesis. Although DNA does not directly determine the synthesis of other macromolecules such as sugars and lipids, it does encode the enzymes that catalyze their synthesis.

Genes are the units of genetic information stored in the sequences of DNA. Each gene contains information about either a single protein (more precisely a single polypeptide) or an RNA molecule (tRNA or rRNA). A gene can be defined as the sum total of base sequences of DNA necessary for the synthesis of a functional polypeptide (or RNA molecule). The regulatory segments of a protein-coding gene are reminiscent of electrical circuits and may be quite complex, containing large numbers of interdependent switches.

Genes are found on both strands of the double-stranded DNA (Fig. 4.2). In eukaryotic DNA, coding regions (exons) are separated by long noncoding sequences (introns). Exons and introns and the regulatory sequences of a gene lie on the same strand, and the start and stop codons are at the 3' and 5' ends of the strand, respectively.

The first step in the synthesis of proteins (polypeptides) is transcription. As was discussed in Chapter 3, a large enzyme complex (RNA polymerase) attaches onto the regulatory segment of a gene immediately upstream from the coding region. The transcription complex then begins to synthesize messenger RNA from nucleotides in the cytoplasm using the gene sequence as template (Fig. 4.3). DNA base A pairs with RNA base U. Base C on DNA pairs with base G on RNA and vice versa. Both exons and introns are transcribed, but introns are then excised by another large enzyme complex and the ends of exons are annealed. In some cases, exons can be recombined (spliced) in multiple ways that each lead to different sequences of amino acids. This process is called *alternative splicing* and accounts for the production of more

FIGURE 4.2 Genes are positioned on both strands of DNA. The start codon of a gene faces the 3' end of the DNA strand to which it belongs. In prokaryotes, genes are clustered in units called operons. (See shaded strips in prokaryotic DNA.) Genes belonging to the same operon lie on the same DNA strand. In eukaryotic genes, coding regions (exons) may be separated by noncoding segments (introns). Each gene has a circuit-regulating gene activity. The DNA segment identified with letter P (promoter) denotes the sequence-regulating gene expression.

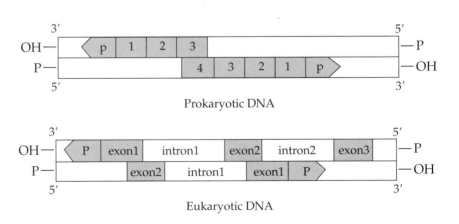

Prokaryotic DNA

Eukaryotic DNA

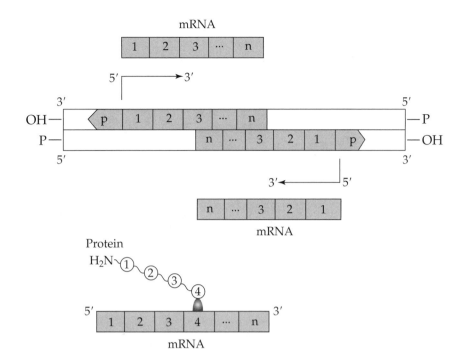

FIGURE 4.3 RNA polymerases use the DNA strand containing the gene as a template to transcribe messenger RNA (mRNA). RNA polymerase moves along DNA in the 3′ to 5′ direction, adding nucleotides to the 3′ end of the growing mRNA chain. The translation from mRNA to polypeptide begins from the 5′ end of mRNA and proceeds from the N-terminal to the C-terminal of the polypeptide.

than one polypeptide by certain eukaryotic genes. The resulting mRNA leaves the nucleus and associates with a ribosome, where translation occurs.

Only 20 out of the hundreds of known amino acids are found in proteins. The amino acids that belong to this subset of 20 vary in size, charge, and reactivity. Like DNA and RNA, polypeptides have an intrinsic sense of direction. The polypeptide grows from its N-terminal amino-acid end to its C-terminal amino acid (Fig. 4.3). The overall process of synthesis of a protein starting with gene transcription and finishing with a complete and functional protein is called gene expression.

The rules of translation from the base sequence of nucleotides to the amino-acid-residue sequences of proteins are universal in living systems. The base sequence of each gene can be partitioned into three-letter codons such as ACT and GTA.

The standard genetic code was described in Chapter 2. Translation from mRNA to amino acid is degenerate in the sense that multiple mRNA codons specify the same amino acid. The genetic code is slightly different for genes encoded by mitochondrial DNA and for a few rare forms of life. Translation of the DNA language into the amino-acid sequence of a protein is illustrated in Fig. 4.4.

One of the biggest challenges of gene informatics today is to identify the sequence segments associated with a gene in an already decoded eukaryotic genome. Long stretches of DNA that begin with a start codon and terminate with a stop codon are likely gene candidates in prokaryotes. In eukaryotes, however, separation of exons by long stretches of introns makes gene recognition more challenging. Pattern-recognition tools used in the prediction of gene sequences on a decoded DNA molecule will be discussed in the next chapter.

FIGURE 4.4 The genetic code given in Chapter 2 relates nucleic acid codons to amino acids. This figure illustrates translation using the codons for serine (S), arginine (R), isoleucine (I), and glycine (G).

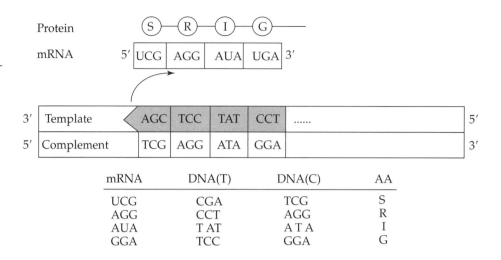

mRNA	DNA(T)	DNA(C)	AA
UCG	CGA	TCG	S
AGG	CCT	AGG	R
AUA	T AT	A T A	I
GGA	TCC	GGA	G

4.3 | Gene Anatomy

Regulation of gene expression is controlled mostly at the transcription level, although other levels of regulation occur at the levels of RNA stability, translational rate, and protein degradation. Regulatory circuits for gene transcription in prokaryotes are relatively simple and will be discussed first.

4.3.1 Prokaryotic Genes

Bacterial DNA contains mostly uninterrupted protein-coding regions in which genes are clustered on DNA in subunits called operons. Genes belonging to the same operon lie on the same DNA strand and encode proteins with related functions. An operon consists of *promoter* (P) and *operator* (O) regulatory sequences, coding regions of each gene, and short sequences separating the coding regions.

Figure 4.5 shows the structure of the bacterial lac operon. Genes belonging to this operon are involved in the digestion of the milk sugar lactose. These genes are lacZ, lacY, and lacA, respectively. The lacZ gene contains 3075 bases (b) and encodes β-galactosidase, the enzyme that splits the disaccharide lactose into the monosaccharides glucose and galactose. The transmembrane protein encoded by the 1254 base-long lacY gene is called *lactose permease*. This protein uses the energy available from the electrochemical gradient across the membrane to pump lactose into the cell. The gene lacA is the shortest (612 b) and the protein it encodes is an enzyme called *thiogalactoside transacetylase*, which is involved in the degradation of small carbon compounds. The regulatory segment of the lac operon is composed of 123 bp. The sequence of 39 bp immediately upstream of lacZ is the *operator*, and the remaining sequence (84 bp) is the *promoter*. As discussed in Chapter 3, RNA polymerase (RNAp) must bind to the promoter in order to initiate transcription. RNAp has low affinity for DNA, except in promoter regions.

When a *repressor* protein binds to the operator region of the operon, it blocks the interaction of the RNAp with the gene-regulatory segment and therefore prevents gene transcription. The repressor protein for the lac operon is called the *lac*

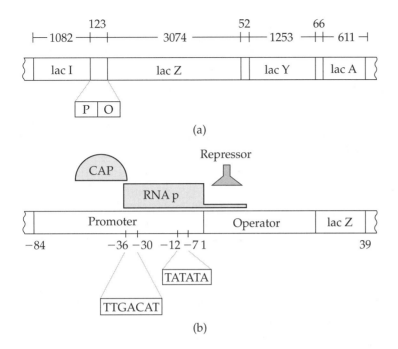

FIGURE 4.5 Prokaryotic genes encoding proteins are clustered into groups (operons) with common transcriptional regulation. The figure shows the schematic diagram of the lac operon. Genes in this operon are involved in the active transport of the milk sugar lactose from the surrounding environment and its subsequent degradation. Various DNA sequences comprising the lac operon, drawn in proportion along the DNA are depicted in (a). The operon contains three genes, lacZ, lacY, and lacA. Binding of regulatory proteins to sites on the control region immediately upstream of the lacZ gene controls the rate of transcription of these three genes (b). The lacI gene encodes the repressor for the lac operon and maps adjacent to the lac regulatory region. The lac repressor binds to the part of the regulatory region called the operator. Binding of repressor to the DNA blocks the binding of RNA polymerase (RNAp) to the adjacent promoter site. RNAp is the protein complex that executes the process of transcription. Activator proteins such as CAP can stimulate RNAp and enhance the transcription rate.

repressor and the gene encoding the lac repressor is lacI. This gene lies immediately upstream of the regulatory sequence and contains 1081 bp (Fig. 4.5). The repressor protein detaches from DNA upon interacting with another molecule, as described later in the chapter. DNA-binding proteins (such as CAP) bind to the promoter region to attract RNAp to the lac operon. Genes belonging to the same operon are *contiguous* in the sense that they are transcribed into a single mRNA molecule and are translated in equal amounts.

The nucleotide sequences of promoters and operators vary considerably from one operon to another in a bacterial genome. These sequences determine the strength with which RNAp is attracted to a particular operon. For example, genes that are highly expressed in a cell have promoters with high affinity for RNAp. Mutations in the promoter or the operator of the lac operon may halt the expression of all three genes in the lac operon. Consequently, a mutation in a regulatory sequence on DNA might lead to marked alterations in gene-expression patterns.

4.3.2 Eukaryotic Genes

The distribution of genes along eukaryotic DNA differs in three aspects from that of prokaryotic DNA:

1. Only a fraction of eukaryotic DNA codes for proteins or RNA molecules. According to the preliminary analysis of the human genome, about 1.1 percent of the genome represents protein-coding genes. Thousands of repeats of simple sequences comprise a significant portion of the DNA. For example, the sequence ATATAAT repeats itself many times in fruit fly DNA.

2. Unlike prokaryotic genes, the coding region of a eukaryotic gene is not necessarily continuous, but is composed of a series of coding regions (exons) that are separated by long noncoding regions (introns). Introns separate the different structural domains of the protein prescribed by the gene and often account for most of the gene. In the two genes associated with breast cancer, BRCA-1 on chromosome 17 and BRCA-2 on chromosome 13, introns constitute more than 98 percent of the genes. BRCA-1 has about 100,000 bp of DNA, but encodes a protein of 1863 amino acids. The BRCA-2 gene contains about 200,000 bp and encodes a protein of 3418 amino acids.

3. Eukaryotic DNA may have multiple sequence regions regulating the expression of just one gene (Fig. 4.6). Experiments involving deletion and insertion of DNA have provided much of the information about the physical location of gene regulatory elements along the DNA. Some of these regulatory elements are positioned immediately upstream of the start of the coding region and others may be as far upstream as 50,000 base pairs. By definition, the promoter is the regulatory element closest to the first exon (coding segment). Regulator sites distant from the first exon are called enhancers.

Eukaryotic proteins that regulate transcription by binding to regulatory sites are called transcription factors. Amino-acid sequences of transcription factors are conserved in eukaryotes. For example, several transcription factors that drive transcription in yeast are also effective in driving transcription in human cells. Many transcription factors are not specific to a given gene, but function as regulatory proteins for multiple genes; they are called general transcription factors. Other transcription factors are specific to particular sets of genes. Specific transcription factors are often regulated by the activity of cell-signaling pathways such that exposure of a cell to an external stimulus can lead to the increased synthesis or activation of these transcription factors. Once activated or increased in concentration these can, in turn, bind directly to gene-specific regulatory elements and regulate transcription of selected genes. Generally, it is thought that the function of these specific transcription factors is to reorganize the DNA in the vicinity of a specific gene in such a way as to make it accessible to the general transcription factors.

The assembly of proteins involved in transcription initiation begins with binding of the general transcription factor TFIID to the sequence called the TATA box at the promoter and concludes with the activation of the RNA polymerase II, the protein complex responsible for transcription of protein-coding genes. Both the distal and the proximal regulatory DNA sequences contain binding sites for a large number of regulatory proteins. These transcription factors typically stimulate transcription. *Activator* proteins that are bound to the *enhancer* transiently bind to RNA polymerase II by looping out the intervening DNA. The folding of DNA enables the proteins bound to distant regulatory sites to physically interact with the proteins bound to the proximal sites (Fig. 4.6).

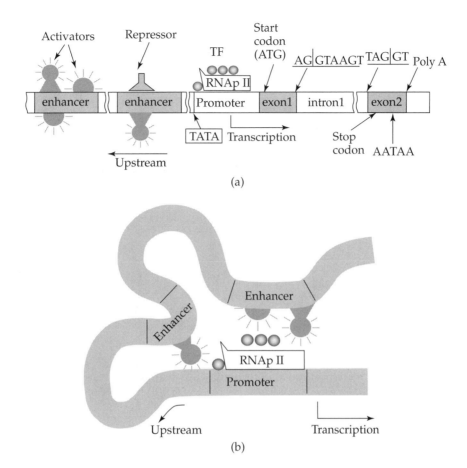

FIGURE 4.6 Eukaryotic genes encoding proteins have complex regulatory circuits. A protein-coding gene is composed of multiple regulatory components (promoters and enhancers), and amino-acid coding regions (exons) separated by noncoding regions (introns). The start codon is placed a few base pairs after the transcription initiation site, and the last exon contains a termination codon. The base sequences that appear at the boundaries between exons and introns are shown in (a). Since such short sequences appear frequently on DNA, advanced pattern-recognition techniques must be used to separate an exon from an intron. Both the exons and introns are transcribed into mRNA. The transcription continues past the stop codon in the last exon onto the so-called poly A sequence. The introns are excised and exons are brought together before mRNA leaves the nucleus and enters the cytoplasm for translation. The polypeptide transcription machine, RNA polymerase II (RNApII), binds to the promoter region immediately upstream from the first exon. However, the binding of polymerase II and its subsequent transcription function depend on the assembly of the so-called general transcription factors (GTF) on the promoter region. One important feature of many promoters is called the TATA box, a six-base sequence composed of T-A and A-T base pairs. The start point of transcription is typically 25 bp downstream from the TATA box. The general transcription factor TFIID binds to the TATA box and along with another regulatory protein (TFIIB) stimulates the interaction of RNApII with the promoter region. The transcription must await the assembly of other general transcription factors at the promoter and subsequent phosphorylation of the polymerase. The general transcription factors are highly conserved in evolution, and factors derived from yeast can be used in human cells to initiate transcription. The rate of transcriptional initiation is also affected by transcription factors that bind to distant regulatory elements (enhancers). Such transcription factors typically act as activators. Some transcription factors act as repressors by preventing the activating transcription factors from binding to DNA. DNA looping might enable transcription factors bound to enhancers to physically interact with the components of the transcription initiation complex assembled at the promoter (b).

FIGURE 4.7 Cytogenic map of human chromosomes (a). This drawing is based on the visual appearance of a chromosome when stained and examined under a microscope. Particularly important are visually distinct regions, called light and dark bands, which give each of the chromosomes a unique appearance. Genes are preferentially distributed on the segments of DNA with relatively high GC concentrations (light bands in the figure). Chromosomes shown in (a) are numbered according to size (sex chromosomes are the exception). The Web site http://genome.ucsc.edu/ allows the user to identify the genes as a function of position along the human chromosomes as illustrated here for chromosome 10 (b).

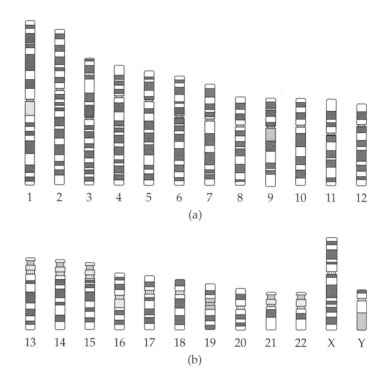

Packaging of DNA into nucleosomes and its extensive folding brings points separated by thousands of base pairs into close proximity. Thus, activators can bind to *enhancers* that are tens of thousands of base pairs from the initiation complex and still greatly stimulate transcription. Most known eukaryotic transcription factors that bind to enhancers are activators. Some of these factors are involved in opening up the DNA at the transcription-initiation site. However, eukaryotic gene transcription is also regulated by the activity of repressor proteins that bind to regulatory elements and repress transcription. Activation of a gene may require the displacement of repressor proteins by appropriate transcriptional activators.

Protein-coding genes are linearly arranged on the chromosomes of multicellular organisms. The approximate locations of thousands of genes on human chromosomes have already been identified (Fig. 4.7) and are presented at the Web site address http://genome.ucsc.edu/. Gene mapping provides significant information about the function of genes and the progression of certain diseases such as cancer. Many of the gene mutations that cause cancer occur in the cell-division-cycle-regulating genes that have already been mapped onto chromosomes. Gene mapping also allows for whole genome comparison of closely related organisms. See the next chapter for examples.

4.4 | Gene Circuits

4.4.1 Boolean Networks

A mathematical language for the regulation of gene expression is currently being developed. One of the simplest mathematical models used in gene-expression analysis is

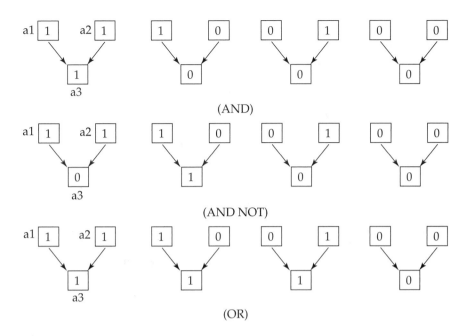

FIGURE 4.8 A three-node Boolean network and the corresponding Boolean logic functions. In this network, the state of node a3 depends on the inputs (states) from node a1 and a2. The rows in the figure indicate the AND, AND NOT, and OR response of node a3 to the four distinct input patterns.

Boolean networks. Boolean networks are logic-based mathematical models in which a node in the network is either on or off. A Boolean network consists of N numbered nodes, a_n, where $1 < n < N$. Figure 4.8 illustrates genetic circuits with three nodes. A node has an associated expression level, which can only take the values 0 or 1.

In a Boolean genetic network, each node represents either a gene or a biological stimulus. Stimuli can be physical factors such as the presence of mechanical stress, high temperature versus low temperature, or ion concentration. The nodal values represent the amount of gene activity (in the case of a gene) or the level of stimulus. The level of activity associated with each gene could potentially refer to mRNA abundance or protein activity. The representation allocates the continuous values of chemical concentrations or activity into two groups: high (1) and low (0). Therefore, in the notation of logic, each node in the network is either in an ON or an OFF state.

The state of a node in a Boolean network is determined by the states of some other nodes in the network. Each element (node) of the Boolean network receives input from a set of other elements in either 0s or 1s. Depending on the input, the element turns ON or OFF. Genes of living organisms interact with each other through protein intermediates and other molecules. These interactions are represented in Boolean networks with directional arrows that connect nodes. The direction of the arrow is from the node that exerts influence to the node being affected. The value each node takes in response to input from other nodes is described by a Boolean logic equation. A logic equation states the conditions for a node to be ON (take on the value of 1). If these conditions are not satisfied, then the node takes on the value 0. Figure 4.8 identifies the definitions of the commonly used logic symbols AND, AND NOT, and OR.

The simplest hypothetical genetic circuit is composed of just two genes, A and B, as shown in Fig. 4.9. The product of each gene in this two-gene network represses the expression of the other gene.

Such a system then has two different steady-state patterns of activity shown as (a) and (b) in Fig. 4.9. In the first steady-state pattern, gene A would be active and

FIGURE 4.9 A Boolean network with two nodes (*A*, *B*). The network structure is such that a node in the "ON" state turns off the other and vice versa. The symbols *a*, *b*, *c*, and *d* represent the four trajectories available to the network. Note that only two of those (*a*, *b*) are acyclic.

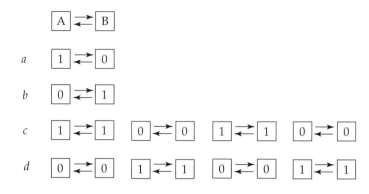

repress gene *B*; in the second pattern, gene *B* would be on and repress gene *A*. The network also has two unsteady trajectories that cycle between two states, (0, 0) and (1, 1). Thus, even a two-node genetic circuit has the potential to exhibit different behavior patterns, depending on the perturbation imposed on the system.

4.4.2 Regulation of Gene Expression in Prokaryotes

The gut bacterium *Escherichia coli* cells (*E. coli*) are good models to study the regulation of gene expression in prokaryotes. *E. coli* cells are available in most biological labs and are easy to grow. The genome of *E. coli* has been decoded, and the functions of most of its genes are known. Experiments with *E. coli* indicate dramatic increases in the levels of appropriate proteins in response to environmental changes. *E. coli* cells might contain as few as 10 copies of infrequently needed proteins and 10^5 copies of highly abundant proteins. The concentrations of proteins are tightly controlled at the gene transcription level. The transcription rate of some proteins varies over a 1000-fold range in response to the supply of nutrients or other environmental perturbations.

The preferred energy and carbon source for *E. coli* (and most organisms) is glucose. *E. coli* grown on glucose express very low levels of catabolic enzymes involved in the breakdown of more complex carbohydrates. For example, in low glucose *E. coli* contain only a few copies of the enzyme β-galactosidase, which is required to transform the milk sugar lactose into galactose and glucose. The concentration of β-galactosidase increases manyfold when *E. coli* are grown in a glucose-poor and lactose-rich environment. *E. coli* grown on glucose can transform lactose into its immediatory allolactose, but cannot split this intermediary into galactose and glucose in the absence of enzyme β-galactosidase.

The gene for β-galactosidase is lacZ, the first gene in the lac operon. In the absence of lactose, the *E. coli* cells contain no allolactose, and the lac represser protein remains bound tightly to the lac operator, preventing the transcription of β-galactosidase and associated genes (Fig. 4.10b).

When *E. coli* are switched to a lactose-rich and glucose-poor environment, they begin to take in lactose and transform it to allolactose. This sugar has strong affinity for the lac repressor protein. Binding of allolactose to the lac repressor protein causes a configurational change in the protein, dramatically weakening its attraction to the lac operator site. Subsequently, the repressor detaches from DNA and the operator site becomes available for the binding of the transcription machinery (Fig. 4.10c).

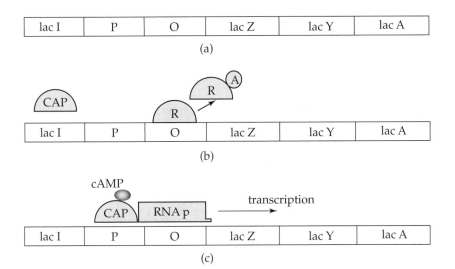

FIGURE 4.10 Regulation of the lac operon in *E. coli*. The lac repressor (R) is bound to the operator unless it forms a complex with allo-lactose (A). Binding of the repressor to the operator blocks transcription. RNA polymerase (RNAp) binds to the operator site and initiates transcription only when the CAP protein binds to the promoter site. CAP binds to DNA only when it forms a complex with the second messenger cAMP.

However, the detachment of the repressor from the operator is not a sufficient condition for the initiation of transcription. A number of operator sites on the DNA compete for the binding of RNA polymerase and those with higher affinity attract these transcription enzyme complexes to their sites more frequently. The affinity of the polymerase to the lac operator increases dramatically (as much as fiftyfold) upon binding of a promoter protein to an adjacent promoter site, as shown in Fig. 4.10c.

The promoter protein in this case is the *catabolic gene activating protein*, also called CAP. CAP stimulates transcription only when it forms a complex with cyclic AMP (cAMP), a signaling molecule involved in the regulation of multiple cell processes. Cyclic AMP is derived from ATP by the sequential hydrolysis of pyrophosphate. Unlike ATP and ADP, cAMP is not involved in metabolic pathways. Generally, the concentration of this signaling molecule in a cell is minute. A rise in concentration to the order of 10^{-8} M can modulate the activities of target proteins.

In bacteria, cAMP serves as a hunger signal, indicating the absence of glucose and leading to the synthesis of enzymes that can exploit other energy sources. This is accomplished by its transcription-enhancing influence on regulatory DNA binding proteins. The binding of CAP–cAMP complex to the promoter sequence of DNA creates an additional interaction site for RNAp and thereby stimulates transcription.

How does the binding of CAP to the promoter of the lac operon depend on the presence of glucose? The concentration of cAMP in the cell varies inversely with the concentration of glucose. In the presence of abundant glucose, cAMP concentration decreases, the CAP–cAMP complex dissociates, and the promoter detaches from DNA, making the lac operator site much less attractive for RNAp. Further transcription of β-galactosidase is repressed. In the absence of glucose and presence of lactose, the repressor detaches from the lac operator and the promoter stimulates the transcription of β-galactosidase. Within a few minutes of the addition of lactose, *E. coli* cells begin synthesizing β-galactosidase.

Regulation of the lac operon is a classical example of feedback control. In feedback control, the product of a later step in a pathway affects the rate of occurrence of an earlier step. Humans and animals use feedback control in common tasks; it is part of the learning process. Suppose that we have three shots to hit a target with a baseball. If the first throw is too short, then we throw with more force in the second trial and so on. Thus, the outcome of the first trial affects the throwing motion of the second trial. Similar feedback principles appear to shape molecular interactions within a cell.

4.4.3 Regulatory Logic of Eukaryotic Genes

Whereas multiple prokaryotic genes are regulated by a single regulatory element on DNA, a single eukaryotic gene can have multiple regulatory units. Eukaryotic genes with complex regulatory circuits often encode gene-specific transcription factors that are involved in cell division and subsequent development. Consider, for example, a simplified version of the regulatory logic of the Endo16 gene of the sea urchin shown in Fig. 4.11. This gene codes a secreted protein that is involved in pattern recognition at late stages of embryogenesis.

The gene has a proximal regulatory site A and distal regulatory sites B and C. The module A has three sites ($A1$, $A2$, $A3$) that directly interact with the basal

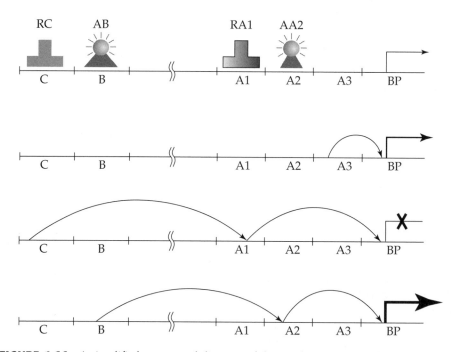

FIGURE 4.11 A simplified conceptual diagram of the regulatory circuit of a typical eukaryotic gene involved in development. The symbols A, B, and C denote, respectively, proximal (A) and distal enhancer (B) and distal repressor (C) clusters of regulatory sites of the gene. The module A contains three regulatory sites, two of which, when occupied with appropriate transcription factors, amplify and relay signals arriving from the distal modules to the basal promoter (BP) of the gene. The transcription is repressed when both C and $A1$ are occupied with their specific transcription factors. When site $A2$ is occupied, it acts as a switch to B; namely, it amplifies the signal coming from B and relays it to BP. In all other cases, the regulatory site $A3$ dominates the signal for transcription. See Yuh et al. (1998) for the regulatory logic of the sea urchin gene Endo16.

promoter (BP). The input signals from the distal modules B and C are interpreted in A, and then the output is transmitted to BP. Let $\gamma = \gamma(t)$ denote the output signal. The output signal takes the values (0), (1), and (2), depending on the rate of transcription. The parameter $\gamma = 0$ when the transcription is repressed. The values $\gamma = 1$ and $\gamma = 2$ correspond, respectively, to the moderate and high rates of transcription. Transcription is blocked when the specific site $A1$ and module C are occupied with appropriate repressive transcription factors. When sites $A1$ and C are not occupied with repressors, the site $A2$ of module A amplifies the enhancing signal coming from B and transmits it to BP. In all other cases, the signal γ is dominated by the intermediate level signal transmitted by the site $A3$. The rate of transcription of this gene depends on the amount of transcription factors specific to regulatory sites A, B, and C. For example, transcription is completely repressed in cells that express repressive factors for C and $A1$. In others, the expression pattern will depend on the concentrations of factors that interact with B, $A2$, and $A3$. The regulatory structure model considered here allows for diverse expression patterns. Which branch of the logic tree gene transcription takes at any point in time depends on the concentration and availability of the specific transcription factors that bind to the regulatory sites. It is also interesting to note that the logic of gene regulation is specified completely by the DNA sequence of the regulatory elements. However, the availability and activity of transcriptional activators and repressors is often regulated epigenetically by cell-signaling pathways.

4.4.4 Time Course of Transcription Regulation

The rate of gene transcription in prokaryotes changes within seconds, following an external input. Cell structure in prokaryotes is simple and allows for rapid movement of ligands that affect gene expression. The time course of eukaryotic gene transcription depends on the concentrations of transcription factors in the nucleus. The following example illustrates how the time course of gene expression depends on protein concentration. Consider a hypothetical genetic circuitry composed of three genes (a, b, c), stays as is two promoters (Pab, Pc), and a termination site T as shown in Fig. 4.12.

The proteins encoded by genes a, b, and c are denoted as A, B, and C, respectively. These proteins exist in negligibly small concentrations in the cell in the absence of external signaling. Repressor proteins Rab, Rb, and Rc remain bound to DNA and block the interaction of RNA polymerase with the transcription initiation sites of genes a, b, and c. The genes a and b share the common promoter Pab, but are separated by the terminator switch T. Transcription of gene a requires entry of a lipid-soluble hormone (H) into the cell. The hormone H binds to the repressor protein Rab and displaces Rab from its binding site on DNA. RNA polymerase then binds to the promoter Pab. The polymerase transcribes gene a, encoding protein A, and transcription stops at the gate T. As the concentration of protein A increases in the nucleus, these proteins begin to interact with Rc and displace Rc from the DNA; thus, the transcription of gene C is initiated. After the concentration of C reaches level Co, T closes and the transcription continues through b to produce signal protein B. In this gene circuit, the promoter Pab and the terminator T act as instant electronic switches responding to threshold concentrations of signaling proteins. The concentrations of signaling proteins in turn, depends on protein production and protein loss.

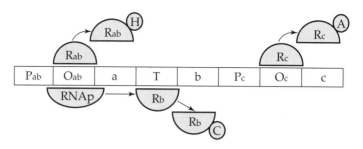

FIGURE 4.12 A hypothetical genetic circuit composed of three genes (*a*, *b*, *c*), two promoters (Pab, Pc), and a termination site *T*. Two of the genes in the circuit, *a* and *b*, share the common promoter Pab (operon), but are separated by the terminator *T*, which controls the transcription of *b*. The transcription of the operon containing *a* and *b* is initiated when a lipid-soluble hormone (H) enters the cell, displaces the repressor from the operator and an RNA polymerase (RNAp) binds at the promoter Pab. The polymerase transcribes gene a encoding protein A. The transcription of gene *c* must wait until a sufficient concentration of *A* is achieved to turn on the promoter for *c* (Pc). Subsequently, the signal protein C controls the terminator switch *T*. After the concentration of C reaches a certain level, *T* closes, and the transcription of *b* continues produce signal protein B.

When the concentrations of controlling signals change and a promoter turns off, a pipeline of already initiated transcripts must be cleared before transcription ceases. Note that in this very simple genetic circuit, each node receives an input signal from one node and transmits output to only one node.

4.5 | Bionetworks

In a cell, the gene circuits discussed previously are part of a large network composed of a large number of proteins and gene-regulatory elements. Metabolic pathways are also examples of bionetworks. In fact, living cells could be idealized as a network of chemicals linked by chemical reactions and enzymes. Electrical engineers have developed methodology for studying the interplay between network topology (connectivity of network nodes) and network robustness against failures and attacks in large networks such as the Internet. The robustness of such networks is typically investigated by removal of randomly selected nodes or node clusters from the network and determining the damage done to the integrity of the network as a whole. Researchers have developed similar methodology for protein and gene networks in cells. These networks are directed networks in the sense that the flow of signals (chemicals) has a well-established sense of direction. Bionetworks can be built at least in theory by considering the binary interactions between the network nodes. Let us illustrate bionetwork construction using the binary data gathered on eight gene products:

$$A \longrightarrow B; B \longrightarrow C; C \longrightarrow D; D \longrightarrow F;$$

$$B \longrightarrow E; E \longrightarrow E'; E' \longrightarrow F; F \longrightarrow G.$$

An inspection of these binary relations leads to the following pathw:

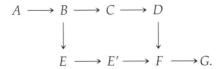

Such binary relations are available for literally thousands of gene products that serve as enzymes in metabolic pathways. Similar binary data are available for many of the signaling pathways that constitute the foundation for communication of cells with other cells and the environment. Using systematic (deductive) logic, researchers that developed the *Kyoto Encyclopedia of Genes and Genomes* (KEGG) have mapped the metabolic pathways for organisms including *E. coli*, yeast, and human. KEGG catalogs virtually all chemical compounds found in living cells and links each compound to a chemical pathway between genes and gene products. One of the most useful forms of data presented by KEGG is the collection of graphical diagrams representing nodal interactions in bionetworks. These diagrams are essentially what engineers call *bond graphs*, subsystems linked together by directed line segments. The reader may find the bionetworks presented in the KEGG Web site http://www.genome.sd.jp/kegg/pathway/map/ useful in the study of metabolic and regulatory pathways. A typical example of a KEGG pathway is shown in Fig. 4.13, which shows the pathway of the synthesis of the aromatic amino acids phenylalanine, tyrosine, and tryptophan in *E. coli* cells. Aromatic amino acids are hydrophobic in nature and are synthesized by plants and microorganisms, but not by humans. The KEGG flow diagram indicates the wiring between the building blocks of the pathway as well as the direction of flow. Elements of the network, the nodes, are comprised of molecules and genes. Each box (node) represents a gene product (enzyme) that mediates a metabolic reaction, and a circle indicates a chemical compound (metabolite).

The number contained in each box is the enzyme classification number of the enzyme (EC). Pressing on boxes and circles leads to detailed information on the corresponding enzymes and compounds. It is interesting that the enzymes encoded by the trp operon of *E. coli* (see Fig. 4.14) appear in the tryptophan pathway shown in Fig. 4.13. Enzymes coded by the genes trpE, trpD, trpC, trpB, and trpA of the trp operon are synthesized sequentially and in equimolar amounts by translation of the polygenic trp mRNA. Translation begins prior to the completion of transcription. Trp mRNA is synthesized in about four minutes and then rapidly degrades. Its short life span of about three minutes enables bacteria to respond quickly to changing needs for tryptophan. *E. coli* cells have the ability to vary the rate of production of its enzymes for tryptophan over a 700-fold range.

The trp operon is regulated as follows. The repressor protein for the trp operon (product of the trp R gene) has low affinity for the operator of the trp operon, but its affinity increases dramatically when bound to a tryptophan molecule (Fig. 4.14). When tryptophan is plentiful, transcription is blocked by the tryptophan-repressor complex. As the level of tryptophan in the cell decreases, repression is lifted and transcription begins. The reader can obtain similar information for all the pathways contained in KEGG. An important aim of large-scale biology is to complete the bond graph diagrams for all genes in a genome. Much progress has occurred in bacteria. The present challenge is to expand the pathways to include the gene circuitry of eukaryotic cells.

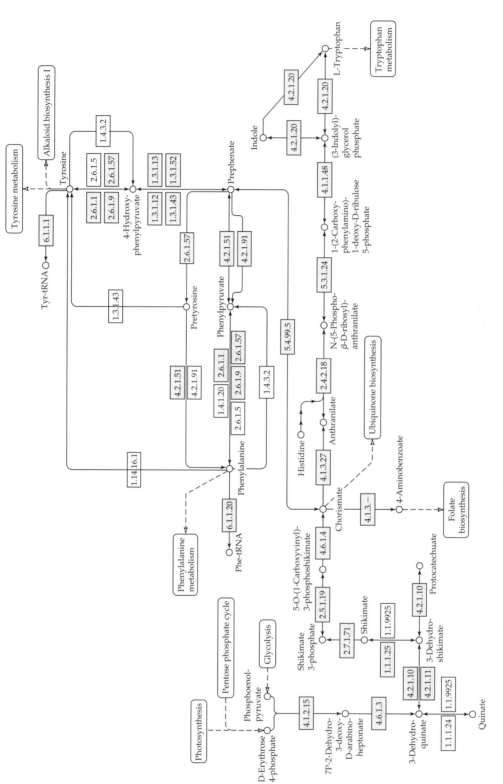

FIGURE 4.13 KEGG metabolic pathway associated with the biosynthesis of aromatic amino acids phenylalanine, tyrosine, and tryptophan. These essential amino acids are all hydrophobic. (Modified from http://www.genome.ad.jp/dbget-bin/getpathway?orgname=hsa&mapno=00380.)

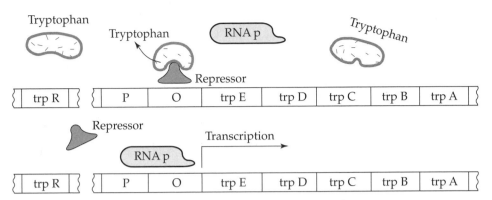

FIGURE 4.14 The trp operon in *E. coli* involved in the biosynthesis of aromatic amino acids phenylalanine, tyrosine, and tryptophan.

4.6 ASSIGNMENTS

4.1   Use the genetic code given in Table 2.1 to correlate the nucleotide sequence of the gene cynR of the *Escherichia coli* with the amino-acid sequence of its product. This gene belongs to the lac operon. The sequence of the gene can be found by searching the Entrez database for the keywords "lac operon." The following nucleotide sequence corresponds to the beginning of the gene cynR: ct accgtgattc atttccgcca acaaccgcgc atttatccaa. The gene is on the sense DNA strand because of inclusion of the phrase "*CDS complement* (...)" at the beginning of the DNA coding.

Ans: The translation of the sequence yields the following amino acid sequence for the protein coded by cynR: MLSRHINY....GNESR.

4.2   The *E. coli* gene cynS lies on the complementary strand of the decoded strand of DNA (cDNA). Parts of the cDNA sequence follow:

   1861 gcgtaacctt atttttaaac catcaggagt tccaccatga ttcagtcaca aattaaccgc;

   2341 aaatatctgc cgaccaaacc gttctgacag.

Determine the amino acid sequence.

Ans: Amino-acid sequence of the protein coded by cynS is as follows:

   MIQSQI ... LPTKPF

4.3   The following DNA sequence is from the coding region of a gene:

   accgtgattcatcccgccaacaaccgctattatac.

The codon for the first amino acid in the sequence could begin with any of the first three letters in the base sequence. Each starting nucleotide defines an open reading frame. List the three possible amino-acid sequences corresponding to the three open reading frames for the given nucleotide sequence. The DNA bases are shown in capital letters A, T, G, and C in most textbooks in biology, but are represented in small letters at the NCBI Genome Web page.

Ans: Amino-Acid Sequence (1) WHVGRLLAII; (2) GTKGGCWRY; (3) ALSRAVVGVNM.

4.4   Suppose that the sequence given in 4.3 belongs to the complementary sequence of a gene. Determine three possible amino-acid sequences compatible with this sequence.

Ans: Amino-Acid Sequence (1) TVIHPANNRYY; (2) P*FIPPTTTAII; (3) RDSSRQQPLLY.

4.5   The first five amino acids of the enzyme β-galactosidase written in single-letter amino-acid representation are as follows: MTMIT. How many different DNA coding sequences give rise to this chain of amino acids? Write down four such nucleotide sequences.

Ans: Fourty-eight different nucleotide sequences give rise to the given amino-acid sequence.

4.6   The lac operon lies on the complementary strand of DNA in *E. coli*. The sequence for lacZ is presented in the Web page http://www.ncbi.nlm.nih.gov:80/entr...uids=1786532&dopt=GeBank. Determine the coding sequence for the first 10 amino acids encoded by this gene. Note that you can reach the website through Entrez by nucleotide search using "lac Operon AND *E. coli.*"

Ans: MTMITDSLAV.

4.7 Determine the DNA sequence separating genes lacZ and lacY in the lac operon of the *Escherichia coli.*

Ans: taat ggatttcctt acgcgaaata cgggcagaca tggcctgcc.

4.8 Using the NCBI Taxonomy home page, determine the genetic code for the genes of vertebrate mitochondria. What are the differences between the vertebrate mitochondrial code and the standard code?

Ans:

AGA	Ter	* (Mitochondria)	Arg	R(Standard)
AGG	Ter	* (Mitochondria)	Arg	R(Standard)
AUA	Met	M (Mitochondria)	Ile	I(Standard)
UGA	Trp	W(Mitochondria)	Ter	*(Standard)

4.9 Suppose that the genome of a hypothetical organism contains equal proportions of bases A, T, G, and C and that these bases are randomly distributed along the DNA. What would be the ratio of the number of serine residues to the number of tryptophan residues in the proteins coded by this organism for each tryptophan?

Ans: 6/1 ratio.

4.10 The mRNAs found in the cells of an organism contain mRNA bases in the following ratio: A: 40%; U: 20%; C: 25%; and G: 15%. Determine the probability that a codon chosen at random is (a) AUG; (b) CGU.

Ans: $P(AUG) = 0.40 \cdot 0.20 \cdot 0.15 = 0.012$.

4.11 The mRNAs found in the cells of an organism contain mRNA bases in the following ratio: A: 40%; U: 20%; C: 25%; and G: 15%. Determine the probability that an amino acid chosen at random in a protein of the organism is (a) glycine; (b) arginine.

Ans:

$$P(G) = P(GGU) + P(GGC) + P(GGA) + P(GGG)$$
$$= 0.15 \cdot 0.15 \cdot 0.20 + 0.15 \cdot 0.15 \cdot 0.25$$
$$+ 0.15 \cdot 0.15 \cdot 0.40 + 0.15 \cdot 0.15 \cdot 0.15 = 0.0225.$$

4.12 Go to the www.ncbi.nlim.nih.gov Web page, and read about the use of the similarity search engine BLAST. What is an accession number? How do you obtain the accession number of a nucleotide sequence? How many different types of input data format are allowed in BLAST? What is the meaning of the parameter E (expectance)? Using the accession mode input of BLAST,

determine the three most similar bacterial genes to *E. coli* lacA gene. What kind of proteins do these genes code for? Which human gene has the most similarity to *E. coli* lacA? What is the protein encoded by this gene?

4.13 Short base sequences of about 10–20 bases are used in gene informatics technology as identification markers for protein-coding genes. Given a sequence containing n bases, determine the probability for finding an exact match to the n-base sequence along a long DNA molecule in which bases are arranged in random. How many times would you expect to encounter an 11-base sequence along such a DNA containing 4 million base pairs?

Ans: $P = \left(\frac{1}{4}\right)^n$ and $E = 4,000,000/P$.

4.14 Use the sequence similarity search engine BLASTp to determine which genes in *Escherichia coli* code proteins containing the amino-acid sequence TEYKTVRGLTRGLM. Determine background information on the products of these genes.

4.15 *Whole Genome Informatic.* Search for the genome of the fruit fly (drosophila melanogaster) in the Genome division of Entrez (www.ncbi.nlm.nih.gov). What is the accession number of the genome? When was the genome decoded? How many chromosomes, base pairs (bp), protein-coding genes, and RNA genes does the fruit fly genome have?

4.16 *Boolean Genetic Circuits.* Let us consider a Boolean network with four nodes. Assume that node 3 receives input from nodes 1 and 2. The response of node 3 for input signals from other nodes is such that

$$x3 = 0 \text{ for } x1 = x2 = 1;$$
$$x3 = 0 \text{ for } x1 = 0 \text{ and } x2 = 1;$$
$$x3 = 1 \text{ for } x1 = 1 \text{ and } x2 = 0;$$
$$x3 = 0 \text{ for } x1 = x2 = 0.$$

The tabulation of outcomes of a node for all possible input-signal combinations such as that presented in this problem is called the truth table of the node. Determine the logic equation for this node.

Ans: $x3: = x1 \text{ AND NOT } x2$.

4.17 *Boolean Genetic Circuits.* In the same Boolean network, node 2 receives input from nodes 0 and 1. The truth table for node 2 is as follows:

$$x2 = 1 \text{ for } x0 = x1 = 1;$$
$$x2 = 0 \text{ for } x0 = 0 \text{ and } x1 = 1;$$

$x2 = 0$ for $x0 = 1$ and $x1 = 0$;

$x2 = 0$ for $x0 = x1 = 0$.

In this network, nodes 0 and 1 are constitutively expressed. Determine the logic equations for nodes 0, 1, and 2.

Ans: $x0: = 1$; $x1: = 1$; $x2: = x0$ AND $x1$.

4.18 A four-node genetic circuit has the following logic functions:

$$x0: = 1; x1: = 1; x2: = x0 \text{ AND } x1;$$
$$x3: = x1 \text{ AND NOT } x2.$$

Determine the steady-state expression of the genes in the circuit.

Ans: $x0 = x1 = x2 = 1$; $x3 = 0$.

4.19 *Inverse Circuit Dynamics.* The wild-type state (steady-state) of a four-node genetic circuit is denoted by $p0$ and has the following values:

$$x1 = x2 = x3 = 1; x4 = 0.$$

The system is then subjected to the following directed perturbations, one at a time: Delete node 1 (set $x1 = 0$), delete node 2 (set $x2 = 0$), delete node 3 (set $x3 = 0$), and high-express node 4 (set $x4 = 1$). The resulting expression matrix generated is as follows:

	$x1$	$x2$	$x3$	$x4$
$p0$	1	1	1	0
$p1$	−	1	0	1
$p2$	1	−	0	0
$p3$	1	1	−	1
$p4$	1	1	1	+

In this matrix, deletions and overexpressions are shown with symbols − and +, respectively. For each node in the genetic circuit a_n, consider all pairs of rows (i, j) in the expression matrix for which the expression levels of a_n differ, excluding the row where the value of a_n was perturbed experimentally. For each such pair, find the set of all other nodes whose expression levels also differ between the two rows. The network itself is self-contained. The nodes that are determined by this procedure constitute the candidate signaling nodes for a_n. Construct candidates for logic functions for each node.

Ans: a_1 and a_2 do not depend on other nodes; therefore, $x1 = x2 = 1$; a_3 may depend on $x1$, $x2$, and $x4$; a_4 may depend on $x1$, $x2$, and $x3$.

4.20 *Inverse Circuit Dynamics.* Design four-node Boolean genetic circuits consistent with the expression profile matrix presented in Problem 4.19.

Ans: Model A: $x1: = 1$, $x2: = 1$; $x3: = x1$ AND $x2$; $x4 = x2$ AND NOT $x3$.

4.21 *Second Messenger.* Discuss the role of cAMP in the life cycle of the simple eucaryote, slime mold (Dictyostelium discoideum). Specify the changes in gene expression in high concentrations of cAMP.

4.22 *Enzyme function and structure.* Using KEGG, determine the function and the pathway of the human enzyme, glyceraldehyde-3-phosphate dehydrogenase, EC number = (1.2.1.12). (See http://www.genome.ad.jp/kegg/kegg2.html.) Alternatively, you could search the NCBI Locus Link resource for the gene coding glyceraldehyde-3-phosphate dehydrogenase. (Start with www.ncbi.nlm.nih.gov/, press entrez, and then press nucleotide.) Once you have found the Locus link report, click on the E.C. number for the gene (it should be 1.2.1.12) then so to the Enzyme Classification entry at the www.exspasy.ch Web server. Follow the link to the Kyoto database (KEGG). Go to the end of the page, click KEGG database, and click on the link for MAP for the glycolysis pathway. From this, switch to the view of the pathways, for human. The enzymes present in humans are highlighted in green. Determine the molecular structure of this enzyme using www.ncbi.nlm.nih.gov/. (Press Entrez, press protein, and press structure, and press view or save.)

4.23 *Enzyme Function.* What is the functional role of the human enzyme alcohol dehydrogenase (NAPD+)? What is its E.C. number? Determine whether the yeast genome also encodes this enzyme. (*Hint:* In the KEGG glycolysis map, click on the E.C. number of the enzyme.)

4.24 Using the KEGG database ligand search tool, determine the numbers of known enzymes that hydrolyze (a) ATP and (b) GTP. Speculate why ATP is used much more frequently than GTP as a readily available chemical energy source. Both dATP and ATP contain triphosphate, but only ATP (not dATP) is used as a readily available energy source. Explain how structure affects function in this case. Note that dATP and ATP are precursors of DNA and RNA, respectively.

4.25 *G Proteins.* An important example of a G protein-linked receptor is the β_2-adrenegic receptor, which binds to the hormones epinephrine and norephinephrine.

Both of these hormones are charged compounds and are released by the adrenal gland in response to fright and heavy exercise. Under these stress conditions, all tissues have an increased need for glucose and fatty acids. Binding of these hormones to β_2 adrenegic receptors on the surface of liver cells cause rapid breakdown of glycogen stored in the liver the release of glucose to the blood stream. Conduct a literature search to identify the difference between β_1- and β_2-adrenegic receptors. In order to seek justification for the notion that this hormone receptor forms seven transmembrane α-helices, use the www.ncbi.nlm.nih.gov/ Web site to identify the sequence of 22–24 hydrophobic amino-acid residues that repeats itself seven times in the amino-acid sequence of this protein. Compare estimated physical length of these repeated sequences with the typical width of a lipid bilayer. Describe the different effects epinephrine has on different cell types.

4.26   Using KEGG metabolic pathways, compare the hydrolysis of lactose in *Escherichia coli* and human. You can find the pathway diagrams at http://www.genome.ad.jp/dbget-bin/get_pathway?org_name=eco&mapno=00052. The enzyme β-galactosidase is shown in these figures with the E.C. symbol 3.2.1.23.

4.27   The mammalian protein NEM-sensitive factor (NSF) is required after vesicle budding, but before vesicle fusion. Search the Homology databases created by the National Library of Medicine (or the KEGG database) to determine NSF's homolog in yeast. What is the function of the homolog yeast protein?

4.28   Molecular genetics is increasingly used in the diagnosis and treatment of hereditary diseases. Point mutations in the epithelial cell-surface adhesion molecule integrin $\alpha6\beta4$ lead to a number of skin abnormalities. Heterologous mutations in the $\beta4$ integrin gene produce the blistering disease epidermolysis bullosa. In one case, the maternal mutation, recorded as R1281W, involved the substitution of a tryptophan codon in place of arginine at amino-acid position 1281 in exon 31. The paternal mutation, identified as R252C, involved the substitution of a codon for arginine for cysteine at amino acid 252 within exon 8. Determine the substitution values for these substitutions. Discuss the consequences of defects in the integrin $\alpha6$ and $\beta4$ subunits in skin diseases such as junctional epidermolysis bullosa.

4.29   Transmembrane proteins are often anchored in the membrane by a stop-transfer sequence, a short stretch of positively charged amino acids, immediately following the hydrophobic signal sequence. Proteins that span the membrane several times often have multiple signals (polytopic), which accounts for their specific topology. Most transmembrane segments of proteins are made up of 20 hydrophobic amino acids. Search for such 20-amino-acid motifs in the amino-acid sequence of the G protein: gkvlskifgn kemrilmlgl daagkttily klklgqsvtt iptvgfnvet vtyknvkfnv wdvggqdkir plwrhyytgt qglifvvdca drdridearq elhriindre mrdaiilifa nkqdlpdamk pheiqeklgl trirdrnwyv qpscatsgdg lyegltwlts nyks.

REFERENCES

Alon U, Surette MG, Barkai N, and Leibler S. "Robustness in bacterial chemotaxis" *Nature* 1999, 397: 168–171.

Baxevanis AD and Oulette BF, editors. *Bioinformatics: A Practical Guide to the Analysis of Genes and Proteins*. 2d Edition, John Wiley and Sons, Inc., 2001.

Chen T, He HL, and Church GM. "Modeling gene expression with differential equations." *Pacific Symposium in Biocomputing* 1999, 99: 101–112.

Clayton JD. "Keeping time with the human genome." *Nature* 2001, 409: 829–831

Edwards JS and Palsson BO. "Metabolic flux balance analysis and the in silico analysis of *Escherichia coli* K-12 gene deletions." *BMC Bioinformatics* 2000, 1: 1–17.

Fleischmann RD et al., "Whole genome random sequencing and assembly of haemophilus influenza Rd." *Science* 1995, 269: 496–512.

Fell D. *Understanding the Control of Metabolism*. Portland Press, 1996.

Galperin et al., "Comparing microbial genomes: How the gene set determines the lifestyle." In *Organization of the Prokaryotic Genome*, R L Charlebois, Ed. Washington DC: American Society of Microbiology, 1999: 91–108.

Heinrich R and Schuster S. *Regulation of Cellular Systems*. New York: Chapman and Hall, 1996.

Ideker TE, Thorsson V, and Karp RM. "Discovery of regulatory interactions through perturbation: inference and experimental design." *Pacific Symposium in Biocomputing*, 2000.

Kanehisa M. *Post-Genome Informatics*. NY: Oxford University Press, 2000.

Karnopp D and Rosenber R. *System Dynamics: A Unified Approach*. New York: John Wiley and Sons, 1975.

Lander ES et al. "Initial sequencing and analysis of the human genome." *Nature* 2001, 409: 860–928.

Lohr D, Venkov P, and Zlatonova J. "Transcriptional regulation in the yeast GAL gene family: a complex genetic network." *FASEB Journal* 1995, 9: 777–787.

McAdams HH and Shapiro L. "Circuit simulation of genetic networks." *Science* 1995, 269(4): 650–656.

Mertzen RM, McCraith SM, Spinelli SL, Torres FM, Fields S, Grayhack EJ, and Phizicky EM. "A biochemical genome approach for identifying genes by the activity of their products." *Science* 1999, 286: 1153–1157.

Mitchell AP. "Control of meiotic gene expression in Saccharomyces cerevisiae." *Microbiol Rev* 1994, 58: 56–70.

Norman TC, Smith DL, Sorger PK, Drees BL, O'Rourke SM, Hugues TR, Roberts CJ, Friend SH, Fields S, and Murray AW. "Genetic selection of peptide inhibitors of biological pathways." *Science* 1999, 285: 591–595.

Schena M, Shalon D, Davis RW, and Brown PO. "Quantitative monitoring of gene expression patterns with a complementary DNA." *Science* 1995, 270: 467–471.

Shoemaker DD. "Experimental annotation of the human genome using microarray technology." *Nature* 2001, 409: 922–927.

Somogyi R and Sniegoski CA. "Modeling the complexity of genetic networks: understanding multigenic and pleitropic regulation." *Complexity* 1996: 45.

Tatusov et al., "A genomic perspective on protein families." *Science* 1997, 278: 631–637.

Tatusuv RL et al., "Metabolism and evolution of Haemophilus influenzae Rd deduced from a whole-genome comparison with *Escherichia coli*." *Curr. Biol.* 1996, 6: 279–291.

Thoma JU. *Simulation by Bondgraphs*. Berlin, Germany: Springer–Verlag, 1990.

Thomas R and D'ari R. *Biological Feedback*. Boca Raton, FL: CRC Books, 1997.

Tsoukalas LH and Uhrig RE. *Fuzzy and Neural Approaches in Engineering*. New York: John Wiley and Sons, Inc., 1997.

Venter CJ. "The sequence of the human genome." *Science* 2001, 291: 1304–1351.

Whitesitt JE. *Boolean Algebra and Its Applications*. Reading, MA: Addison–Wesley Publishing, 1962.

Winzeller et al., "Saccharomyces deletion project." *Science* 1999, 285: 901–906.

Yuh C–H, Bolouri H, and Davidson EH. "Genomic cis-regulatory logic: experimental and computational analysis of a sea urchin gene." *Science* 1998, 279: 1998–2000.

Genomics: The Technology behind the Human Genome Project

5.1 | Introduction

The genome can be defined as the collection of DNA molecules that carries the hereditary information of the organism. Each cell in an organism, with the exception of postmeiotic germ cells (eggs and sperm), contains an identical copy of the genome. The field of genomics is the study of the sequence, content, and history of the genome. Our knowledge of genome evolution from ancient ancestor precursors is incomplete, but it is clear that random mutations in the sequence of DNA and DNA duplication have played a significant role in the evolution of genomes. The concept of sequence similarity has been instrumental in the comparison of genomes and in constructing a tree of life indicating kinship relationships among species.

Recent advances in biotechnology allow sequencing of even the most complex genomes and thereby enable the comparison of genomes across species as well as within members of the same species. The technology used in genome sequencing includes enzymatic techniques for cutting DNA into pieces and combining DNA from different sources (recombinant DNA), separating DNA fragments according to size (electrophoresis), and making copies of DNA using the cells' machinery for DNA replication (cloning) and the polymerase chain reaction (PCR). These and other methodologies can also be used to perturb genomes. Such genomic perturbations include gene mutation, insertion, and deletion and have been instrumental in the creation of genetically engineered agricultural products and in the development of living factories for producing desired human proteins.

This chapter begins with an introduction to genomics and provides a brief history of the evolution of genomic DNA. Sequence similarity is used extensively as a tool to uncover kinship between genes. Finally, the conceptual plan used to sequence the human genome is presented and the basic biotechnology associated with genomic decoding is outlined. Genomics is of particular interest to engineers and computer scientists because of its extensive use of cutting-edge technology such as nanotech sensors, microrobotics, informatics, and pattern-recognition methodology.

5.2 | History of the Genome

Biologists have long been interested in constructing a timeline (history) and kinship for the millions of life forms on Earth. The only figure in Darwin's book *Origin of Species* is that of an evolutionary tree, describing the hierarchical structure relating species to their most recent common ancestors. Until the molecular biology revolution, biologists constructed evolutionary trees from morphological comparisons of contemporary species and used limited fossil data to create a time line for the vertical descent of species. However, such trees were not comprehensive enough to include most species. Moreover, the morphological data used were subject to interpretation. Consequently, differences and commonality in cellular organization are often not conclusive enough to reliably categorize organisms into families and subfamilies.

More recently, biologists have turned to the use of gene-sequence similarity as a criterion to uncover the hierarchical kinship relations between organisms. Genes considered for comparison are those that are common to all organisms. Such genes have similar, but often not identical, nucleotide sequences in different organisms. Family trees of organisms that are based on gene similarity are called gene trees. The most comprehensive classification of living species is based on sequence similarity analysis of the gene for 16S ribosomal RNA, which specifies a component of the machinery that translates the nucleotide sequence of the gene into a protein. The choice of an rRNA gene to construct a family tree of life is appropriate because indirect evidence points to RNA as the first emerging type of macromolecule at the initial stages of life. RNA could function as a template for both DNA (reverse transcription) and protein (translation), and it has an enzymatic role in fundamental cellular activities.

Figure 5.1 shows the tree of life drawn on the basis of 16S rRNA. The rRNA-based life tree partitions living organisms into three branches protruding from the trunk: bacteria, archaea, and eukaryota. These branches differ in their organization of DNA and details of their translation machinery. There are also structural differences among the organisms belonging to different branches. These differences include the chemical composition of lipid membranes as well as the presence and absence of cellular organelles. Bacteria and archaea both lack nuclei and together form prokaryotes. Archaea includes microbial species that grow at 95°C in the highly acidic conditions found in hot sulfur beds. The extreme resistance of the enzymes produced by archaea to heat and acid make them highly attractive to biotechnology companies for their potential use in industry.

Eukaryotes are distinguished from other forms of life not only by their DNA organization, but also by the presence of nuclei, other organelles, and a cytoskeleton. Genes for cytoskeletal proteins such as actin, myosin, and microtubules are only found in eukaryotes. The word eukaryote means "true nuclei." Eukaryotes include humans and other animals, plants and fungi, and a rich variety of other organisms including algae and amoebae.

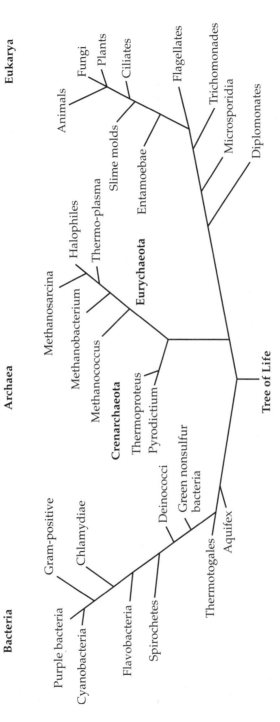

FIGURE 5.1 The universal tree of life based on the sequence similarity of the 16 S rRNA. The tree divides organisms into three domains: Eubacteria (or bacteria), Archaea, and Eukaryota.

The rRNA-based tree of life shown in Figure 5.1 depicts the universal ancestor cells as the trunk and the descendents as branches and leaves (present day life forms). In the deepest branches of the tree lie organisms that live at high temperatures such as those found near volcanos, sulfur springs, or the volcanic vents that gash the ocean floor. Similar physical conditions were thought to exist at the early stages of life on Earth. Biologists have long aspired to derive information on the genome of universal ancestor cells by running the tree of evolution backward going from its leaves down to the point where all its branches coalesce into a single trunk. Working back to the ancestor using the sequence of DNA letters is similar to the construction of the words of ancient mother tongues by linguists. However, there are difficulties in predicting the trunk from its known branches and leaves.

For example, the recently decoded genomes of ancient organisms living in extreme conditions did not reveal major genomic differences from those that live in more moderate conditions. Moreover, tree construction based on molecular similarity was dependent on the molecule or molecule clusters chosen for comparison. In some cases, tree variation could be explained by the possibility that early organisms more or less freely exchanged genes and gene clusters independent of reproduction.

In animals, genes are passed vertically from parent to child but single-celled creatures have the capacity to engulf each other to occasionally create an organism with a composite genome. Gene sequence similarities suggest that both archaea and eukaryotes might have acquired metabolic genes from bacteria in lateral gene traffic through a process called *horizontal transfer*. It is feasible that a community of ancestor cells used the same genetic code and frequently exchanged genes by horizontal transfer. At a certain stage of complexity, an emerging cell structure might have limited a cell's ability to exchange genes, and the ancestral pool split into the three domains seen today.

5.3 Sequence Similarity and Homologous Genes

This section introduces the reader to quantitative measures of sequence similarity and how these measures are used in exploring kinship between genes and across species.

5.3.1 Mutations

DNA undergoes mutation due to imperfections in replication, repair, and quality-control processes. External environmental factors such as radiation and chemical insult could be partially responsible for imperfections in DNA replication. A mutation is any change in the base sequence of a gene or noncoding DNA segment. A *point mutation* is a change affecting a single nucleotide in a gene. The point mutations in the following sequences are marked in ***bold***:

$$3'\ldots \text{ATGGGCTGCACGGAT}\ldots 5'$$
$$3'\ldots \text{ATGC GCTG}T\text{ACGGAT}\ldots 5'$$

Point mutations may cause a corresponding change in the amino-acid sequence of the protein that the gene produces. For example, the first mutation transforms the codon GGC to CGC and, in doing so, alters the coded amino acid from proline (P) to alanine (A). The other mutation in the sequence, however, does not affect the amino-acid sequence of the coded protein. Codons TGC and TGT both correspond to the amino acid threonine (T).

Point mutations in the coding regions frequently lead to shifts between synonymous codons coding the same amino acid. Such mutations do not alter the amino-acid sequence of a protein. Point mutations that cause changes in the amino-acid sequence may not always result in drastic changes in the function of the gene product. The amino-acid sequence coded by a gene might show slight variations among the members of the same species. *Alleles* of a gene include all the distinct (slightly varying) sequences of the same gene found in a given species. Point mutations occur much more frequently in noncoding regions of the genome than the coding regions because gene mutations that lead to nonfunctional proteins do not survive the forces of evolution.

A *rearrangement mutation* affects a large region of DNA. Insertions of additional material or deletions of a stretch of the gene are among the examples of observed DNA rearrangements. The following is an example of rearrangement mutation:

$$3'\ldots \text{ATGCGCTGCACG}\mathbf{\mathit{GAT}}\ldots 5'$$
$$3'\ldots \text{ATGCGC}\mathbf{\mathit{GAT}}\text{TGCA\_G}\ldots 5'$$

In this example, the order of codons shifted and one nucleotide in the second sequence was deleted. Rearrangement mutations that occur in the sequence of a gene may prevent gene expression or result in a gene product that is unrecognizable by the cell. In some instances, mutated genes may survive and contribute to the diversity of species.

5.3.2 Sequence Similarity and Homology

Two sequences are called *homologous* if their level of similarity implies that they are derived from a common ancestor. How do we determine whether two sequences are close, but inexact, copies of an ancestor gene? How do we evaluate the degree of similarity of two nucleotide (or amino-acid) sequences? Let us expand on these questions by considering a simple nucleotide sequence such as GTAATCG and imposing a series of point mutations on it:

GTAATCG (ancestor sequence)
GT**C**ATCG (GT**T**ATCG) (GT**G**ATCG) GTAATCG
GT**C**ATCG GTAATCG (GTA**G**TCG) (GTA**T**TCG) GTA**C**TCG

The point mutations are shown in bold in the preceding sequences. The sequences in parentheses identify those mutations that either kill the host, or prevent it from reproducing, and are therefore not carried over to the population. Only the sequences in the last row are known, and a common ancestor needs to be inferred by observations of sequence similarity. Let us align the sequences in the bottom row two at a time and score their similarity by assigning a positive integer value, say, 3, for each matching base pair and a minus integer value, say, −5, for each nonmatching element.

Then the similarity score (S) for the pairs shown would be as follows:

$$S = 13 \text{ for GTCATCG and GTAATCG;}$$
$$S = 5 \text{ for GTCATCG and GTACTCG;}$$
$$S = 13 \text{ for GTAATCG and GTACTCG.}$$

The sequence shared among the pairs with the highest S value is the predicted ancestor sequence GTAATCG. The other two sequences can be obtained from this common ancestor by a single substitution and are said to be homologous sequences to the ancestor sequence.

Sequence comparison for assessing similarity requires sequence alignment in order to achieve maximal levels of identity. *Global alignment* of pairs of genes (or proteins) potentially related by common ancestry is defined as an alignment throughout their lengths where all bases are aligned with another base or a gap. The following pairs of sequences constitute examples of global alignment:

<table>
<tr><td>GTAATCG</td><td>GTAATCG</td></tr>
<tr><td>GTACG_ _</td><td>GTA_ _CG</td></tr>
</table>

Another mode of alignment is *local alignment*, where one does not need to align all the bases in all the sequences. See, for example,

GTCTACG<u>GTAATCG</u>AGGTCATCGGTGGTCGGTGTCG
GTAAG:

Local alignments may provide information on sequence motifs of proteins found at the sites of interaction with other proteins. For example, sequences of the catalytic sites of enzymes with similar function are conserved across species. Multiple local-alignment options exist when two or more fairly long DNA segments are compared. In such cases, the alignment that gives the highest sequence alignment similarity score is considered as the best possible alignment.

Suppose, for example, that one considers the similarity of two sequences of different length, say, GTAATCG and GTACG. How does one align these two sequences to give the best sequence similarity index value? An alignment provides a mapping of residues in one sequence onto those of another. In alignment, each base in a sequence is used only once for comparison. Let us use the symbol "_" to denote a gap in the sequence. We then have the following possible modes of alignments:

<table>
<tr><td>GTAATCG;</td><td>GTAATCG;</td><td>GTAATCG;</td></tr>
<tr><td>GTACG_ _;</td><td>_ _GTACG;</td><td>_GTACG_.</td></tr>
</table>

Assuming that a gap is worth (-6), a substitution (-5), and matching sequence (3), we calculate the following similarity scores: $S = -13, -27,$ and -29. Thus, the alignment that appears first in the row of alignments highlights the similarity between the two sequences best.

The choice of scoring system used in evaluating similarity is somewhat arbitrary. A commonly used scoring formula is as follows: Add one for each matching base, subtract one for each mismatch (substitution), subtract five as gap opening penalty, and subtract one for each gap in the alignment. Consider, then, the following alignment:

<table>
<tr><td>GTAACTGCTAGA_ _;</td></tr>
<tr><td>GTAC_ _ GC_ _ GTCG.</td></tr>
</table>

Using the aforementioned scoring system, we find that

$$S = 6(1) + 3(-1) + 3(-5) + 6(-1) = -18.$$

The similarity score S, even when it accounts for insertions or deletions, is not sufficient when comparing degrees of similarity of pairs of sequences of significantly different lengths. Suppose, for example, one has the task of matching two sequences, one of 9 bases and the other of 17 bases, along a DNA sequence of 124 bases. The closest matching sequences in both cases to yield an S score of 5. Which sequence is more closely matching to its counterpart on the sequence of 124 bases? Matching of the longer sequence is more successful, because the longer the segment, the lower is the probability of a higher score.

For example, if one were to search for homologous sequences for ACGT and GTCATCTACGU along a bacterial DNA, thousands of identical matches would have been found for the first sequence, but few for the latter. This can be illustrated by a simple calculation involving probabilities. Consider a large pool of randomly generated nucleotide sequences of length r. The probability P of selecting a particular random sequence B is, then,

$$P(b) = 1/(4^r).$$

Let us consider next another randomly generated sequence of length m. Let us identify this longer sequence with BB. Sequence BB presents $(m - r)$ continuous sequences of length r as possible matches for the sequence B. These sequences overlap with neighbors in all, but the nucleotides at each end. The expected number of times the sequence B is encountered in the longer sequence BB is then given by the equation

$$E(b) = (m - r)/(4^r).$$

This equation shows that the number of possible hits decreases rapidly with increasing sequence length r. Thus, the higher the sequence length to match, the harder is the challenge.

A comparison of amino-acid sequences of proteins also yields information about their origin and evolution. Methods used to compare amino-acid sequences take into account the fact that some amino-acid mutations hardly affect the structure or function of a protein. The most prevalent replacements occur between amino acids with similar side chains such as substitutions within the following groups: (G, A); (A, S); (S, T); (I, V, L). Amino-acid replacements involving chemically similar amino acids appear more often on DNA than expected for random mutations. In such cases, the physical and chemical properties of the original residue are conserved. Researchers developed substitution matrices that put a numerical value for each possible substitution among the 20 amino acids found in proteins. In substitution matrix tables, punishment for replacements between chemically similar amino-acid pairs is smaller than that with nonsimilar pairs. The substitution matrix also assigns values for amino-acid matches between two sequences. A tryptophan match between two sequences is scored highest. Tryptophan is not chemically similar to any of the other 19 amino acids found in proteins, and its presence in protein sequences is largely conserved during evolution.

5.3.3 Ortholog and Paralog Genes

In many eukaryotes, DNA contains segments in which at least one protein-coding gene and its slightly different copies are clustered. Some of these inexact copies also code for proteins. DNA sequences that are thought to have arisen by duplication of an ancestral gene are referred to as duplicated protein-coding genes. *Paralog* genes are those genes that appear in more than one copy in a given organism by a duplication event. Products of paralog genes may diverge in function during the course of evolution. *Orthologous* families of genes consist of genes in different species that arose from a common ancestral gene during speciation. If two genes are orthologs, they can be traced by descent to the gene pool of a common ancestor. Ortholog genes perform similar functions in different organisms.

How does one establish whether two genes are orthologs (or paralogs)? For a given gene from one genome, the gene from another genome with the highest sequence similarity is the ortholog if the best hit is highly significant statistically. Paralogs are genes that belong to the same organism with highly significant sequence similarity. Thus, the first step is to establish gene similarity beyond what would be expected by random coincidence.

The National Center for Biotechnology Information (NCBI) provides a publicly accessible database search engine called BLAST for similarity alignment. BLAST stands for "The Basic Local Alignment Search Tool." The Web site http://www.ncbi.nlm.nih.gov?Education/blasttutorial.html provides a detailed background on BLAST. Given a sequence, BLAST searches in its large database for matches or close-matches. Because the number of matches depends on the sequence length, the program not only provides the similarity score S, but also the so-called *expect* value (E), the number of hits one can expect to see just by chance when searching a database of particular size. The lower the value of E, the higher is the probability that the match has not occurred by chance. The E values resulting from a BLAST search has a well-defined statistical interpretation, making real matches easier to distinguish from random background hits. BLAST can compare an amino-acid or nucleotide sequence against a library of sequences of amino acids and nucleotides, and such comparisons often lead to important insights into gene and protein function.

The Web site http://www.ncbi.nlm.nih.gov/COG/ presents in-depth information about orthologous gene families gathered by comparing protein sequences encoded in complete genomes. The information is presented in the database entitled Clusters of Orthologous Groups of Proteins or COGs. Each COG includes proteins from different species that are thought to be orthologs and therefore have the same function. At the beginning of 2001, of the known 4292 *Escherichia coli* genes, 2752 had already been placed in COG families. Each COG contains proteins with similar sequence and function. Research is underway to map known human genes into COGs. Preliminary studies have indicated the presence of 2758 orthologs between human and fly and 2031 between human and worm.

Orthologous families of proteins may contain multiple paralogs from different species. Such protein families in one species contain several to hundreds of members. In the human, there are more than 200 members of the protein kinase family. Protein kinases phosphorylate target proteins and effect a structural or functional change. Proteins involved in cell adhesion and in the regulation of cell division also comprise large families in vertebrates. In fact, protein families constitute as much as half of the protein-coding DNA in vertebrate genomes.

5.4 | Decoding the Human Genome

The revolution in New Biology and the onset of the Genomics Era was largely technique driven, advances in understanding often coming after the development of a new technique or approach. In several instances, the new approach was itself based on a serendipitous discovery. For example, the first DNA sequence-specific restriction endonuclease was discovered following the accidental finding that the bacterium Haemophilus influenzae rapidly broke down foreign phage DNA. There are now over 150 sequence-specific endonucleases available commercially. The use of restriction enzymes led to important advances in DNA sequencing and recombinant DNA methodology. Similarly, the discovery that bacteria living in hot springs have DNA polymerases that work well at extremely high temperatures made the polymerase chain reaction (PCR) that has revolutionized molecular genetics feasible for large-scale use. This chapter introduces some of the important techniques that form the underpinning of the new biology. It is not intended to be a complete listing of techniques, nor an in-depth guide to the use of them. Rather, it is intended to give the reader a fundamental understanding of how these procedures work and what they can be used for.

5.4.1 Nucleotide Sequencing

The nucleotide sequences of thousands of mRNA-coding genes were determined long before whole genome sequencing began. mRNA can be purified and complementary DNA (cDNA) made by reverse transcriptase. Sequencing the DNA then led to the elucidation of the nucleotide sequence of many individual genes. However, genomic DNA is much longer than a cDNA fragment containing a gene, and sequencing long pieces of DNA continues to be a challenging task. Currently available sequencers can only sequence DNA pieces less than 1000 base-pairs long. To sequence longer pieces of DNA, scientists came up with a creative approach: Cut many copies of the DNA randomly to generate a set of smaller, but overlapping DNA segments, sequence these segments, and develop a computer program to integrate the sequencing data to reconstruct the sequence of the original DNA with no prior knowledge of where the pieces originally came from. This method is called the *shotgun* procedure. Shotgun sequence assembly is in fact an inverse problem: Given a set of overlapping sequences randomly sampled from a target sequence, reconstruct the order and the position of those sequences in the target. This is accomplished using computational power to match the end point sequences of overlapping DNA fragments.

Shotgun sequencing technique has been successfully employed in sequencing the genomes of a large number of prokaryotes. However, the technique needed to be expanded for sequencing the chromosomes of multicellular organisms. Vertebrate chromosomes are much longer than prokaryotic DNA and contain long stretches of simple repeats. Therefore, the publicly funded International Human Genome Sequencing

Hierarchical shotgun sequencing

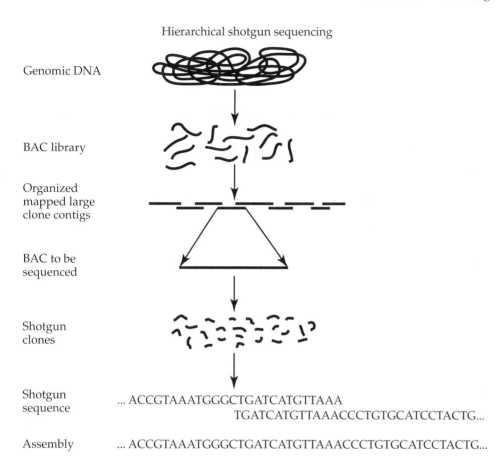

Genomic DNA

BAC library

Organized
mapped large
clone contigs

BAC to be
sequenced

Shotgun
clones

Shotgun
sequence

... ACCGTAAATGGGCTGATCATGTTAAA
        TGATCATGTTAAACCCTGTGCATCCTACTG...

Assembly   ... ACCGTAAATGGGCTGATCATGTTAAACCCTGTGCATCCTACTG...

FIGURE 5.2   Schematic representation of the hierarchical shotgun sequencing strategy that was used by the International Human Genome Sequencing Consortium in sequencing the human genome. A library of DNA-segment clones is constructed by fragmenting the target genome and cloning it into large fragment cloning vectors (BAC library). The genomic DNA fragments are then organized into a physical map and individual clones are selected, cut, and sequenced. Finally, the clone sequences are assembled to reconstruct the sequence of the genome. (Modified from Lander et al., *Nature* 2001, 409: 860–921.)

Consortium adapted a hierarchical shotgun methodology to sequence the human genome. Eric S. Lander and collaborators described the human genome sequence procedure adapted by the consortium in a research article published in *Nature* on 15 February 2001. The steps in the initial sequencing of the human genome are shown in Fig. 5.2. These steps can be summarized as follows in the subsequent sections.

5.4.2 Obtain and Purify DNA

DNA was obtained from multiple donors from diverse racial and ethnic backgrounds. Standard procedures and ready-to-use kits are available for purification of DNA from whole blood, bone marrow, lymphocytes, and cultured cells. The use of DNA from several individuals enables researchers to uncover patterns of DNA-sequence differences among the genomes of different individuals. In the human genome, nucleotide types vary at well-defined positions along DNA with about 1 percent frequency. These variations are referred as single nucleotide polymorphisms or SNPs. The correlation between SNPs and predisposition to cancer and other diseases is a subject of intense study (www.hgvbase.cgb.ki.se).

5.4.3 Create a Set of Overlapping DNA Fragments

As discussed in the next section, chromosomal DNA was cut into relatively large pieces using enzymes called restriction endonucleases. Using a technique called electrophoresis, researchers chose a subset of overlapping fragments of approximately the same size (about 150,000 bp) for further treatment and analysis. The set covered the length of the DNA many fold and the common sequences at the ends of overlapping pieces used to integrate sequencing data obtained for each fragment using the shotgun technique.

5.4.4 Clone Copies of the Selected DNA Fragments

Sequencing requires many copies of the DNA fragments included in the set of overlapping sequences. These copies are obtained by a procedure called cloning. Cloning requires insertion of the DNA into host cells and inducing the DNA-replication machinery of the host to make copies of the inserted DNA. Cloning procedures lead to a DNA library composed of different colonies (clones) of cultured host cells. Each cell in a clone carries an identical copy of the foreign DNA insert and therefore can be visualized as a supplier of the DNA segments needed for further processing.

5.4.5 Map-Cloned DNA Segments along Chromosomes

Since human DNA contains very long stretches of repeated units, it is important for data-integration purposes to know the spatial position of the cloned DNA fragments along the chromosomes. Therefore, clones carrying different, but overlapping, fragments of DNA must be placed (mapped) along the length of DNA. So-called segmentation fingerprints are obtained for each cloned DNA, and the common features in these fingerprints are used to organize the overlapping DNA pieces along the chromosomes. This procedure is called DNA mapping and will be described in the next section.

5.4.6 Use Shotgun Sequencing for Each Cloned DNA Segment

Consortium researchers used shotgun sequencing on ordered overlapping DNA clones. The insert DNA in each clone is cut, and multiple copies of the fragments made using the polymerase chain reaction (PCR) technique discussed in the next section. These copies are then sequenced starting at both ends. The sequences are then integrated using computational pattern-recognition techniques that match DNA fragments by sequence at both ends. The sequences of the overlapping DNA segments (~150,000 kb) were similarly matched to obtain a draft copy of the human genome. The computational effort involved in putting together hundreds of thousands of sequences into a unified sequence is a testament to the creativity and collaboration of biologists, engineers, physicists, and computer scientists working in the field of genome sequencing.

5.5 | Cutting and Sizing DNA

5.5.1 Cutting DNA with Restriction Enzymes

The age of recombinant DNA technology was ushered in by the discovery of sequence-specific (restriction) DNA endonucleases. Until the early 1970s, the only enzymes known to digest DNA (DNAses) cut DNA into thousands of random pieces. Restriction endonucleases cut DNA at specific sequences resulting in the generation of fragments of predictable size and with a known sequence at each end. An example of a restriction enzyme cleavage sequence is shown in (Fig. 5.3). Cleavage sequences range from four to eight base pairs and are present with variable frequency throughout the genome. Some restriction enzymes cut the DNA in a staggered fashion that results in fragments with short single-stranded tails on each end. These "sticky ends" have a tendency to reanneal or recombine with their complementary sequence, a process that can be accelerated and strengthened by enzymes known as DNA ligases. In the absence of a complementary "sticky end" from another DNA, the cut DNA will simply recombine with itself. However, investigators soon learned that they could mix DNA from two different sources that had both been cut with the same restriction enzyme and generate recombinant fragments that contained sequences from both sources. Other restriction enzymes cut the DNA symmetrically and produce

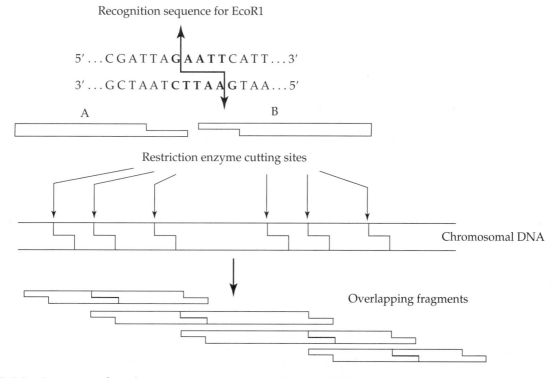

FIGURE 5.3 Generation of overlapping DNA segments. Double-stranded DNA can be cut at the recognition sites by enzymes such as EcoR1. This recognition sequence occurs on average about 4000 bp in a typical genome. However, partial digestion (see text) yields larger fragments. The cut pieces anneal resulting in different combinations of DNA segments. The recombinant DNA can then be cloned in host cells.

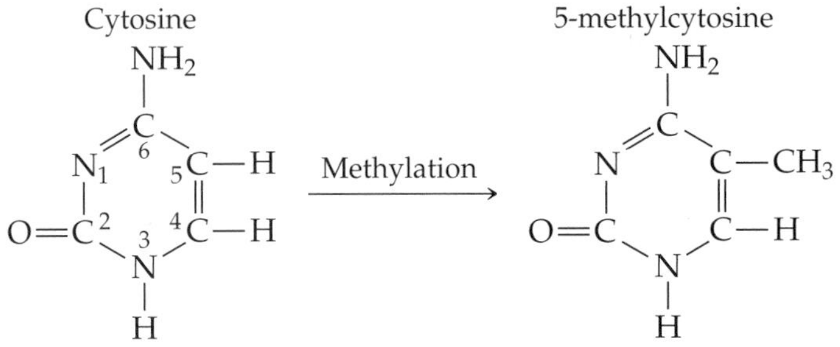

Cytosine   5-methylcytosine

NH_2 NH_2

C C

N_1 6 5C—H   Methylation →   N   C—CH_3

$O=C^2$ 3 4C—H $O=C$ C—H

N N

H   H

FIGURE 5.4 Methylation of a cytosine base in the DNA helix. Methylation means addition of methyl group CH_3 to nucleotide bases. Bacteria produce restriction enzymes that cleave and degrade phage (bacterial virus) DNA. Methyl groups inserted at recognition sites block the restriction enzymes from cutting the bacterial DNA. Methylation is a covalent modification and in vertebrates is an indicator that distinguishes active genes from those that are not. DNA methylation helps turn off genes. Binding of proteins to DNA that recognizes methyl C shuts down a gene completely. The covalent modification of C has no effect on base pairing.

"blunt ends." Although blunt ends cannot be recombined directly, small oligonucleotides can be ligated to them to make almost any combination of DNA ends. The ability to cut and splice DNA at will made possible major advances in DNA sequencing and cloning. Although the PCR has now replaced restriction endonucleases for some applications (as discussed later), they are still regularly used in all molecular biology laboratories for routine subcloning and diagnostic purposes.

In the course of genomic sequencing, copies of chromosomes are broken into shorter overlapping pieces using partial digestion of genomic DNA with restriction enzymes. At low levels of restriction enzymes, only the most accessible sites are cut and the DNA is fragmented into quite large pieces.

For example, the enzyme EcoRI, made by *E. coli*, cuts double-stranded DNA only where it finds the sequence GAATTC. Different species of bacteria make different restriction enzymes. A natural biological function of these enzymes is to protect bacteria by attacking viral and other foreign DNA. The bacterial DNA is protected from the cutting action of the enzyme by methylation at A or C residue (Fig. 5.4).

Some restriction enzymes (rare cutters) cut the DNA very infrequently, generating a small number of very large fragments (several thousand to a million base pairs (bp)). Most enzymes cut DNA more frequently, thus generating a large number of small fragments (less than a hundred to more than a thousand bp). On average, restriction enzymes with four-base recognition sites will yield pieces 256 bases long, six-base recognition sites will yield pieces 4000 bases long, and eight-base recognition sites will yield pieces 64,000 bases long. Since hundreds of different restriction enzymes have been characterized, DNA can be cut into many different small fragments.

5.5.2 Electrophoresis: Separating DNA Fragments According to Size

DNA fragments can be separated from each other by electrophoresis, the core technology used to separate complex molecular mixtures from biological samples. Gel electrophoresis produces high-resolution separation of DNA molecules; even fragments

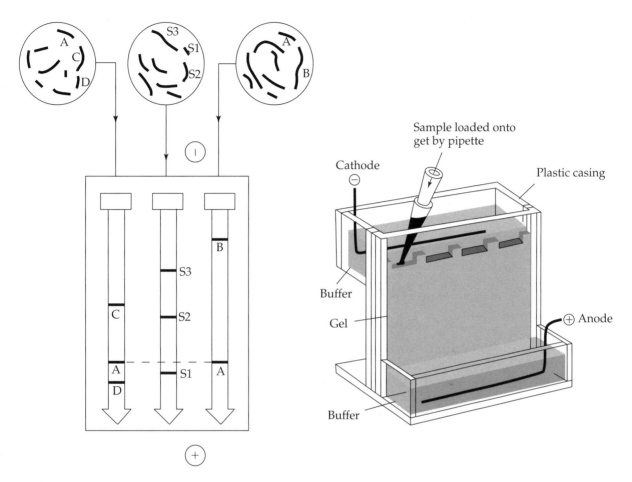

FIGURE 5.5   A typical electrophoretic procedure used to separate fragments of DNA according to size. Generally, the sample of particles (molecules, ions) is run in a support matrix in an electrical field. The matrix acts as a sieve by retarding the movement of large macromolecules while allowing smaller molecules to migrate more freely. The matrix also inhibits convective mixing caused by heating. After a period of migration, DNA fragments of different sizes are positioned according to their size on the matrix. The fragments can then be visualized by staining the gel with the fluorescent DNA chelator, ethidium bromide.

that differ in size by only a single nucleotide can be resolved. Electrophoresis is a technique based on the migration of charged molecules in solution in response to an electric field (Fig. 5.5). Electrophoresis offers a powerful means of separating proteins and nucleic acids because a molecule with a net charge migrates in an electric field. The velocity v of the molecule depends on the electrical field strength (E), the net charge on the protein (z), and the frictional coefficient f according to the following equation:

$$v = (Ez)/f.$$

The electrical force Ez driving the charged particle toward the oppositely charged electrode is opposed by the viscous drag force fv. Since the movement is creepingly slow, Newton's second law reduces to the condition of force balance, and hence the preceding equation governing the motion of a charged particle in a gel matrix. The frictional coefficient f depends on both the mass and shape of the molecule and the viscosity of the medium.

Separation of DNA fragments of different sizes by gel electrophoresis takes about an hour. The lengths of molecules (fragments) separated by mobility are inferred by running molecules of known size on a different lane of the same gel. The DNA is extracted by cutting out the appropriate band on the gel and passing an electrical current through it in a process known as electroelution. The negatively charged DNA fragments migrate away from the gel and toward the positively charged pole. For DNA fragments less than 500 nucleotides, polyacrylamide gels can differentiate nucleotides that differ in length by a single nucleotide. These gels have small pore sizes. The larger pore-size agarose gels are used in the electrophoresis of larger DNA molecules.

5.6 | Making Multiple Copies

5.6.1 Plasmids and Cloning

Plasmids are small circular DNA particles that infect bacteria such as the gut bacterium *E. coli*. Cutting plasmid DNA with a restriction enzyme such as EcoR1 generates sticky ends and allows for the insertion and ligation of any DNA that has also been cut with EcoR1. These recombinant plasmids can then be used to infect bacteria and "clone" the inserted DNA. As the bacteria grow exponentially, the original DNA inserted into the plasmid is multiplied manyfold. Because the plasmid infection process is not very efficient, those bacteria that have incorporated the plasmid must be selected.

To do this, the plasmid is engineered to contain a gene that provides antibiotic resistance to the infected bacterium. Simply growing the bacterial culture in the antibiotic allows only those bacteria that have incorporated the plasmid to survive and propagate. The use of selection pressure is an important tool that is used over and over again in molecular and cell biology.

The simplest cloning plasmids or "vectors" do not allow the inserted gene to be expressed (transcribed or translated). Other plasmids known as expression plasmids contain promoter sequences that allow the inserted gene to be expressed in bacteria and in mammalian cells. Such plasmids can be used to generate large amounts of recombinant protein in bacterial cells and to transfect cultured mammalian cells.

Hundreds of different cloning and expression plasmids are now commercially available. Many of these have regions adjacent to the insertion site that contain sites for multiple restriction enzymes. These multiple cloning sites allow for the convenient insertion of DNA sequences using any one of a number of restriction enzymes. Other vectors incorporate a "tag" that is in frame with the inserted gene. The resulting fusion protein can then be easily detected or purified, using reagents that are specific to the tag.

One use of cloning vectors is in the generation of cDNA or genomic libraries that contain DNA that is representative of a particular organism or tissue. Until recently, the basic procedure for cloning an unknown gene consisted of the following steps: A protein with the desired properties was purified and a partial amino-acid sequence determined. Oligonucleotides corresponding to the amino-acid sequence were used to screen a cDNA library and isolate the gene of interest. However, the availability of the complete sequence of the human genome, gene chips, and advances in structural, computational, and bioinformatic technologies will soon render library screening obsolete.

By fragmenting DNA of any origin (human, animal, or plant) and inserting it in the DNA of rapidly reproducing foreign cells, billions of copies of a single gene or DNA segment can be produced in a very short time.

5.6.2 Recombinant DNA

The first step in cloning is the joining of the DNA of interest to another piece of DNA, usually derived from a cloning *vector* (plasmid) that can be replicated within a bacterial cell. The resultant DNA incorporates fragments from two different DNA molecules. DNA combined from two or more resources is called recombinant DNA. To create recombinant DNA from two sources, DNA molecules from each source are treated with the same restriction endonuclease (such as EcoR1 or BamHI), which cuts the same site on both molecules (Fig. 5.6). The ends of the cut are called sticky ends because they are able to base pair with any DNA molecule containing the complementary sequence. When these pieces are mixed together, DNA with complementary sticky ends pairs and the recombinant DNA is sealed by the enzyme DNA ligase.

5.6.3 Cloning Vectors

A cloning vector is a circular DNA that has the necessary sequences to induce its own replication in the host cell. The DNA fragment to be amplified can be combined with a cloning vector (e.g., a plasmid) that is itself a replication unit (replicon). A necessary condition for DNA to be a replicon is the possession of a specific sequence called the origin of replication. As discussed in Chapter 3, DNA polymerase, the primary molecular machine for DNA replication, binds to DNA at the origin of replication and begins to replicate the DNA, which includes the recombinant DNA.

Vectors used in cloning vary according to the size of the DNA to be cloned. A YAC (yeast artificial chromosome) consists of the portion of a yeast chromosome that is needed for replication within a yeast cell. Up to 1.5 million nucleotides of foreign DNA can be cloned using YACs, and they are used to clone very large pieces of DNA. A similar vector, called a BAC (bacterial artificial chromosome), can be used to clone approximately 100,000 to 150,000 nucleotides of foreign DNA. BAC vectors were used to create an overlapping set of clones in the sequencing of the human genome.

Bacterial plasmids and prokaryotic and eukaryotic viruses are also used as cloning vectors. Viruses are modified by DNA deletion so that they cannot kill and lyse the host cells. Bacterial plasmids are double-stranded circular DNA molecules of 1–200 kb in size that replicate and are inherited independently of the bacterial chromosome. Some plasmids replicate at a much higher rate than genomic DNA reaching a copy number as high as 700 per cell. Commercially available plasmid

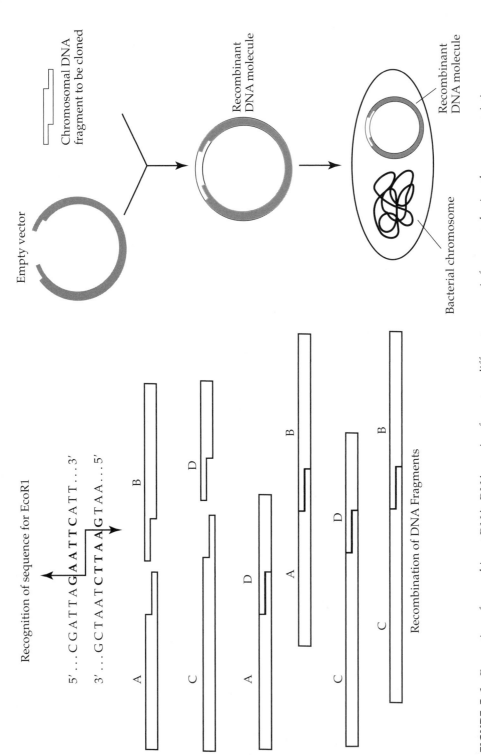

FIGURE 5.6 Formation of recombinant DNA. DNA coming from two different sources is fragmented using the same restriction enzyme. Matching sticky ends bring fragments from different sources together. These pieces are covalently linked with the help of the enzyme DNA ligase. The figure on the right shows the insertion of a DNA fragment into an open circular DNA (an empty vector) and thus the formation of a cloning vector.

vectors are small circular molecules of double-stranded DNA derived from larger plasmids. Plasmid vectors can be used to clone up to 5000 bp of foreign DNA, which is large enough to carry a few prokaryote genes. Bacteriophage lambda, which infects *E. coli*, can carry up to 20,000 bp of DNA. When a viral vector is used, the normal replication mechanisms of the virus can produce 10^{12} copies of the inserted DNA along with viral DNA. When the host cell is a plant cell, a commonly used vector is a plasmid found in the plant-infecting bacterium *Agrobacterium tumefacians*.

5.6.4 Presenting Vectors to Host Cells

A necessary step in cloning is the entry of the cloning vector into a host cell. Vectors derived from naturally infecting viruses enter host cells directly. Entry of other vectors must be facilitated. The reason for this is that both the vectors (recombinant DNA molecules) and the plasma membrane of the host cell are negatively charged and therefore repel each other. That is, the reason viral vectors are the vector of choice in the experimental gene-therapy research involving humans and animals.

The charge repulsion between the vector and the host cell can be alleviated using calcium (Ca^{2+}) salts. The salts reduce the charge repulsion and thereby increase the permeability of the bacterial cell wall to DNA. Bacteria that have been treated this way are termed "competent" for transformation. The entry of plasmid vectors into competent bacteria is usually achieved by exposing the bacteria to the plasmid DNA for 30 min on ice and then "shocking" them for 45 s at 42°C. Other methods used to insert vectors into host cells include mechanical methods of insertion such as injection with a pipette and electroporation, in which cells are exposed to rapid pulses of high-voltage current for temporarily increasing host cell-membrane permeability to DNA. Current methods of inserting vectors into cells are not yet very efficient. Only a small fraction of cells, when presented with cloning vectors will absorb them. However, more effective new methods of vector insertion into host cells that involve coating DNA with lipid are becoming common.

5.6.5 Selecting Cells That Contain Insert DNA

Cells that are used as host cells in cloning range from bacterial and yeast cells to plant and mammalian cells. Yeast cells not only have the machinery for protein modification as human cells do, but they are also easy to grow in the laboratory and have a rather short life cycle (two to eight hours). After the host cells (bacteria or yeast) are introduced to cloning vectors, they are grown on a solid medium. Because of the concentrations of cells dispersed on the solid medium, cells that survive will divide and grow into distinct colonies (Fig. 5.7). As noted previously, when cloning vectors are introduced to host cells in a cell suspension, only a fraction will take up vector DNA. Moreover, since the method of making a DNA insert vector is far from perfect, not all vectors taken up by cells carry the desired DNA.

A variety of markers are used to differentiate cell colonies that contain insert DNA from those that do not. Plasmids usually contain genes for resistance to antibiotics (Fig. 5.7). Some of the artificial chromosomes used as vectors also use antibiotic genes as selectable markers, and some include genes that get inactivated only after successful insertion of DNA. Simply growing the host cells in the appropriate antibiotic allows only those cells that have successfully incorporated the vector

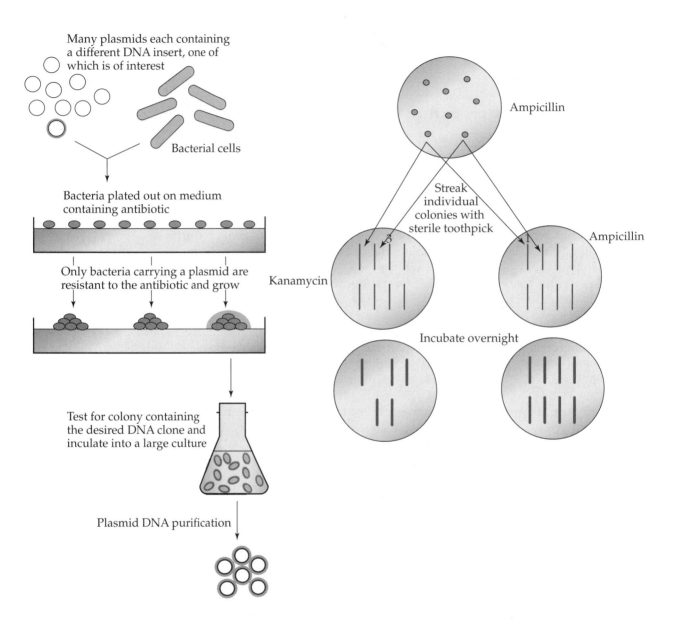

FIGURE 5.7 Separation of cell clones containing insert DNA from others that do not contain the desired DNA fragment. The figure on the left indicates the procedure with plasmids used as cloning vectors. The figure on the right indicates the use of antibiotics as a marker for differentiating cell colonies with insert DNA from the rest of the colonies.

containing the antibiotic-resistance gene to survive. Surviving colonies that contain the desired recombinant DNA can be isolated from the solid medium and grown in large amounts in liquid culture. Procedures for the purification of the circular plasmid DNA from these cells are well established. Generally, the identity of the recombinant DNA is confirmed by restriction enzyme analysis and sequencing. Using the cloning procedures previously described, a large number of human genes have been cloned in *E. coli* or yeast, making it possible to produce unlimited amounts of human proteins such as insulin for diabetes and erythropoietin for treating anemia.

5.6.6 Polymerase Chain Reaction (PCR)

Although restriction enzymes and plasmid cloning techniques are still in routine use, the advent of the polymerase chain reaction (PCR) has revolutionized the study and analysis of genes. Although DNA polymerases use single-stranded DNA as a template for DNA synthesis, they require a small section of double-stranded DNA to "prime" the reaction. This property led to the use of synthetic oligonucleotide primers that could direct the DNA polymerase to a section of single-stranded DNA that is complementary to the primer. Given the appropriate nucleotides, DNA polymerase can then replicate the single-stranded DNA to produce a double strand. Heating the double strand to approximately 90°C separates the two strands, which can each act as a template for new synthesis provided a primer is provided for each strand. The addition of new DNA polymerase allows the reaction to be repeated for many cycles resulting in the amplification of the original DNA template manyfold. However, most DNA polymerases are inactivated at high temperatures, and the need to add fresh enzyme at each step hindered the automation of the PCR, which was not immediately widely adapted.

The discovery of heat stable Taq DNA polymerase allowed for a single addition of the enzyme at the beginning of the reaction and led to the automation of the PCR using sophisticated heating blocks called thermal cyclers. By definition, PCR is a process based on a specialized polymerase enzyme, which can synthesize a complementary strand to a template DNA strand in a mixture containing the four DNA bases and two DNA fragments (primers, each about 20 bases long) flanking the target sequence (Fig. 5.8). The mixture is heated to about 95°C to separate the strands of double-stranded DNA containing the target sequence and then cooled to 37°C to allow the primers to find and bind to their complementary sequences on the separated strands. The mixture is heated to 72°C to enable the polymerase-mediated extension of the primers into new complementary strands. Repeated heating and cooling cycles multiply the target DNA exponentially, since each new double strand separates to become two templates for further synthesis. In about one hour, 20 PCR cycles can amplify the target DNA a millionfold. The steps in the PCR are outlined in Fig. 5.8.

The PCR can be used to amplify DNA from RNA provided that a reverse transcriptase step precedes the PCR reaction. This form of PCR is called RT-PCR. The PCR can be used not only to amplify DNA, but also to introduce mutations, in cloning, and has found a particular utility in DNA sequencing. (See next section.) The availability of the complete sequence of the human genome, and that of several other organisms, facilitates the design of PCR primers that can be used to detect, amplify, and modify any region of any gene. The PCR can amplify a desired DNA

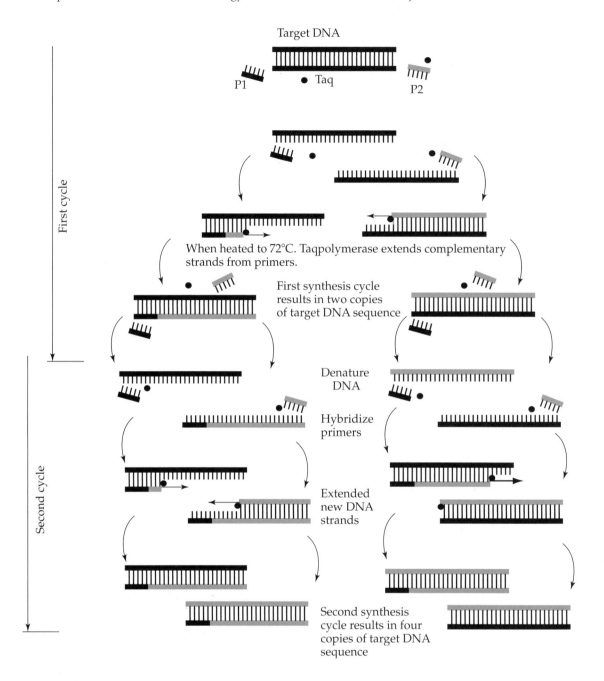

FIGURE 5.8 DNA amplification using the polymerase chain reaction. In PCR, the reaction mixture contains a target DNA sequence to be amplified; two primers, P1 and P2; Taq polymerase (a heat-stable DNA-replication enzyme); and the nucleotide building blocks dATP, dTTP, dCTP, and dGTP. The latter are used as building blocks of the newly synthesized DNA. The reaction mixture is heated to 95°C to denature the target double-stranded DNA into a single-stranded form. Subsequent cooling to 37°C allows the primers to hybridize to the complementary sequences in the target DNA. When heated to 72°C, Taq polymerase extends complementary strands from the primers. The first synthesis cycle results in two copies of the target DNA sequence. The number of copies of DNA increases geometrically with each DNA-synthesis cycle.

sequence of any origin (virus, bacteria, plant, or human) hundreds of millions of times in a matter of hours, a task that would have required several days with recombinant technology (cloning). PCR is especially valuable because the reaction is highly specific, easily automated, and capable of amplifying minute amounts of sample. For these reasons, PCR has also had a major impact on clinical medicine, genetic disease diagnostics, forensic science, and evolutionary biology.

5.7 | Sequencing

5.7.1 Genomic Libraries and Fingerprinting

As noted earlier in this chapter, biologists think of the genome of an organism as a library. The human genome is composed of 24 distinct volumes (22 numbered chromosomes, along with X and Y chromosomes) containing all the human genes. When a genome is broken into smaller pieces, each of these smaller fragments still represent a volume in the library. The libraries used in genome sequencing are sets of host cell cultures containing overlapping DNA fragments encompassing an entire genome. Also available are chromosome-specific libraries, which consist of cell populations containing fragments derived from a source DNA enriched for a particular chromosome. In decoding the human genome, BAC libraries were established using partial digestion of genomic DNA with restriction enzymes. This library represented sixty-five-fold coverage (redundant sampling) of the genome. Individual recombinant colonies in the library are suppliers of the DNA fragments needed for sequencing. Cloned DNA fragments can be isolated and their nucleotide sequences determined in a subsequent shotgun procedure. Cloning procedures provide unlimited material for experimental study.

However, this collection of overlapping clones of a chromosomal library has no obvious order indicating the original positions of the cloned pieces on the uncut chromosome. How do we map these DNA fragments (~150,000 bp) back onto the chromosome? To do this, a process called fluorescence in situ hybridization (FISH) is used. The DNA from the chromosomal BAC or YAC libraries is labeled with a fluorescent marker and allowed to hybridize to preparations of metaphase chromosomes. An experienced cytogeneticist can easily recognize individual chromosomes by their size and shape allowing them to assign the fluorescent signal to a particular region of a known chromosome. Mapping utilizes a collection of fluorescent marker sequences. Over the years, biologists built a collection of sequences (>10 bp) that hybridize to chromosomes at well-defined positions. When such a marker hybridizes to a DNA fragment under consideration, it signals the position of the fragment along the physical length of the DNA that constitutes the chromosome.

For finer mapping, libraries are ordered by obtaining a typical signature (fingerprint) for each clone using restriction digestion and subsequent electrophoresis. In creating a library of ordered BAC clones for the human genome, each clone was cut using the restriction enzyme HindIII. The size distribution of the resulting fragments was obtained by agarose gel electrophoresis. Patterns of restriction fragments observed on the gel (size distribution) provided a fingerprint for each BAC. Fingerprinting was based on the observation that overlapping clones shared at least one restriction site and had fragments of common size. Occurrence of repetitive elements in the adjoining DNA fragments was then used as criteria to determine overlapping DNA segments and create an ordered set of continuous clones (contigs) along the length of the DNA. Each BAC in the ordered set could then be separately sequenced. One of the other procedures used to map genes on chromosomes is radiation hybrid mapping, which uses PCR to amplify the sequence of interest and compares it to the known localization of marker sequences.

5.7.2 Sequencing DNA Fragments

Current sequencing protocols are based on the dideoxy sequence method, which yields less than 750 bp of sequence per reaction. Consequently, a large number of reads are required to cover a eukaryotic genome. The Maxam–Gilbert and Sanger sequencing approaches are most widely used. These methods differ primarily in the way the nested DNA fragments are produced. Both methods work because gel electrophoresis produces very high-resolution separations of DNA molecules; even fragments that differ in size by only a single nucleotide can be resolved. All steps in these sequencing methods are now automated.

Sanger sequencing is also called the chain-termination method. An enzymatic procedure is used to synthesize DNA chains of varying length in four different reactions, stopping the DNA replication at positions occupied by one of the four bases and then determining the resulting fragment lengths. Each sequencing reaction tube identified in Fig. 5.9 as T, C, G, and A contains the DNA template to be sequenced, a primer sequence, and a DNA polymerase to initiate synthesis of a new strand of DNA at the point where the primer is hybridized to the template. The medium also contains the four building blocks needed to extend the temple. In addition, each of the tubes shown in Fig. 5.9 contains one type of base analog, which terminates the growing chain wherever it is incorporated. Tube A has the analog for A, tube C has the analog for C, and so on. After an incubation period, each tube will have a collection of DNA fragments with a terminating base analog at one end. The fragments of varying length are then separated by electrophoresis, and the positions of the nucleotides are analyzed to determine sequence. Shorter fragments move faster and appear at the bottom of the gel. The sequence is read from bottom to top. In modern automated sequencing, the nucleotide analogues are each labeled with a different fluorescent probe and the fluorescent signals detected by the automated reader.

5.7.3 Assembly of the Draft Genome Sequence

Computer scientists working on the assembly of the draft human genome from the sequenced segments wrote software for the alignment of sequenced fragments, detection of overlaps, and other factors. The draft sequence has 94 percent coverage of the whole genome. The work conducted by the International Consortium and the privately owned Celera enabled the mapping of thousands of genes on chromosomes. In the draft document, the total number of genes in the human genome is estimated to be between 30,000 and 40,000 genes. The functions of more than half of the predicted human genes are as yet unknown. Humans have about three times as many genes as the fruit fly and worm, but because of alternative splicing of mRNA transcripts, the same gene can often yield several different protein products.

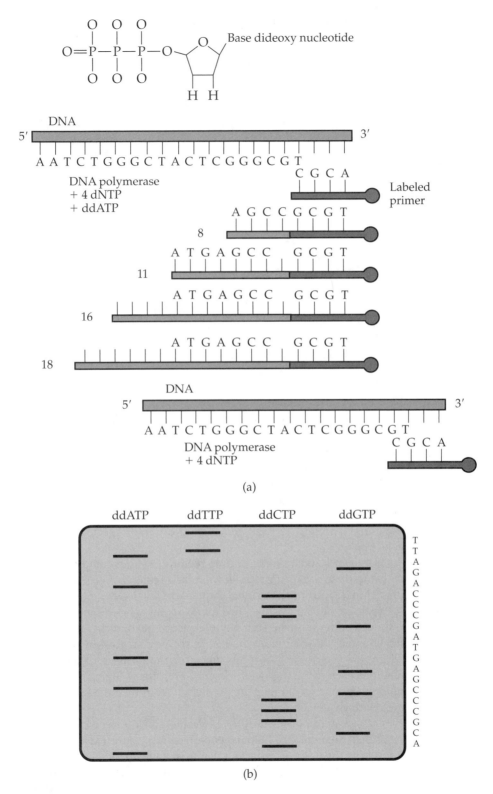

FIGURE 5.9 Termination of DNA replication upon binding of an analog to base A results in DNA pieces with varying lengths (a). The resulting fragments are loaded onto lane A of the acrylamide-sequencing gel (b). The sequence is read from bottom to top.

The human genome project has also uncovered about 1.4 million nucleotide locations on the human genome where single-base DNA differences (single nucleotide polymorphism) may occur among different individuals. Some gene polymorphisms are associated with particular traits such as susceptibility to certain diseases and response to treatment. For this reason, a great deal of effort is devoted to the generation of human SNP maps, which can be used to uncover patterns underlying such common diseases as cardiovascular disease, diabetes, arthritis, and cancer.

5.8 | Gene Annotation

One important aim of decoding genomes is to predict the location, sequence, and function of the genes belonging to that organism. Prediction of gene sequence in prokaryotes is relatively simple. Prokaryotic DNA is organized into clusters of genes called operons. As discussed earlier, the expression of genes belonging to the same operon is regulated by common regulatory sequences. In these organisms, it is relatively straightforward to identify most genes by the presence of long open reading frames (ORFs). Prokaryotic protein-coding regions consist of contiguous ORFs. Statistical principles indicate that an ORF (ATG to stop) longer than 300 bp would occur randomly every 36 kb on a single strand of idealized DNA with equal concentration of all four nucleotides. The lengths of protein-coding genes correspond on average to a 1000-bp ORF, indicating the implausibility of random distribution of nucleotides along DNA. A simple algorithm retaining the longest of the overlapping ORFs and applying a size threshold (for instance 300 bp) already detects most real genes. *ORF Finder*, a graphical analysis tool, can find all open reading frames in a decoded genome within a short period of time. Gene predictions gain additional strength if the predicted gene sequence exhibits significant homology (similarity) with genes from different species already deposited in the gene bank.

In plants and animals, coding regions of genes comprise only a tiny fraction of DNA. In these organisms, the short coding regions of genes (exons) are separated by much longer introns. Key signal sequences called splice sites occur at the boundaries between introns and exons. The exons in eukaryotic genes are flanked by AG and GT nucleotides, but the splice-site sequence beyond the nucleotides immediately at the boundary is otherwise variable. In the human genome project, scientists laid out all known genes along chromosomes for the purpose of pattern recognition using methods such as neural networks and even fuzzy logic. Despite these efforts, computer programs for direct gene prediction from the human or mouse DNA sequence are relatively inaccurate.

As gene transcripts (mRNA molecules) mirror the coding regions of genes, computational prediction of human genes must rely on the existing library of cellular gene transcripts. As discussed earlier, mRNA is made up only of exons because intronic sequences are spliced out in the nucleus. Standard techniques have evolved to isolate and purify mRNA from cells. Each mRNA molecule can then be used as a template to build a complementary DNA (cDNA) using the enzyme reverse transcriptase.

The sequence of a cDNA is therefore a direct reflection of the coding regions of a gene. The genomic data mining software BLAST can be used to evaluate local similarity between a genomic DNA segment and cDNA. The best local alignment between cDNA and genomic DNA identifies the locations where exons terminate and introns begin.

5.8.1 Homolog Gene Databases

Candidate gene sequences are compared for homology with known gene sequences across genomes to determine the existence of orthologs. Currently, most genes present in the COG database are those that belong to archaea and bacteria. The yeast Saccharomyces cerevisiae is the only eukaryote represented in the COG. For the majority of COGs, the protein function is either known from direct experiments, mostly in *E. coli* or yeast, or can be confidently inferred on the basis of significant sequence similarity to functionally characterized proteins from other species.

Another source of information for gene prediction is the identification of the regions of the genome rich in the nucleotide bases G and C (G + C). These regions are relatively dense with genes. Regions of DNA rich in the dinucleotide composition CG, also known as CpG islands, are often located upstream of the transcription start site. Such statistical information, along with homology studies and pattern-recognition techniques are used in annotation of genes. A researcher experienced in human gene annotation of uncharacterized consecutive segments of the human genome may use a graphical gene-annotation assistant such as GeneBander in assembling information from a large number of sequence tests for gene prediction.

The preceding discussion clearly shows that the bottleneck in genomics is no longer in sequencing genomes, but lies in their annotation. Gene annotation is a dynamic process; scientists will want to improve on the original annotations of the previously decoded genomes as the annotation techniques improve with time. As more genome databases become available, there is a clear need for characterization to be carried out using a controlled set of shared vocabulary terms. The goal of the recently established Gene Ontology™ Consortium is to produce a dynamic controlled vocabulary that can be applied to all organisms even as knowledge of gene and protein roles in cells is accumulating and changing. Such standardization will make New Biology accessible to experts in engineering and physics and that could impact on the future of this most exciting science.

5.8.2 Genetic Maps

A genetic map is a chromosome map of a species that shows the position of its known genes and markers relative to each other, rather than as specific physical points on each chromosome. A genetic marker is a segment of DNA with an identifiable physical location on a chromosome and whose inheritance can be followed. A marker can be a gene, or it can be some section of DNA with no known function. Because DNA segments that lie near each other on a chromosome tend to be inherited together, markers are often used as indirect ways of tracking the inheritance pattern of a gene that has not yet been identified, but whose approximate location is known.

The DNA from different, but related, species can be compared to highlight the most interesting regions of a genome. In mammals, moderate resolution maps generated by restriction enzymes show increasing similarity with decreasing evolutionary distance. All mammals have roughly the same number of nucleotides in their genomes. Many of the genes belonging to different mammal species can be traced to common ancestor genes. For example, there is a 95–98 percent similarity between related genes in humans and apes.

The rough drafts of the mouse and human genomes allow mapping of the human genome onto the mouse genome and vice versa. Human and mouse genomes are composed of about 150 segments that have considerable sequence similarity, but these segments are integrated in different ways to create the chromosomes. This is illustrated in Fig. 5.10.

Figure 5.10 shows the full complement of human chromosomes, cut schematically, into about 100 pieces, then reassembled into a reasonable approximation of the mouse genome. The numbers alongside the mouse chromosomes indicate the human chromosomes containing homologous segments. This piecewise similarity between the mouse and human genomes means that insights into mouse genetics are likely to illuminate human genetics as well. Some mouse and human gene products

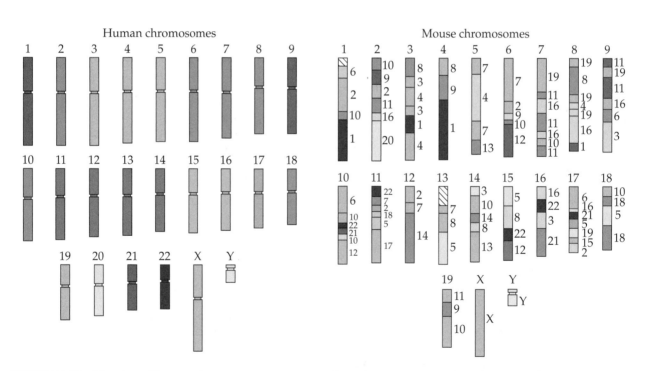

FIGURE 5.10 Mapping of human chromosome segments onto homologous segments along the mouse chromosomes. Numbers identifying sections of mouse chromosomes refer to the identity of human chromosomes with similar sections. (Modified from http://www.ornl.gov/hgmis/publicat/97pr/05gmous.html.)

are almost identical, whereas others are nearly unrecognizable as close relatives. The variation in the similarity of gene products may be because some nucleotide changes are "neutral" and do not yield a significantly altered protein. Others, but probably only a relatively small percentage, would introduce changes that could substantially alter the function of a protein.

5.9 | Plants and Animals with Modified Genomes

Scientists have used recombinant DNA technology to perturb the genomes of multicellular organisms such as fruit fly, mouse, and sheep. The perturbations involve gene insertion, deletion, substitution, and modification. Gene insertion into organisms is frequently employed to find efficient ways of producing desired proteins for medical purposes. The standard technique for making *transgenic* animals entails microinjecting a genetic construct—a DNA sequence incorporating a desired gene—into a large number of fertilized eggs. A few of them take up the introduced DNA so that the resulting offspring express it. These animals are then bred to pass on the transgene. Transgenic mice have been used as model animals to gain a better understanding of genetic deficiencies found in the human. The cells of transgenic animals contain copies of the perturbed genome and can be used to analyze the function of the modified genes. For example, it is possible to create mouse white blood cells lacking certain surface receptors and compare their response to normal cells when exposed to immune challenges.

Transgenic animals that carry the same inserted DNA construct do not necessarily possess identical genomes. *Cloning*, on the other hand, creates animals and plants with identical genomes. However, the ancestor cell may have a genomic modification imposed on it using recombinant DNA technology. Cloning and transgenic biotechnology has already made significant advances in designing agricultural products that resist attack from harmful microorganisms and agents. An example of such a modified organism is BT corn, which was engineered to be resistant to pests by the insertion of a bacterial gene.

Recently, some scientists began cloning animals. Animal cloning is based on nuclear transfer, which involves the use of two cells. The recipient cell is normally an unfertilized egg taken from an animal soon after ovulation. The chromosomes of the recipient egg cell are removed by suction using a fine pipette. Then, typically, the donor cell, complete with its nucleus, is fused with the recipient egg. In most successful cloning experiments, the donor cell is also an embryonic cell. Some fused cells start to develop like a normal embryo and will develop to term if implanted into the uterus of a surrogate mother (Fig. 5.11).

Although the cloning of Dolly from a mammary-derived culture showed the feasibility of cloning from an adult cell, this procedure for cloning animals fails much more frequently than it succeeds. The methodology for nuclear transfer is not yet efficient. The fusion of an adult cell with an egg cell lacking DNA yields positive results only when special measures are taken to synchronize the cycles of the donor and recipient cells. Donor cells are made quiescent by reducing the concentration of nutrients in their culture medium. Pulses of electric current are delivered to the egg to encourage the

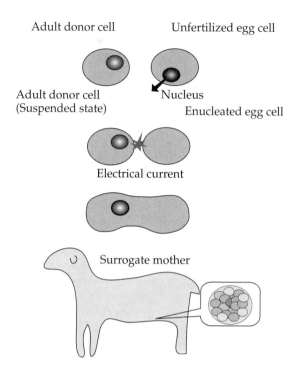

Adult donor cell   Unfertilized egg cell

Adult donor cell   Nucleus
(Suspended state)   Enucleated egg cell

Electrical current

Surrogate mother

FIGURE 5.11   Major steps involved in the cloning of a sheep.

cells to fuse and to mimic the stimulation normally provided by a sperm. Even with these measures, the success rate for developing an embryo is rather small.

Some cloned cattle and sheep are unusually large. The number of embryos surviving to adulthood is smaller when donor cells are taken from animals at a more advanced developmental stage. All the cloning studies described so far show a consistent pattern of deaths during embryonic and fetal development, with laboratories reporting only 1 to 2 percent of embryos surviving to become live offspring. These low statistics may reflect the complexity of the genetic reprogramming of adult donor DNA needed if healthy offspring are to be born. This issue is discussed in more detail in Chapter 8.

5.10   ASSIGNMENTS

5.1 Some evidence suggests that genomic DNA evolved by combinatorial fusion of circular DNA of about 300–400 bp long. The known optimal range for DNA closure is about 350 bp, and a large number of already identified proteins have lengths that are multiples of about 125 amino acids. According to the combinatorial fusion theory, the genes and noncoding sequences existed in the form of autonomously replicating DNA rings, very much like the present day plasmids that infect bacteria. The circular DNA rings started to fuse, forming larger DNA molecules consisting of at least several unit genes. Conduct a literature search on the observational facts that may support the combinatorial fusion theory.

5.2   Conduct a literature search about the rRNA-based universal tree of life and the assumptions used in its construction.

5.3   What are the biological features that put archaea and eukaryota closer to one another than to bacteria?

5.4   Symbiosis is the process in which unrelated organisms come together and form a stable association. An example is the symbiotic relation between coelenterates and dinoflagellate algea in coral reefs. Through this mechanism, complementary metabolic capabilities, life cycles, and competencies of different organisms come together for greater efficiency. Symbiosis provides an example of genetic networks consisting of multiple genomes. Illustrate symbiosis using an example involving bacteria and human cells and another involving bees and flowers.

5.5   When two populations of the same eukaryotic species become reproductively isolated, each carries multiple allelic lineages inherited from the ancestral population for a period of time. Eventually, these allelic lineages are lost because of so-called genetic drift or selection until each isolated population carries only the descendants of a single ancestral lineage. Describe genetic drift and how it impacts cell culture technology.

5.6   The restriction enzyme Hpa I is from Hemophilus parainfluenzae and cuts DNA in the middle of the sequence CAATTG; therefore, it does not create DNA fragments with sticky ends. In contrast, restriction enzymes Eco RI, Hind III, and Pst I all produce DNA fragments with sticky ends. What are DNA nucleotide sequences recognized by these enzymes and which organisms produce them?

5.7   For DNA fragments less than 500 nucleotides, polyacrylamide gels can differentiate nucleotides that differ in length by a single nucleotide. On the other hand, pulse-field agarose gel electrophoresis is used to separate extremely long DNA molecules such as bacterial or yeast chromosomes. Mammalian chromosomes must be cut into a few large pieces using restriction enzymes selected to recognize sequences that rarely occur. Conduct a literature search to explain the difference

between polyacrylamide and agarose gel. Why do researchers use a pulse electrical field in place of a constant field in separating large DNA molecules?

5.8   An effective method of determining the amino-acid sequence of a protein is to determine the nucleotide sequence of its gene. For this purpose, complementary DNA is obtained and then sequenced. In principle, there are six different reading frames in which a DNA sequence can be translated into protein (three on each strand). The correct reading frame is generally the one lacking frequent stop codons. For a random nucleotide sequence, a stop signal for protein synthesis will be encountered on average about once every 21 amino acids. Reading frames that actually encode a protein are long stretches without a stop codon. Illustrate this pattern by considering the Eco RI gene of *E. coli*.

5.9   Genetic maps can have different resolution. The finest resolution genomic map is the nucleotide sequence of the genome. Moderate resolution maps typically are restriction site maps. The software DNAMAN provides easy-to-use tools to produce restriction maps. Knowing the DNA-sequence information, one can generate restriction maps involving 180 of the most frequently used restriction enzymes. The software can also be used to reconstruct a restriction map provided that the distribution of all fragment sizes in single and double digestion is known. Conduct a literature search to compare the restriction site maps of some mammals including human.

5.10   A DNA library that includes only genes transcribed in a particular tissue (or cell type) can be made from complementary DNA or cDNA. cDNA is made using mRNA as a template for the enzyme reverse transcriptase. Most eukaryotic mRNAs have a string of A's at their 3′ end (poly A tail). The first step of cDNA production is to extract mRNA from a tissue and allow the mRNA to hybridize with a molecule consisting of a string of T residues. After the hybrid forms, the poly T molecule serves as a primer for reverse transcriptase. A collection of cDNAs from a particular tissue is called a cDNA library. Note that cDNA clones lack intron sequences. Why is a cDNA library a useful tool in gene annotation? Describe why a number of biotech companies are working on the total transcript content of various human cell types in health and disease.

5.11 Nucleotide base-pairing rules can be employed to stop the translation of certain mRNAs by producing an RNA molecule that is complementary to mRNA. This complementary molecule, called antisense RNA, binds to the mRNA and effectively takes it out of circulation for protein synthesis. What are the potential uses of antisense RNA in biological research?

5.12 Conduct a literature survey to generate examples of molecular biotechnology used to improve the nutritional properties of crops.

5.13 What are the key distinguishing features of the computer system ARACHNE, whole genome shotgun assembler? Describe its methodology.

5.14 Why do DNA polymerases used in PCR come from hyperthermophilic organisms that survive temperatures up to 95°C?

Ans. In each PCR cycle, DNA is heated above 90°C and then cooled to about 55°C. The heating step destroys most polymerases unless they were isolated from high-temperature-loving organisms.

5.15 In situ hybridization is a commonly used technique for identification of the rare colonies in a library that contain the DNA fragment of interest. The method takes advantage of the specificity of the base-pairing interactions between two complementary nucleic-acid molecules. Conduct a literature search to describe this method.

5.16 Thousands of human genes produce noncoding RNAs as their ultimate product. These products include transfer RNAs (tRNAs), adapters that translate the triplet nucleic-acid code of RNA into the amino-acid sequence of proteins. Also in this group of genes are the genes that code ribosomal RNAs (rRNAs) that are central to the translation machinery. The RNA genes do not have translated open reading frames (ORFs), and basically there is no straightforward way of predicting an RNA gene. One option is to identify genomic sequences that are homologous to known RNA genes, using BLAST N. Use this program to identify homologs in the yeast genome to the tRNA genes for Phe in *E. coli*. How many tRNA genes are there in *E. coli* and in yeast?

5.17 To learn more about the sequence patterns of human genes (exons and introns), researchers align cDNA sequences of known genes in the RefSeq database to the draft human genome sequence. Such genomic alignments could be used to study exon–intron sequence structure. cDNA alignments on the human genome indicate considerable variation in overall gene size and intron size. Conduct a literature search to quantify the distribution of intron size in the human genome.

5.18 Dr. Mark J. Daly and colleagues at the Whitehead Institute in Cambridge MA made a startling discovery about the SNPs in the human genome toward the end of 2001. Find the reference for the primary article concerning this discovery. What is the hypothesis put forward by Dr. Daly and his colleagues about haplotypes? What is HapMap? Why is it more difficult to uncover the genetic foundations of diseases like cancer than diseases such as cystic fibrosis?

5.19 In a research article published in Nature, scientists presented the results of the initial sequencing and comparative analysis of the mouse genome. Provide the full reference for this article. What is syteny and what percentage of mouse and human genomes can be partitioned into corresponding regions of conserved synteny. What is meant by genome landscape and why do the authors of this article believe that genomes contain not just recipes for proteins but many additional features? What is the origin of the mouse genome and why it is considered to contain the keys for future medical treatments the in human? Describe briefly the RNA genes found in mouse and their relation to human genes.

REFERENCES

Batzoglou S, Jaffe DB, Stanley K, Butler J, Gnerre S, Mauceli E, Berger B, Mesirov JP, and Lander ES. "ARACHNE: A whole-genome shotgun assembler." *Genome Res* 2002, 12: 1777–1789.

DeRuggiero J, Brown JR, Bogert AP, and Robb FT. "DNA repair systems in archaea: mementos from the last universal common ancestor?" *J Mol Evol* 1999, 49: 474–484.

Doolittle WF. "The nature of the universal ancestor and the evolution of the proteome." *Curr Opin Struct Biol* 2000, 10: 355–358.

Glansdorff N. "Microreview: about the last common ancestor, the universal life-tree and lateral gene transfer: a reappraisal." *Mol Microbiol* 2000, 38: 177–185.

Lander ES et al., "Initial sequencing and analysis of the human genome." *Nature* 2001, 409: 860–921.

O'Brien SJ, Menotti-Raymond M, Murphy WJ, Nash WG, et al., "The promise of comparative genomics in mammals." *Science* 1999, 286: 458–462.

Tatusov RL, Koonin EV, and Lipman DJ. "A genomic perspective on protein families." *Science* 1997, 278: 631–637.

Tatusev RL, Natale DA, Garkatsev IV, Tatusova TA, Shankavaram UT, Rao BS, Kiryatun B, Galperin MY, Fedorava ND, and Koonin EV. "The COG database: new developments in phylogenetic classification of proteins from complete genomes." *Nucleic Acids Res* 2001, 29: 22–28.

Venter JC et al., "The sequence of the human genome." *Science* 2001, 291: 1304–1351.

Woese C. "The universal ancestor." *Proc Natl Acad Sci USA* 1998, 95: 6854–6899.

CHAPTER 6

Cell Adhesion and Communication

6.1 | Introduction

In Chapter 3, we discussed the structures and activities that are common to all individual cells and that are, to a large extent, autonomous—that is, independent of other cells or their products. The activities of free-living cells are largely regulated by environmental factors such as nutrient availability and sunlight. During the course of evolution, cells acquired the ability to adhere to one another to form multicellular structures ranging in complexity from simple sponges and slime molds to humans. The development of multicellular structures was accompanied by the development of new levels of cell regulation. Cells within multicellular collectives are able to exert an influence on their neighbors either by direct contact or by the release of short-range signaling molecules. In more complex organisms, collections of like cells (organs and tissues) can exert long-range effects on distant tissues. An example of this type of signaling, common in vertebrates, is the influence of hormones produced by one organ (e.g., the pituitary gland) on the function of other organs such as the adrenal gland or gonads. In the first part of this chapter, we discuss the various modes of cell signaling and the resulting cell response. In later sections, we concentrate on the network of proteins that are involved in cell communication.

6.2 | Modes of Cell Communication

Cells communicate with the environment by binding ligands such as growth factors and hormones to specific cell-surface receptors and by adhesion to one another or to the acellular matrix surrounding the cells (extracellular matrix). The input signals that cells receive from their environment are diverse and depend upon the composition and activity of ligands bound to cell-surface receptors, the composition of the extracellular matrix, and the mode of cell adhesion (dynamic versus static) as well as the external physical forces imposed on them. Cells respond to these signals in a finite number of distinct ways, which include cell growth and division, cell differentiation, cell movement, and apoptosis (programmed cell death).

In many cases, the response of the cell to input signals involves the activation of sets of transcription factors and alterations in gene expression. The proteins that transduce the external signals into a cell response form complex networks. These networks are largely in the form of somewhat interconnected parallel pathways that begin on the cell surface, pass through the cytoplasm, and conclude with regulation of gene expression in the nucleus. Clearly, mapping of the input function space on to the cell response space is not one-to-one and remains to be systematically investigated.

The extent of connectivity of these signal transduction pathways is an area of great interest to biologists and some mathematical models suggest that the pathways must be loosely connected for cells to respond to the wide range of input signals by falling into a small number of stable patterns, or phenotypes. The decision-making logic of some nodes in these pathways is understood. However, most nodes have not been studied in this context. The identification of the key elements (nodes) of signal transduction pathways is extremely important from a medical standpoint because these nodes are potential targets of drugs for many diseases.

6.2.1 Cell Adhesion

Nobel Laureate Gerald Edelman, in his remarkable examination of how genes can define animal shape, pattern, and phenotypic function, develops the hypothesis that

> *morphogenesis and morphologic evolution depend upon a special set of molecular regulatory mechanisms mediated at the cell surface. These morphoregulatory mechanisms link the mechanochemical events that affect cell shape, division and motion to particular patterns of gene expression. They do so by means of sets of cell surface molecules that mediate one form or another of cell adhesion, and that interact directly and indirectly with the cytoskeleton*

It has been 25 years since Edelman first suggested that modulation of cell-surface adhesion is the transduction mechanism mediating the connection between the genetic and epigenetic processes that regulate tissue form and function.

Since that time, several classes of cell–cell and cell–substrate adhesion molecules (CAMs and SAMs, respectively) have been identified as candidates for such morphoregulators. The functional expression of these molecules is of central importance in virtually all developmental, morphogenic, and neoplastic processes. For example, several aspects of the development and function of the immune and nervous systems are profoundly influenced by the expression of particular sets of CAMs and SAMs. CAMs and SAMs are also involved in the expression of other molecules associated with cell junctions. In the later sections of this chapter, we discuss cell adhesion and molecules that mediate adhesion with particular emphasis on integrins and cadherins.

6.2.2 Soluble Ligand-Receptor-Mediated Communication

Cell–cell signaling carried out by soluble ligands is a prominent mode of cell–cell communication in animals. Ligand signaling is grouped into three categories. In *endocrine* signaling, *hormones* act on target cells distant from their site of synthesis. In animals, blood transports hormones from their sites of release to their targets. In *paracrine* signaling, the signaling molecules released by a cell only affect cells in close proximity. The conduction of an electrical impulse from one nerve cell to another and to a muscle cell by *neurotransmitters* and *neurohormones* is an example of paracrine signaling. In *autocrine* signaling, cells respond to substances (growth factors) that they themselves release, utilizing the signal amplification component of the pathway to alter growth and differentiation. Tumor cells commonly overproduce and release growth factors that stimulate uncontrolled growth, leading to the formation of a tumor. Common elements of all these signaling modes are the signal-receiving proteins or receptors, transducers and *effectors*. In most cases, through a cascade of binary protein interactions, the signal is amplified so that even minute concentrations of hormones can cause significant alterations in cell response.

In an earlier chapter, we have seen how a small lipid-soluble hormone affected the transcription of a number of genes. (See Fig. 4.18.) Steroids and thyroid hormones are lipid soluble. They pass across the plasma membrane of the cell and the nuclear envelope and directly interact with receptors that, in their ligand-bound form, act to release repressors and recruit transcription factors to specific DNA-regulatory elements, which in turn activate transcription of target genes. In other cases, lipid-soluble hormones bind to large protein aggregates anchored in the cytoplasm. Binding of the hormone releases an activator transcription factor from the cytoplasmic anchor. The protein activator–hormone complex then enters the nucleus and binds to appropriate regulatory (response) elements of DNA associated with target genes. For further information on lipid-soluble hormones, see the problem on retinoic acid signaling at the end of the chapter. Later in the chapter, we will focus on signaling through lipid-insoluble ligands.

6.3 | Extracellular Matrix (ECM) and Integrins

6.3.1 Extracellular Matrix

Cells in a multicellular organism are in contact with other cells or with extracellular matrix (ECM). This matrix is of fundamental importance for the integrity and function of all organs. There are many examples of cultured cells acquiring tissue-specific organization and function only when grown on the appropriate ECM. This is an area of active research for scientists interested in organ replacement. As shown in Fig. 6.1, there are two fundamental types of ECM. The first of these is a three-dimensional matrix that completely surrounds the cells and is found commonly in tissues such as bone, cartilage, and tendon, as well as surrounding the epithelial components of glandular organs such as the breast or prostate glands. The second type of ECM is a two-dimensional basal lamina that interacts with the basal surface of all epithelia. In this case, the epithelial cells form sheets that contact one another at their lateral borders and contact the two-dimensional extracellular matrix at their

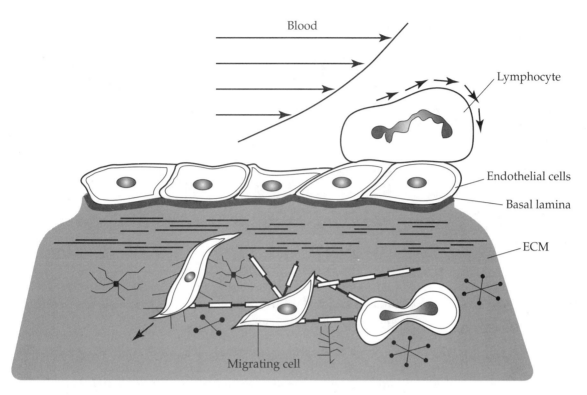

FIGURE 6.1   Modes of cell–ECM interactions. Endothelial cells are deposited on a two-dimensional basal lamina surface, whereas migrating fibroblasts interact with a three-dimensional matrix. In both cases, cell–ECM interactions lead to the imposition of physical force onto cells. This figure also illustrates the interaction of a circulating white blood cell with the endothelial cells by cell–cell adhesion. The arrows indicate the direction of blood flow and the forces exerted on the cells.

basal surface. In many cases, the two-dimensional basal lamina is in turn in contact with a three-dimensional ECM in which other cells are suspended.

Both two-dimensional and three-dimensional ECMs are made up of four major components:

1. *A structural component, one or more members of the collagen family.* Fibrous collagens such as type I collagen are commonly found in three-dimensional matrices and nonfibrous collagens such as type IV collagen in two-dimensional matrices.

2. *An adhesive component responsible for interacting with cells and with other components of the ECM.* Fibronectin is the most common type of adhesive molecule found in three-dimensional ECM and a member of the laminin family in two-dimensional ECM.

3. *Space fillers, such as sulfated and hydrated glycosominoglycans and proteoglycans.* Heparin and heparan sulfate are examples of these. These molecules vary in their carbohydrate, sulfate, and linker protein content from tissue to tissue.

FIGURE 6.2 Epithelial cell adhesion to a basal lamina is an important determinant of cell fate. Cells in contact with the ECM and with neighboring epithelial cells are limited in their ability to proliferate. Cells that lose contact with other cells while adhering to ECM can proliferate (a), whereas loss of attachment to ECM results in apoptosis (b). Many tumor cells do not undergo apoptosis when detached from the ECM and have subverted the role of cell adhesion in the regulation of cell proliferation. (Modified from Giancotti and Ruoslahti, *Science* 1999, 285: 1028–1032.)

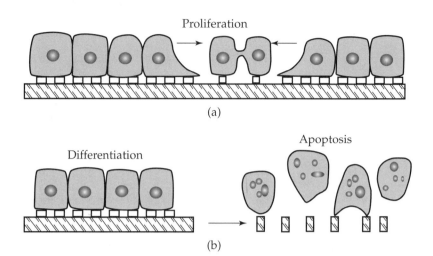

4. *Elastic components such as elastin.* ECM such as that found in the lungs is continually exposed to stretching forces. To accommodate this, the content of elastin is very high in the lung.

In addition to these integral components of the ECM, other molecules, many with an affinity for one of the structural components, are found in the ECM. These include heparin-binding growth factors as well as enzymes involved in matrix remodeling, such as collagenases.

In the normal cell, adhesion (or anchorage) to the ECM is required for cell proliferation (Fig. 6.2a). Cells that are in contact with an appropriate ECM will grow and divide until they come into close contact with other cells and detachment from the ECM leads to apoptosis in normal cells (Fig. 6.2b). In contrast, this anchorage requirement for cell growth is lost in many types of cancer cells. The composition of the ECM constitutes an important signal that determines cell fate. For example, myoblasts proliferate on fibronectin, but stop growing on laminin and begin to form tubelike structures. Mammary epithelial cells also differentiate when deposited on laminin.

The physical connections between neighboring epithelial cells and between epithelial cells and the extracellular matrix are illustrated in Fig. 6.3. Integrins and cadherins are the prominent families of cell-adhesion receptors that form contacts with ECM and with other cells, respectively. Integrins act as receptors for many ECM components, whereas cadherins generally interact with the same cadherin family member on a neighboring cell.

6.3.2 Integrins

The integrins are a supergene family of adhesion molecules that promote cell–cell and, more importantly, cell–ECM interactions. They are calcium-binding, integral membrane glycoproteins that are indirectly linked to the actin based submembrane cytoskeleton. (See Fig. 6.4.) In addition to attaching cells to the ECM, integrins influence the morphology, migratory behavior, and differentiated status of cells. The integrins often act in concert with cell–cell adhesion molecules to influence cellular behavior. For example, neurite outgrowth involves both types of adhesion molecules.

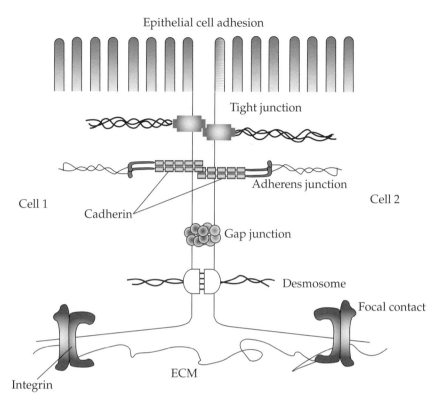

Epithelial cell adhesion

Tight junction

Adherens junction

Cell 1

Cell 2

Cadherin

Gap junction

Desmosome

Focal contact

Integrin

ECM

FIGURE 6.3 Molecular structures (junctions) that connect epithelial cells with other epithelial cells and with the ECM. As discussed in the text, these bridges are composed of multiple proteins. Cadherins are important in cell–cell adhesion and integrins in cell–matrix adhesion.

Each integrin is composed of two, noncovalently linked, dissimilar subunits, termed α and β (Fig. 6.4). In most cases, the extracellular domain of the integrin heterodimer interacts primarily with a region of its ligand that contains the tripeptide arginine-glycine-aspartic acid (RGD). Specificity for different RGD-containing ligands is imparted by other regions of the protein and by the combination of α and β integrins interacting with the matrix. The cytoplasmic tails of integrin subunits are generally short and always devoid of enzymatic features. Therefore, as shown in Fig. 6.4, integrins transduce signals to the cell interior by associating with adapter proteins that connect the integrin to the cytoskeleton, cytoplasmic kinases, and transmembrane growth factor receptors. There are at least eight β and 14 α subunits that have been identified. Each β subunit can interact with any one of multiple α subunits to give rise to a variety of heterodimeric integrins with differing ligand specificities. The integrins have been divided provisionally into subfamilies based on their β subunit composition. For example, the β1 subfamily consists of seven members, each of which contains the β1 subunit and a unique α subunit. This subfamily contains integrins that are known to interact with the extracellular matrix components: laminin, fibronectin, tenascin, and collagen. Cells are capable of displaying multiple β1 integrins on their surfaces at any given time. These integrins, in turn, act together to both promote interactions between cells and between cells and various components of the ECM. Cell communication through integrins influences cellular behavior, leading to changes in gene expression, alteration in intracellular pH, and rearrangement of the cytoskeleton. All cells do not express all integrin genes; the diversity of integrin heterodimers allows specific cell types to interact with specific ligands and can promote cell-type-specific functions.

FIGURE 6.4 Schematic diagram of the integrin family of transmembrane adhesion molecules. Each integrin is composed of an α subunit and a β subunit. Integrins interact with one of many ECM components often by recognizing the tripeptide motif RGD in the substrate (e.g., laminin or fibronectin). The cytoplasmic tails of these subunits are generally short and devoid of enzymatic features. Therefore, integrins transduce signals to the cell interior by associating with adapter proteins (FAK, CAS, Pax, Tal, and Vin) that connect the integrin to the cytoskeleton, cytoplasmic kinases, and transmembrane growth factor receptors. Integrin binding to ECM promotes integrin clustering and association with the cytoskeleton. (Modified from Giancotti and Ruoslahti, *Science* 1999, 285: 1028–1032.)

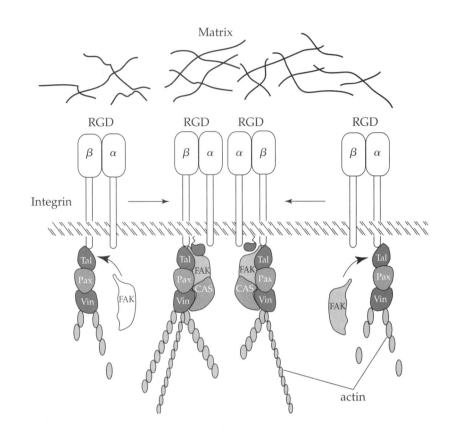

Integrin binding to the ECM generates a mechanochemical cascade that transmits signals into the cell. In addition to this outside–in signaling, cells can also regulate the ability of integrins to actually bind ECM (inside–out signaling). As β1-containing integrins bind to the ECM, they rapidly cluster in the plane of the membrane and associate with a number of adapter proteins and kinases, resulting in the assembly and reorganization of actin filaments (Fig. 6.4). We discuss the integration of adhesion and growth factor signaling pathways later in this chapter. (See Fig. 6.13–6.15.)

The stabilization of nascent integrin clusters by these actin filaments in turn stimulates more clustering to result in the formation of large aggregates on each side of the membrane. Such large aggregates can be observed by electron and fluorescent microscopy and are termed focal adhesions. (See Fig. 6.3.) In this way, integrins link the ECM and the actin cytoskeleton to form strong mechanical structures that in conjunction with other cytoskeletal elements provide the cell with a network of tension-coupled struts and crossbeams, similar to structures based on tensegrity. In cells that are motile and deformable, these cytoskeletal structures must by necessity be able to accommodate and drive shape change by local reorganization without destroying the whole network. In addition to the formation of focal adhesions and the resulting assembly and reorganization of the cytoskeleton, integrin clustering activates various tyrosine and serine kinases, which in turn link cell adhesion to the signaling pathways that regulate cell growth and differentiation. (See Fig. 6.13.)

6.4 | Cell–Cell Adhesion and Cadherins

Cell–cell adhesion molecules (CAM) are the transmitters of signals between cells in direct contact. Of the cell–cell adhesion molecules, the cadherins are the best studied and most important; therefore, we will restrict our discussion to them. Unlike the integrins, cadherins promote cell adhesion principally through homophilic interactions, meaning cadherins will bind to cadherins of the same type. In differentiated epithelial collectives, cadherins have been localized to the membrane domains of adherens junctions where they are thought to be associated with at least three other proteins known as catenins (Fig. 6.5). The catenins have been proposed to function as links between the cytoplasmic domains of cadherins and the submembrane actin-based cytoskeleton.

Cadherins comprise a rapidly expanding superfamily of membrane glycoproteins. The first cadherins to be discovered are now known as the "classical" or type I cadherins. These include E (Epithelial)-cadherin, N (Neural)-cadherin, and P (Placental)-cadherin. Classical cadherins are composed of five extracellular domains of 110 amino acids (known as cadherin domains), a single transmembrane domain, and two cytoplasmic domains. The structure of a classical cadherin is shown in Fig. 6.5. Calcium-binding motifs are situated in the extracellular domains. The first extracellular domain (designated as the EC1 domain) of each classical cadherin contains the cell adhesion recognition (CAR) sequence, His-Ala-Val (HAV). The cytoplasmic

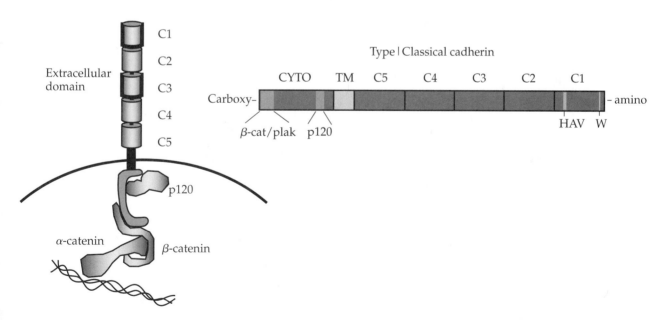

FIGURE 6.5   Structure of a classical cadherin. A classical cadherin consists of five extracellular (EC) domains and two cytoplasmic (CP) domains. The extracellular domains contain calcium-binding motifs, which when bound to calcium confer protease resistance and structural rigidity. The cell adhesion recognition sequence HAV and a conserved tryptophan (W) are located in the first EC domain. The CP domains interact principally with p120, β-catenin (β) or plakoglobin, which also interacts with α-catenin (α) linking the cadherin to the actin cytoskeleton.

domain of the classical cadherins is the most highly conserved region. This domain of the cadherin interacts with the cytoskeleton via its associations with catenins.

The exact mechanism by which classical cadherins interact with each other is a subject of intense investigation. Researchers have performed three-dimensional structure analysis of E- and N-cadherin extracellular domain fragments. These studies revealed the presence of both *cis* (same cell) and *trans* (apposing cells) interactions between cadherin molecules and provoked differing models of "cadherin dimerization" to account for the two kinds of interactions. The crystal structure of the EC1 domain of N-cadherin introduced the "strand dimer" model. This model predicts that each cadherin binds to a neighboring molecule on the same cell, thus forming a strand dimer (Fig. 6.6). The strand dimers on opposing cells then intercalate to form a zipperlike conformation. Although other studies provide evidence for the existence of cadherin *cis* dimers, they do not report the formation of the high-order zipper structures predicted by the "strand dimer" model.

Structural studies demonstrate that the alanine residue (A) of the classical cadherin CAR sequence, HAV, that is found towards the end of the EC1 domain (see Fig. 6.5) interacts with a tryptophan residue (W) located at the beginning of this domain. Calcium ions play an important role in the structural biology of the

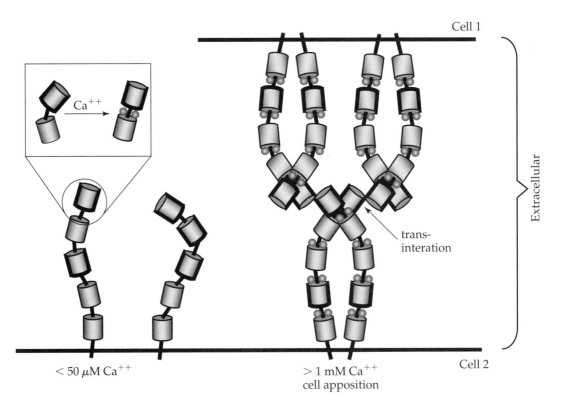

FIGURE 6.6 The cadherin extracellular domain undergoes a steric shift in the presence of calcium that orders the C1 domain and allows *cis-* and *trans-* interactions. (Modified from Pertz et al., *Embo J*. 1999, 18: 1738–47.)

adhesive interface, binding to sites between the cadherin domains and causing changes in conformation. Binding of calcium leads to a more rigid extracellular structure in which the W interacts with the hydrophobic cavity of HAV. Mutation of either W or HAV abolishes adhesion by preventing *trans* interactions between cadherins. Thus, in the presence of calcium, cadherins are competent to form adhesive *trans* interactions. Classical cadherins are typically homophilic adhesion molecules, although several studies have shown that they are capable of forming heterophilic complexes with one another under certain circumstances. Although the antiparallel, *trans* cadherin–cadherin interactions are crucial to calcium-dependent adhesion, like integrins, the formation of stable cadherin-based contacts between cells also depends upon the clustering of cadherins within the cell membrane (*cis* interactions). The juxta-membrane region of the cytoplasmic domain has been implicated to play a role in this lateral clustering, perhaps by interaction with p120.

Under physiological conditions, classical cadherins are associated with their cytoplasmic binding proteins, catenins (including the tyrosine kinase substrate p120), and α-, β-, and γ-catenin/plakoglobin catenins (Fig. 6.5). Remarkably, the classical cadherins cannot promote strong cell adhesion unless they are complexed with the catenins and the actin cytoskeleton. In epithelial cells, E-cadherin complexes cluster together to form the strongly adhesive adherens junction, which forms a belt around cells (Fig. 6.3). Other cadherin family members, desmoglein and desmocollin, cluster together to create another form of junction called the desmosome. Unlike adherens junctions, desmosomes are linked to the intermediate filament (keratin) cytoskeleton rather than actin. Desmosomes form spot junctions rather than belts. When cells are brought into close proximity by cadherin-based junctions, other junctions that do not contain cadherins are formed. Tight junctions form a tight seal at the apical surface of epithelial cells and are important in secretory and absorptive epithelia. Gap junctions are formed from subunits of molecules called connexins and allow the passage of small molecules from one cell to another. Gap junctions are important for cells to act in concert with one another, as for example, in the synchronized contraction required to eject secretions such as milk or digestive enzymes. Furthermore, both cadherins and integrins are subject to complex regulation and these adhesion molecules are regulated by and are themselves regulators of intracellular signaling cascades. The topology of these protein networks is the subject of discussion of a later section.

In addition to their role in promoting cell–cell adhesion through the formation of junctional complexes, members of the cadherin family are responsible for the *specificity* of cell–cell adhesion. This is important during development, particularly in the demarcation of boundaries in the brain. In the developing brain, individual segments called rhombomeres are delineated by the expression of particular cadherins. In a hypothetical example, cells in rhombomere 1, which express cadherin-1, can interact with one another, but not with cells in rhombomere 2, which express cadherin 2. In this case, differential expression of cadherins demarcates boundaries between specific areas of the brain. Similar events take place in other developing organs and are also thought to be important in cancer metastasis. In this case, inappropriate cadherin expression could allow tumor cells to interact with other cells in the body and allow them to invade and metastasize.

6.5 | Examples from Signal Transduction Pathways

6.5.1 Water-Soluble Signaling Molecules

Water-soluble signaling molecules (peptides and charged molecules) cannot diffuse through the cell membrane. Instead, they bind to specific receptors on the cell surface. Binding of hormone to the receptors is the input signal that is transduced to the cell interior (Fig. 6.7). Hormone binding alters the physical configuration and chemical properties of the receptors, clustering them together, and enabling them to interact with cytoplasmic proteins that act as transducers (Fig. 6.8).

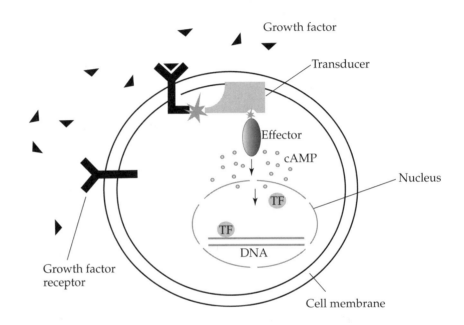

FIGURE 6.7 Schematic diagram of the steps in cell–cell signaling through binding of a charged hormone to a cell-surface receptor belonging to the family of receptors called G protein-linked receptors. cAMP: cyclic AMP; TF: transcription factor.

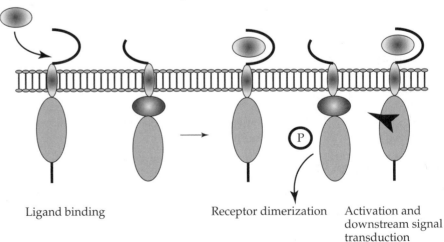

FIGURE 6.8 Ligand binding induces aggregation and in many cases cross-phosphorylation of receptors in the plane of the cell membrane. As with integrins and cadherins, this aggregation in turn results in the recruitment of adaptor molecules and subsequent transduction of the signal.

The effector proteins can be kinases, enzymes that alter the function and structure of proteins through phosphorylation or enzymes that catalyze the formation of small signaling molecules called second messengers. This cascade of events amplifies the external signal, results in changes in gene expression, and culminates in altered cellular processes. Different cell types may have different sets of receptors for the same ligand and react differently to the same signal. Moreover, the binding of a specific ligand to a specific receptor may trigger different responses in different cell types. In the following examples, we illustrate the principles of signal transduction using three different pathways.

6.5.2 Signaling by G-Protein-Linked Receptors

In this section, we focus on signaling mediated by water-soluble hormones such as epinephrine (aka adrenaline), a charged molecule released by the adrenal gland in times of fright or heavy exercise. These hormones bind to transmembrane signal receptors that belong to a protein family called G-protein-linked receptors (Fig. 6.9). Proteins in this family have seven stretches (loops) of transmembrane α-helices. The loop between the α helices 5 and 6 faces the cytosol and is the site of interaction with the so-called G proteins (Fig. 6.9).

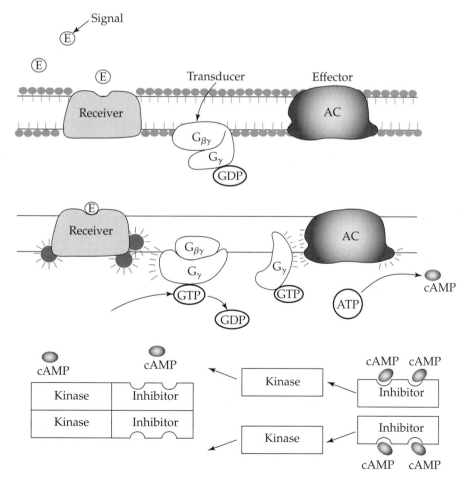

FIGURE 6.9 G-protein-mediated signaling by polypeptide hormones such as epinephrine. The top row shows the inactive G-protein-linked receptor (β2-adrenergic receptor), G protein (transducer), and the effector protein complex, in this case adenylate cyclase. Binding of epinephrine (E) to β2-adrenergic receptor activates the receptor, which in turn activates the G protein, causing it to disassemble. The α subunit of the G protein then interacts with adenylate cyclase, turning the enzyme on (middle row). The cAMP molecules produced by adenylate cyclase bind to the regulatory units associated with the cAMP-dependent protein kinase, PKA. The complex separates into four units and the kinase subunits go on to phosphorylate proteins that influence gene expression.

An example of G-protein-linked receptor is the $\beta 2$-adrenergic receptor (BAR2) expressed by liver cells and other tissues. Binding of hormone to a specific extracellular site on the G-protein-linked receptor activates and alters the physical configuration of this transmembrane receptor. The receptor–hormone complex then interacts with G proteins encountered during movement within the plane of the plasma membrane. The G proteins are peripheral membrane proteins about half the size of G-protein-linked receptors (Fig. 6.9).

G proteins are associated with the plasma membrane of the cell, but are not integral membrane proteins. In many cases, G proteins are modified post-translationally by a lipid molecule. This modification enables them to be highly mobile with the plane of the cell membrane. The G-protein-coupled receptors associate with a complex that is composed of three G-protein molecules called the α, β, and γ subunits. Gα has a single binding site for the guanine nucleotide GTP and its hydrolysis product GDP (Fig. 6.10). The molecule is inactive when bound by GDP. Interaction with a hormone–G-protein-linked receptor complex activates the G protein, resulting in the detachment of GDP. Almost instantaneously, a GTP molecule attaches to the vacant site and subsequently the α subunit detaches from the $\beta\gamma$ subunits. The α subunit diffuses in the plane of the membrane and binds to an effector protein complex (Fig. 6.10). In the case of epinephrine-induced signaling the effector is the enzyme adenylate cyclase, which is a transmembrane protein containing four α-helices that span the plasma membrane. Adenylate cyclase catalyzes the formation of cAMP when it is attached to Gα. The catalytic activity of the enzyme ceases upon detachment of Gα. There are at least two types of G proteins in eukaryotic cells: Gs triggers an effector enzyme to catalyze the formation of second messenger, as described earlier, and Gi inactivates effector enzymes, halting the production of second messenger.

A single hormone–receptor complex activates a large number of G proteins. Each activated G protein transduces the signal to one copy of an effector enzyme, and the effector enzyme catalyzes the production of thousands of copies of the second messenger while bound to an active α subunit. This amplification is an important feature of this form of signal tranduction because hormone concentration in the blood is very small even at peak concentration. As a result, only a few receptors may have a ligand bound to them.

Thus, it is important that a hormone-bound receptor can activate multiple G proteins and that each effector protein can catalyze the formation of thousands of second messenger molecules. Once the hormone concentration in the blood decreases toward zero, G-protein-linked receptors can no longer activate G proteins and eventually the production of second messenger comes to a halt.

The activity of many enzymes and protein kinases depends on the concentration of second messengers such as Ca^{++} and cAMP. In the absence of signaling, the concentration of a second messenger is small and it increases sharply and rapidly upon signaling. A rise in concentration of a second messenger leads to tissue-specific responses because the activity of many enzymes depends on these ligands. A second messenger can modify the rates of enzyme-catalyzed reactions as well as activate or inactivate kinases. Some protein kinases are present in the cytoplasm in an inactivated form in association with inhibitor proteins as shown in Fig. 6.9. The kinase complex shown in the figure is composed of two identical regulatory units (inhibitors) and two catalytic units (kinases).

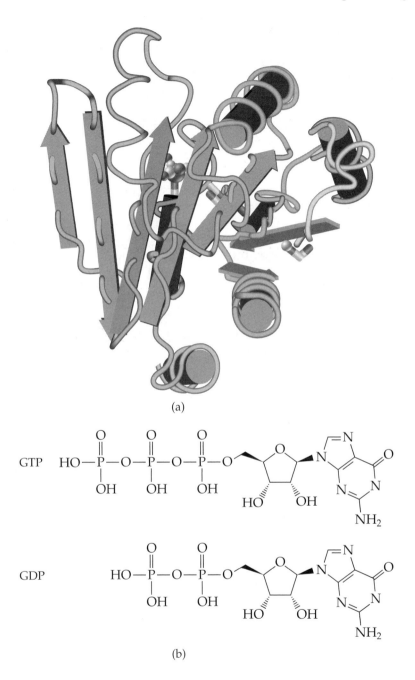

(a)

(b)

FIGURE 6.10   The α subunit of the human G protein (a) is an enzyme that hydrolyzes GTP into GDP (b). Structure of the α subunit was obtained from the http://www.ncbi.nlm.nih.gov/ using the Entrez search for human G protein (small G Protein Arf6-Gdp with protein classicification number 1E0SA).

   In the absence of cAMP, the regulatory units block the enzymatic activity of the catalytic units. Upon binding of cAMP to the regulatory units, the enzyme dissociates. The free catalytic subunits then are enabled to phosphorylate their target proteins, altering the proteins' structure and function. Proteins phosphorylated as a result of activation of G-protein-coupled receptors include transcription factors that modulate gene expression.

G-protein-mediated signaling is tightly regulated in animals. Consider, for example, the effect of the hormone epinephrine on the human body. Adrenergic receptors that belong to the family of G-protein-linked receptors interact with this hormone.

The family includes $\alpha 1$-, $\alpha 2$-, $\beta 1$-, and $\beta 2$-adrenergic receptors. α-adrenergic receptors are found on smooth muscle cells lining the blood vessels of the skin, intestines, and kidneys. Epinephrine binding to these receptors causes the arteries in these organs to constrict and thereby decrease their blood supply. $\beta 1$-adrenergic receptors are expressed on the surface of heart muscle cells. Epinephrine bound to these receptors increases the contraction rate of the heart. $\beta 2$-adrenergic receptors are expressed on the surface of liver cells. The different responses of heart, liver, and smooth muscle cells to the same hormone help supply energy for the rapid movement of the arms and legs in response to bodily stress.

It is interesting to note that upon binding of epinephrine, both $\beta 1$- and $\beta 2$-adrenergic receptors activate G proteins that subsequently stimulate adenylate cyclase, the enzyme that catalyzes the production of cyclic AMP. The increased concentrations of this second messenger cause different effects in liver and heart cells, presumably because these cells constitutively express different sets of genes.

6.5.3 Signal Transduction by Tyrosine Kinase Receptors

Another family of cell-surface receptors is the protein tyrosine kinase (RTK) family (Fig. 6.11), which act as receptors for soluble growth factors such as epidermal growth factor (EGF) and fibroblast growth factor (FGF). In many circumstances, cell proliferation is dependent upon activation of receptor tyrosine kinases by ligand. Mutations in RTKs have been linked to an increasing number of inherited human disease syndromes, including dwarfism, venous malformation, and heritable cancer susceptibility. ErbB2, an RTK that is overexpressed in breast cancers, is the target of the new breast cancer drug herceptin. RTKs possess intrinsic kinase activity that results in the transfer of the high-energy γ phosphate of ATP to hydroxyl groups on target proteins. Receptor tyrosine kinases undergo ligand-dependent dimerization, which activates their intrinsic protein tyrosine kinase response and results in autophosphorylation.

In response to specific extracellular ligands, RTKs play important roles in a variety of processes including growth, migration, survival, and differentiation. RTKs consist of an extracellular ligand-binding domain, which is connected to the cytoplasmic domain by a single transmembrane helix (Fig. 6.11). The cytoplasmic domain contains a highly conserved kinase domain and additional regulatory domains.

Like integrins and cadherins, nearly all RTKs are activated by dimerization or the formation of higher order oligomers. At any given time, inactive RTK monomers on the cell surface exist in equilibrium with inactive and active dimers. The presence of ligand stabilizes the active dimer form of the RTK and allows more efficient transphosphorylation of tyrosine residues in the activation loop of the catalytic domain. This in turn allows the activation loop to open up, allowing access to more ATP and substrates, which further enables phosphorylation of the receptor itself and of other substrates.

In addition to the ability of the activated RTK to directly phosphorylate a wide variety of cellular substrates, autophosphorylation controls the recruitment of a

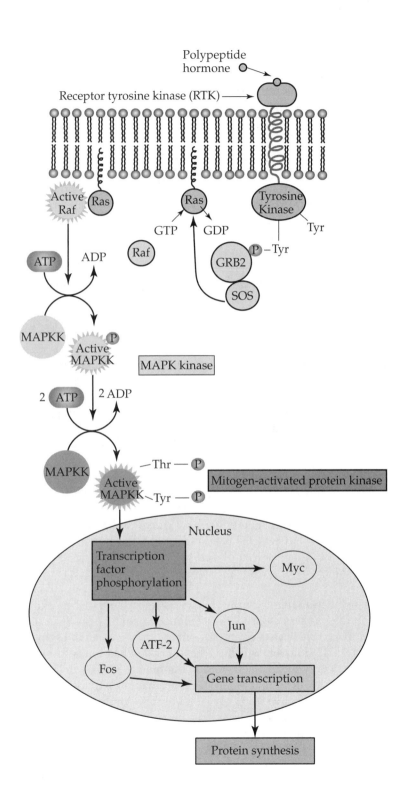

FIGURE 6.11 Schematic representing a simplified RTK pathway. Not shown is the dimerization of the lig-and-bound receptor, which results in its autophosphorylation. Adapter proteins such as the SH2-containing Grb2 interact with a phosphorylated Y residue on the cytoplasmic domain of the RTK. Grb2, in turn, interacts with the guanine nucleotide exchange factor SOS, which can then activate the small G protein Ras. Activated Ras then recruits Raf, which results in activation of the MAP kinase pathway and the regulation of gene expression.

number of signaling proteins. Tyrosine phosphorylation of residues outside the catalytic domain creates sites that bind the modular domains of signaling proteins. Depending on their modular composition, other domains of the recruited or docked signaling proteins can then interact with other target proteins and even the plasma membrane itself. In this way, a complex of signaling proteins is recruited to the activated RTK, some of these interact directly with the activated receptor and others are recruited by the receptor-binding docking proteins.

A large number of different modular docking proteins and their target protein exist, and different cell types and tissues express different classes of these molecules. Consequently, the same RTK can elicit quite different responses in different cells depending upon the combination of docking and signaling proteins that are recruited to the complex. In other words, in cell type 1 an activated RTK may recruit an SH2-containing docking protein A. Protein A, by virtue of its unique modular domain composition then couples with kinase B, which results in phosphorylation and activation of kinase C, which in turn leads to transcription of gene profile E. In cell 2, the same RTK is activated, but SH2 domain-containing protein F (not A) is expressed. Protein F has a different modular composition than protein A and recruits kinase G, which in turn phosphorylates kinase H to lead to gene profile J. A similar argument can be used to explain how two different RTKs can lead to the same response in the same cell. In this scenario, RTK1 and RTK2 have quite different ligand specificities, but similar docking protein-binding specificities. That is they both recruit protein A upon activation and ultimately lead to activation of gene profile E. In reality, the situation is considerably more complex. For example, most RTKs have several sites for docking proteins. Each of these sites can be regulated by phosphorylation or some other conformational change, such as binding of another protein. Control of RTK signal transduction is also exerted by autoinhibitory loops, short- vs. long-term responses, and targeted degradation or endocytosis of key intermediate proteins.

We have already discussed G-protein-coupled receptors. It turns out that RTKs also result in the activation of G proteins. In the case of RTKs, the G protein is a member of the small G protein Ras family that is usually suspended from the inner leaflet of the plasma membrane by virtue of a lipid tail. As with other G proteins, Ras exists in either a GTP bound or a GDP bound state (Fig. 6.11). When activated, the GTP is hydrolyzed to GDP, the resulting conformation change driving some sort of chemical reaction. This leaves the Ras in the inactive GDP bound form.

In order to be available for other reactions, the GDP must be exchanged for GTP, a reaction that is carried out by the exchange factor Sos. Sos interacts with the adaptor protein Grb2 through a modular domain known as an SH3 domain. Grb2 also possesses an SH2 domain, which we have just learned binds to tyrosine phosphorylated RTKs. Consequently, activated RTKs bind to Grb2, which in turn binds to Sos bringing it to the membrane where it can exchange GDP for GTP on Ras. The activated Ras can then interact with any of a number of effector proteins to stimulate many intracellular processes. Perhaps the best-studied example of a Ras-activated pathway is the MAP kinase pathway (Fig. 6.11). RTK/Ras activation of the MAP kinase cascade results in the activation of specific transcriptional programs and is important in the regulation of metabolic processes, migration, cell proliferation, and differentiation. There are at least a hundred different RTKs, which can be organized into supergene families.

6.5.4 Signal Transduction by TGFβ Family Receptors

Members of the TGFβ family are important in the control of embryogenesis and in the regulation of cell growth and the deposition of extracellular matrix materials in the adult. Nearly 30 members of the family have been described to date and all of them bind to and signal through members of the TGFβ-receptor family of serine/threonine kinases. As is the case with adhesion molecules and RTKs, ligand binding results in the interaction of two members of the receptor family (Fig. 6.12). Upon ligand binding, type II receptors interact with and phosphorylate type I receptors, which in turn phosphoryate an effector protein known as a receptor (R) Smad.

Unphosphorylated R-Smads exist in the cytoplasm associated with a cytoplasmic anchor protein called SARA. Upon receptor activation, R-Smad is phosphorylated and binding to SARA is diminished thereby exposing a nuclear import signal on the R-Smad. Phosphorylated R-Smad is then able to interact with another member of the Smad family known as a co-Smad. The R-Smad–co-Smad complex can then enter the nucleus and regulate transcription of specific sets of genes.

This mode of signal transduction is much more direct than that of either G-protein-coupled receptors or RTKs in that the activated receptor directly activates

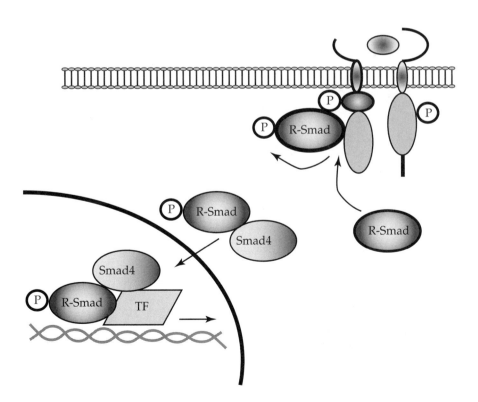

FIGURE 6.12 Signal transduction by TGFβ family receptors. Upon ligand binding, type II receptors interact with and phosphorylate type I receptors, which in turn phosphoryate an effector protein known as a receptor (R) Smad. Phosphorylated R-Smad is then able to interact with another member of the Smad family known as a co-Smad. The R-Smad–co-Smad complex can then enter the nucleus and regulate transcription of specific sets of genes.

the transcription factor responsible for altering the transcriptional profile. However, complexity is introduced at several other levels. For example, access of ligands to receptors is carefully controlled by a number of soluble proteins that bind TGFβ factors and prevent receptor binding. Although only one co-Smad has been identified (Smad 4) there are a number of R-Smads that are specific to particular receptors. Finally, the ability of the R-Smad–co-Smad complex to regulate transcription is itself regulated by a number of co-factors and other transcriptional activators and repressors. In this way, signal transduction pathways that regulate these Smad modifiers can affect the response of a given cell to TGFβ family receptor activation.

A good example of this is the interaction of Smads with the transcription factor, TCF. TCF activity is regulated by the wnt signal transduction pathway that is discussed in more detail later. Smad4 can directly interact with TCF and cooperatively regulate the transcription of genes that have both Smad- and TCF-binding sites in their promoters. Consequently, cells that are exposed to both wnt and TGFβ signals can alter the transcription of different sets of genes from those exposed to either ligand alone.

6.6 Crosstalk between Signal Transduction Pathways

The protein molecules involved in integrin-mediated signal transduction are illustrated in Fig. 6.13 as a function of their position with respect to cell membrane. The

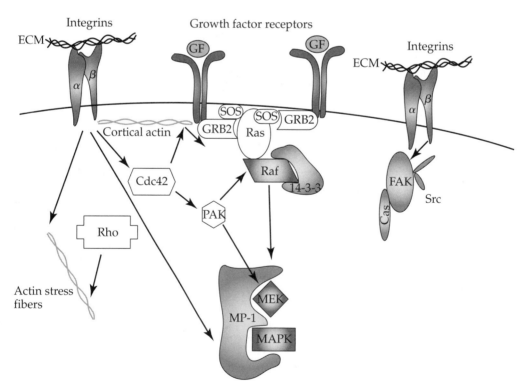

FIGURE 6.13 Schematic of integrins and growth factor receptors and the protein molecules associated with cell-division cycle.

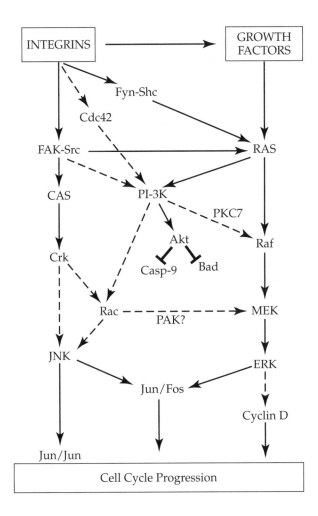

FIGURE 6.14 Portion of the cell-division-cycle network of proteins that begins with integrins and growth factor receptors. Nodes in the network identify the proteins involved and arrows point to the direction of flow of the signaling information. The diagram is simplified in the sense that some of the adapter proteins are not shown, and the phosphorylation events are not underscored. (Modified from Giancotti and Ruoslahti, *Science* 1999, 285: 1028–1032.)

figure also illustrates the connectivity of signal transduction pathways that begin with integrins and growth factors.

The protein network involved in cell-cycle progression receives input at the cell membrane from cell-adhesion molecules and bound growth factors (Fig. 6.14). The arrows in Fig. 6.14 indicate the direction of the progression of the signals from each node (protein element) of the network to other nodes. This combinatorial control of cell division by cell adhesion and soluble growth factors is discussed in more detail in Chapter 7.

6.6.1 Wnt/β-Catenin Signaling Pathway: An Example of Combinatorial Regulation of Intracellular Signaling by Cell Adhesion and a Soluble Ligand

Our final example of a signal transduction pathway is the wnt pathway. (See Fig. 6.15 for a representation of the homologous vertebrate wnt and fruit fly

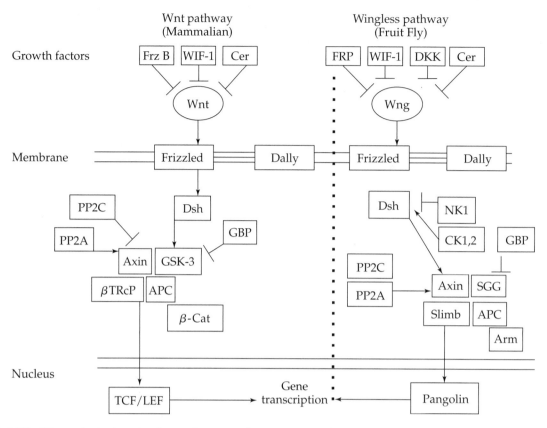

FIGURE 6.15   Wnt and wingless signal transduction pathways. See text for details.

wingless pathways.) Members of the wnt family of growth factors are important in embryogenesis and in cancer. An interesting aspect of this pathway is the important role played by β-catenin, which we have already discussed in the context of cadherin-based adhesion. Wnt ligand interacts with its receptor, a seven-pass transmembrane serine/threonine kinase known as frizzled. Activated frizzled recruits the dishevelled protein, which in turn regulates the activity of a kinase GSK-3. GSK-3 then regulates the assembly of a complex of proteins, which includes the tumor-suppressor gene, APC, and a scaffold protein, axin. In the absence of a wnt signal, this complex results in the phosphorylation and subsequent ubiquitination and degradation of β-catenin. Consequently, cytoplasmic levels of β-catenin are very low in cells that are not stimulated by wnt. When exposed to a wnt signal, this complex is no longer able to target β-catenin for degradation, which is then free to interact with the transcription factor TCF and activate gene transcription. β-catenin/TCF-regulated genes are important for axis development in the embryo and are particularly relevant in colon cancer.

The ability of cadherins to bind β-catenin means that the cytoplasmic pool of β-catenin is susceptible to sequestration to the cell membrane by cadherins. In this way, it is possible to envisage a combinatorial mode of regulation of a signaling pathway by cell-adhesion molecules (Fig. 6.16). β-catenin is important in many

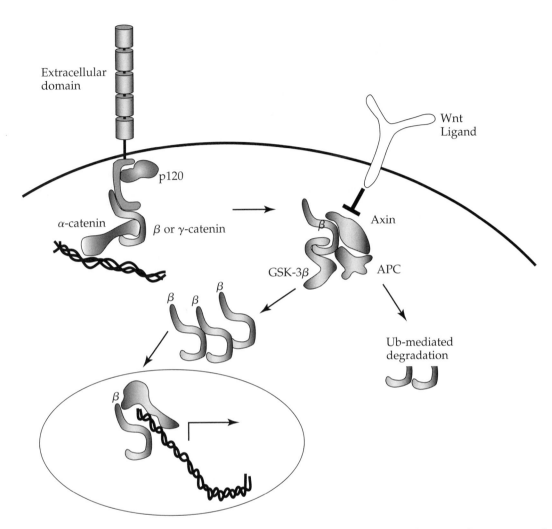

FIGURE 6.16   The dual functions of β-catenin in cell adhesion and the wnt signal transduction pathway. See text for details.

different developmental and disease states and is a key molecule involved in the integration of adhesion and growth factor signaling. A recent publication showed that β-catenin is also important in the regulation of brain size (Chen and Walsh, 2002). In this study, the authors showed that mice making more β-catenin than normal developed much larger brains with a folding pattern characteristic of human brains.

6.6.2 Signal Transduction and Genetic Regulatory Networks Give Cells the Properties of Complex Adaptive Systems

In this chapter, we have emphasized that cells respond to and interact with the environment via an extraordinarily complex network of extranuclear signaling pathways coupled with the equally complex network of genetic regulatory circuits discussed in Chapter 4. The components of these pathways can be regulated by concentration (altered protein expression) or by transient and reversible posttranslational modifications (phosphorylation and degradation).

Additional complexity arises as a result of interactions among signaling and genetic regulatory circuits. More than 200 unknown genes with kinase domains exist in the sequence tag database indicating that the pathways so far elucidated represent the tip of the iceberg. Remarkably, the result of the activity of these networks is the "emergence" of ordered properties such as a differentiated cell phenotype, organ specific structures and memory. In a single cell type, perturbation of these networks (by ligand receptor interactions, for example) leads to a restricted number of output phenotypes. The cell may divide, die (apoptosis), move, differentiate, or produce secreted molecules destined to influence neighboring or distant cells. How cells are able to interpret the biochemical output of the network appropriately is the major unanswered question in cell biology today. The most common approach to this study is the identification and characterization of the individual components of a particular pathway. Examples of this include the human genome project, detection of genetic alterations in disease states such as cancer and the dissection of the various kinase-regulated signaling pathways. Clearly, in order to completely understand a phenomenon it is important to know the components in as much detail as possible. Nevertheless, it is possible to derive thematic information using the information that is currently available. For example, it is not necessary to know every word in a language to speak it or write poetry. Very few of us have a vocabulary that exceeds 10 percent of the words in the *Oxford English Dictionary*. Additional language sophistication arises as a result of alternative interpretation of the same words and phrases by different individuals and societies.

Generally, one cannot interpret studies of a single linear pathway in isolation. Rather, one should consider the influence of the experimental manipulation upon the network as a whole. An example of this is the homeobox "language" discussed in Chapter 8. This influence may not be linear or predictable using conventional logic. That is, a small perturbation of a component which is part of a signaling pathway amidst a highly connected parallel processing network is likely to lead to consequences that cannot be predicted based on a knowledge of all the individual components. These are characteristics of nonlinear dynamic systems (chaotic systems). The more highly connected the network the more likely the system is to tend toward nondeterministic chaos; that is, its behavior cannot readily be predicted. For example, if each individual pathway connects with all others it will be impossible to predict the outcome.

In cells, it is unlikely that individual signaling and genetic regulatory circuits all communicate with one another. Perhaps some "nodes" represent pathways that do not connect with others, and deleting the gene for these nodes null leads to dead ends. Some circuits (pathways) may be highly connected and can influence and be influenced by other pathways. For example, one could imagine that spill over from one pathway to another may occur in cases of over-expression of certain components (amplification of a growth factor receptor for example). This may well lead to consequences that are an indirect result of inappropriate stimulation of an alternative pathway.

An increasing amount of evidence points to the connectivity of signaling and genetic regulatory circuits. Consequently, combinatorial control of intracellular signal transduction from different growth factors, receptors as well as adhesion molecules, results in a network of reactions, the output of which can be both fine tuned and robust at the same time.

6.7 ASSIGNMENTS

6.1 White blood cells roll along blood venules, adhere to the vascular endothelium and subsequently transmigrate into the surrounding tissue at the sites of inflammation. Rolling is rotation of the cell with little slip at the contact region and is characterized by quickly occurring and reversible cell–substrate interactions. Some cancer cells use similar rolling mechanisms in blood vessels during metastasis. Conduct a literature search to identify the surface receptors involved in cell rolling.

6.2 Integrin $\alpha6\beta4$ is implicated in epithelial-layer closure during wound healing. Conduct a literature search on the structure and function of this integrin.

6.3 Discuss the role of cadherins in sorting cells according to type in a suspension of human tissue cells.

6.4 Research literature indicates that fluid shear stress induces alterations in gene expression in endothelial cells that are adherent to a substrate. Discuss how cells could sense fluid shear stress.

6.5 Conduct a literature search to find out why mutation of either W residue or HAV region abolishes cadherin-mediated adhesion.

6.6 Describe the signaling proteins recruited by the autophosphorylation of the RTK receptors.

6.7 Discuss the wnt signal transduction pathways using the KEGG diagram as a foundation. What are the major proteins involved in this pathway? What is the significance of β-catenin in the wnt pathway? What is the mechanism that tightly controls the concentration of this protein in cell nucleus and cytoplasm?

6.8 Normal epithelial cells have receptors for TGFβ. Activation of this receptor leads to the expression of growth-arrest genes and other target genes. Describe why cells lacking TGFβ receptor undergo uncontrolled growth.

6.9 Discuss a second-messenger signaling pathway of yeast using the KEGG database. Elaborate on the different responses of the cell to different levels of concentrations of the nutrients present.

6.10 What are the proteins that connect integrin-mediated signaling to growth factor-induced signaling? Discuss the potential consequences of mutations that result in the dysfunction of the connecting proteins.

REFERENCES

Aplin AE, Howe AK, and Juliano RL. "Cell adhesion molecules, signal transduction and cell growth." *Curr Opin Cell Biol* 1999, 6: 737–744.

Blume-Jensen P and Hunter T. "Oncogenic kinase signalling." *Nature* 2001, 411: 355–365.

Bourne HR. "How receptors talk to trimeric G proteins." *Curr Opin Cell Biol* 1997, 2: 134–142.

Chen and Walsh. "The regulation of cerebral cortical size by control of cell cycle exit in neural precursors." *Science* 2002, 297: 365–369.

Giancotti FP and Ruoslahti, E. "Integrin-signaling." *Science* 1999, 285: 1028–1032.

Gottardi CJ and Gumbiner BM. "Adhesion signaling: how beta-catenin interacts with its partners." *Curr Biol* 2001, 11: R792–R794.

Hartman JL 4th, Garvik B, and Hartwell L. "Principles for the buffering of genetic variation." *Science* 2001, 291: 1001–1004.

Huelsken J and Birchmeier W. "New aspects of Wnt signaling pathways in higher vertebrates." *Curr Opin Genet Dev* 2001, 5: 547–553.

Massague J. "How cells read TGF-beta signals." *Nat Rev Mol Cell Biol* 2000, 3: 169–178.

Nagafuchi A. "Molecular architecture of adherens junctions." *Curr Opin Cell Biol* 2001, 5: 600–603.

CHAPTER 7

Cell Division and Its Regulation

7.1 | Introduction

The first part of this chapter is concerned with the mechanics of protein–gene interactions that occur during the cell cycle. Living organisms develop, reproduce, and remodel, and cell division is a fundamental step in these processes. Cells multiply by dividing into two daughter cells. Before division, the nuclear DNA makes two copies of itself, and these copies along with cytoplasm and its associated organelles are equally distributed to the two new daughter cells. The rate of cell division in prokaryotic cells (such as bacteria) as well as in unicellular organisms (such as yeast) is largely determined by the availability of nutrients and the release of toxic wastes that limit growth. The growth of these cells in bioreactors is important to chemical engineers that use them to produce recombinant proteins or other agents. Understanding how the cell cycle is regulated in these organisms is essential to maximizing production.

Cell division in multicellular organisms is regulated by much more complex signals in order to maintain the intricate organization and function of the body. Some cells in the adult animal such as neurons and muscle cells perform a set of special differentiated functions and do not divide after birth. Understanding why neurons do not divide in the adult is important to bioengineers in the area of nervous-tissue regeneration and transplantation. Cell division is particularly important during embryological development and in renewing tissues such as the gastrointestinal tract and skin to replace the millions of cells that die every day. In all these situations, cell division is subject to extensive controls. Uncontrolled cell division in the embryo leads to profound embryological deformity and death, whereas cell division that escapes regulation in the postembryonic state leads to cancer. In this chapter, we focus on mitotic division of body or somatic cells. The meiotic division of germ cells is considered in the next chapter in which we focus on embryonic development.

7.2 | Modes of Cell Division

The products of the cell division should be two daughter cells, each possessing the required organelles and faithful replicas of the chromosomal template of the parent cell. The stages through which cells pass between one division and the next is referred as the cell cycle. Reduced to its simplest form, the cell cycle is comprised of two principle events: replication of chromosomes and segregation of replicated chromosomes. Both events require the involvement of complex protein machinery. Recent advances in high-resolution imaging and the use of fluorescent probes have uncovered the physical events that occur in different phases of cell division. These events are described in the sections that follow.

7.2.1 Cell Division in Prokaryotes

The simplest mode of cell division is observed in prokaryotes. These unicellular organisms lack nuclei and have a single chromosome in the form of a double-stranded circular DNA. The chromosome lies in a diffuse nucleiod region in the central part of the cell. The chromosome, which in its extended form, is several hundred times longer than the cell, attaches at one point to the plasma membrane and is compacted and folded with the help of DNA-binding proteins (Fig. 7.1).

Before cell division, DNA in a prokaryotic cell replicates, resulting in two identical copies, which then move rapidly towards opposite poles of the cell and attach to the plasma membrane. After the completion of DNA replication and subsequent chromosome separation, a wall forms and divides the cell into two compartments. Most prokaryotic cells divide every 30 min in the presence of nutrients. Many of the genes that mediate chromosome segregation in prokaryotes have orthologs in eukaryotes, indicating the universal nature of the cell-division-cycle machinery. More about prokaryotic cell division can be found in recent reviews of the subject including that by RB Jensen and L Shapiro (1999).

7.2.2 Cell Division in Eukaryotes

Eukaryotic cells also divide following DNA replication. Most animal cells divide much less frequently than prokaryotes, and some cells in multicellular organisms do not divide at all. Gross examination of animal cells growing in culture reveals two populations with distinct morphologies. A relatively small percentage of the cells (~10–20 percent) are rounded with some of them containing condensed chromosomes. These cells are actively undergoing division (mitosis) and are in the so-called M phase of the cell cycle. The physical events that occur during the multistage M phase include separation of duplicated chromosomes as well as partition of organelles into daughter cells.

The cells that are not dividing appear flatter in the culture dish. They are more firmly attached to the bottom of the dish and exhibit size variations. These cells are between cycles of mitosis (M) and are said to be in interphase. Interphase can be divided further into three phases, based on the DNA content of the cell. The period of active DNA synthesis is called the S phase, and the gap phases between M and S and

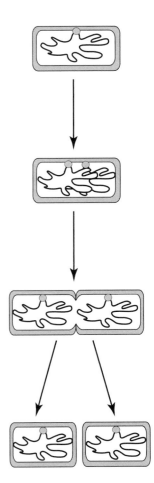

FIGURE 7.1   Schematic of a prokaryote and the circular DNA contained within. The DNA is about 1.7-mm long, whereas the cylindrical cell has a diameter of about 1 μm and length of 4 μm. Proteins that bind to DNA guide this long molecule pack into the small space by folding on itself many times. Cell division follows this DNA replication.

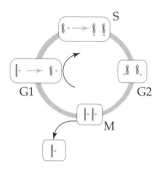

FIGURE 7.2   Subdivisions of the cell cycle. The period between the formation of a cell and its division to two new cells can be divided into four segments: a period of active DNA synthesis (S), a period of active cell division (M), a gap between S phase and M phase (G2), and a gap between M and S phases (G1). The DNA is shown as a tangled thread in the rectangle boxes corresponds to interphase (G1, S, and G2). DNA is represented by a vertical bar in M. A circle in a rectangular box associated with a phase represents the centrosome. In most cultured cells, the G1 and S phases take about 10 hours each. M phase takes less than an hour, and G2 phase approximately 4 hours.

S and M are called G1 and G2 phases, respectively (Fig. 7.2). For diploid cells with n pairs of chromosomes, G1 cells have $2n$ DNA molecules, cells in S phase are between $2n$ and $4n$, and cells in G2/M have $4n$ molecules of DNA. The physical and molecular events that occur during the four distinct phases of cell cycle are the subject of this chapter. Division of germ cells that lead to sperm and eggs (meiosis) will be covered in the next chapter.

7.3 | Molecular Basis of Cell Division

The cell cycle is regulated by a series of biochemical steps that translate an external stimulus into a response by the cell. To do this, the cell integrates a large amount of extrinsic (environmental) and intrinsic information. Factors extrinsic to the cell include the presence or absence of chemical nutrients, spatial clues, differentiation inducers, and growth factors (Fig. 7.3). The cell processes these signals in order to select an outcome from a set of finite outcomes, including staying in G1 (G0) phase, dividing, or undergoing apoptosis (programmed cell death). If normal G1 cells are too crowded on the culture plate, not attached to the substratum, or not provided with the appropriate nutrients and growth factors, they cannot enter the DNA-replication phase (S). On the other hand, the presence of too much of one input, or the appearance of too much input, might allow cells to proceed even if other input conditions are not met. The outcome for the identical set of external stimuli is not the same for all cells of an organism, but depends strongly on the cell type (Fig. 7.3).

7.3.1 External Signals

Growth factors are extracellular peptides that bind to surface receptors, which are often tyrosine kinases. (See Chapter 6.) Growth factor binding initiates a signal transduction pathway that leads to expression of genes required at early stages of

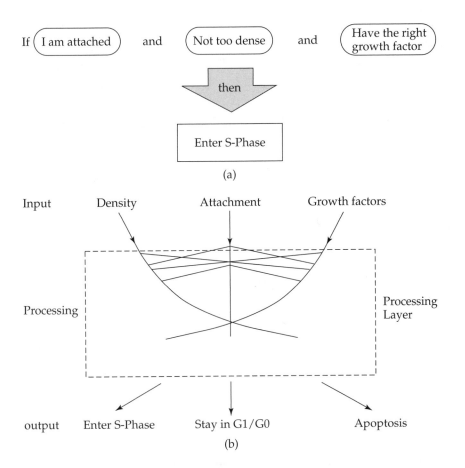

FIGURE 7.3 Decision-making process of an animal cell during the transition from G1 to S phase. The process is represented as a Boolean (a) and neural network (b).

cell division. When normal eukaryotic cells are cultured in vitro, nutrients and growth factors must be provided by the culture medium for cells to undergo cycles of cell division and volumetric growth. Simply reducing serum levels in the medium will prevent most cells from cycling through the phases of cell division. Growth factor-induced cell division requires prolonged exposure to growth factors until the so-called restriction point is reached and growth factors are no longer necessary. Growth factor simulation leads to a rapid increase in the expression of more than a hundred different genes, setting the stage for preparation to enter the S phase.

Cell-cycle events also require signals provided by surrounding cells and the extracellular matrix. Contact-induced signals impact the decision of a cell to divide. Many cells cannot grow unless they are attached to the substratum, and normal cells do not enter S phase when they are very crowded. Signals due to integrin-mediated adhesion are transduced into the nucleus using pathways downstream of receptor tyrosine kinases. Thus, the signal tranduction pathways that begin with growth factors are connected downstream with those beginning with adhesion receptors (Fig. 7.4). The connectivity of these pathways enables cellular response to a soluble mitogen (growth factor) to be altered by changes in cell adhesion (Fig. 6.13). Cancer cells are able to grow in the absence of attachment because a component of the processing layer that normally transmits the "I am anchored" signal is mutated and constitutively activated,

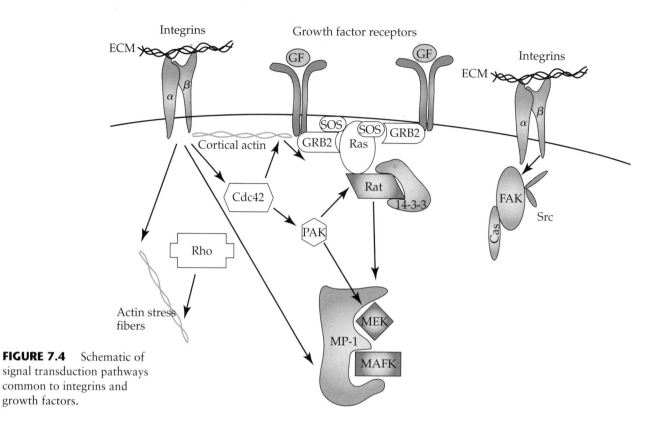

FIGURE 7.4 Schematic of signal transduction pathways common to integrins and growth factors.

even in the absence of anchorage (Fig. 7.3). In other words, it misleads the processor. Some cancer cells can grow in suspension and in very low levels of growth factors. In these cases, a component of the processing layer that is shared by both inputs, or is downstream of an intersection of the inputs, is constitutively activated leading the cell to believe that "I am anchored and exposed to growth factors." One emerging theme from recent research is that cell-adhesion molecules regulate the actin cytoskeleton and may thereby indirectly alter the phosphorylation states of kinases in the cell-cycle regulation pathways.

7.3.2 Cyclins and Cyclin-Dependent Kinases

At the molecular level, cell-cycle progression is associated with cyclical appearance of proteins that belong to the cyclin family. The expression of hundreds of genes, many of which encode transcription factors, changes during the cell cycle. The abundance/activity of cell-cycle-associated proteins also varies dramatically with time depending on the phase of the cycle (Fig. 7.5). Orderly progression through the cell cycle requires a sequential activation and inactivation of different cyclin-dependent kinases also called Cdks (Fig. 7.6). The name cyclin arose because the concentrations of proteins that belong to this family rise and fall in phase with the cell cycle. Binding of an appropriate cyclin to Cdk is required for its activation.

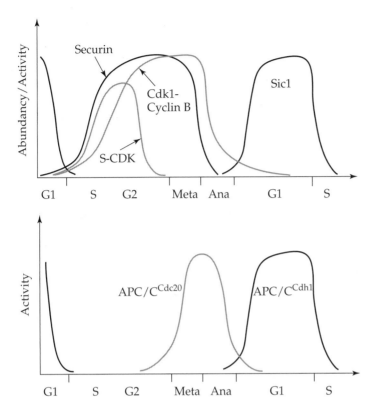

FIGURE 7.5 The cyclical appearance of some cell-cycle-associated proteins during the various phases of the yeast cell cycle.

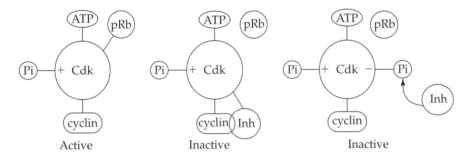

Active   Inactive   Inactive

FIGURE 7.6 Schematic of protein complexes formed by cyclin-dependent kinases. The activation of cyclin-dependent protein kinase depends not only on the binding of an appropriate cyclin, but also on its state of phosphorylation, as well as the presence of inhibitor proteins (Inh). First, a phosphate group (Pi) must be added to the positive (+) regulator site. This is a specific threonine side chain that activates the protein. Secondly, a phosphate elsewhere in the negative (−) regulation site of protein (which is covalently bound to a specific tyrosine side chain) must be removed. Binding of an inhibitor may halt activity of the Cdk even if the cyclin remains bound. Only one of the three states of the Cdk complex shown in the figure can activate a target protein such as Retinoblastoma protein (pRb). In this way, a Cdk acts much like a computer chip. It has a binary output (activate or not activate) based on multiple input signals (binding of cyclin, inhibitor, state of phosphorylation, and so on).

Activated *cyclin-dependent protein kinases* (Cdk) propel the transition from one phase of the cell cycle to another. Their mode of action is phosphorylation, and the overall effect is the release of crucial transcription factors needed for the expression of cell-division-cycle genes. Cyclin–cdk complexes act sequentially and are deactivated rapidly upon completion of their task. Both Cdks and cyclin families have multiple members. Cyclins and Cdks appear to have homologs in all species belonging to the domain eukaryota and the number of cyclins increases with increasing complexity of the organism. For example, the number of cyclins that mediate yeast cell division is smaller than that of human.

Although Cdks function only when bound to their corresponding cyclin, binding of cyclin to its appropriate Cdk does not necessarily result in the activation of Cdk. Cdk activity is tightly controlled by a set of activators and inhibitors. This is evidenced by the observation that Cdk–cyclin complexes are found in the cytoplasm and the nucleus, but they are active only in the nucleus, where their targets are located. Switching off the activity of Cdk–cyclin complexes is carried out by a number of protein inhibitors such as p27 or p21. These inhibitors block Cdk activity by binding to the cyclin–Cdk complex. In some situations, inactivation involves the action of kinases that inactivate Cdks by phosphorylating them at a negative regulatory site (Fig. 7.6).

7.3.3 Targeted Protein Degradation

Degradation of selective proteins marked for destruction generates directionality in the cell cycle. A key event in the activation of a cyclin–Cdk complex is the targeted destruction of the *inhibitor* proteins that normally prevent its premature activation. This introduces a recurrent theme in cell-cycle regulation that involves the cyclical degradation of important regulatory proteins. Many proteins are post-transcriptionally modified in a way that marks them for destruction. This modification involves the covalent attachment of multiple copies of a small protein called *ubiquitin* by a process called *ubiquitination* (Fig. 7.7).

Large protein complexes called SCF and APC mediate the degradation reactions in the transition from G1 to S and in the M phase, respectively. In general terms, one can think of the APC and SCF as master regulators that function to reset the cell-cycle clock by getting rid of components that are not required for or may inhibit the subsequent stage of the cycle.

Ubiquitinated proteins are degraded rapidly by the *proteosome*, a multiprotein particle that contains proteolytic enzymes. Cyclins are also degraded rapidly upon completion of task in a process involving ubiquitination. The proteosome itself can be thought of as a sort of "black hole"; anything that is polyubiquitinated will be directed to it, and anything that enters it will be destroyed. Consequently, the factors and protein complexes that regulate protein ubiquitination are very important in the control of the cell cycle.

In the case of proteins that are involved in G1/S progression, phosphorylation is an important element in the targeting of these proteins for ubiquitination. Once phosphorylated, the proteins are recognized by a member of a family of protein complexes called the SCF and promptly targeted for destruction. In the case of the APC, the targeted proteins are recognized, not by phosphorylation, but by a stretch of amino acids in the primary structure called the destruction box.

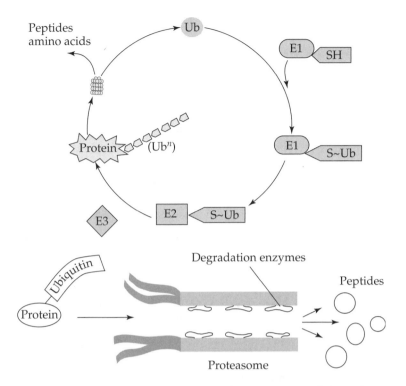

FIGURE 7.7   Degradation of proteins marked by ubiquitination. Cells have the necessary machinery to rapidly degrade unassembled, damaged, or misfolded proteins. The same machinery is also used to confer short half-lives on certain normal proteins whose concentrations affect the cell cycle and other important processes. Ubiquitination of proteins is carried out by a series of enzymes called E1, E2, and E3. Ubiquitin is first activated by E1 and transferred to an ubiquitin-conjugating enzyme (E2). The activation and transfer of ubiquitin is carried out by a thioester cascade. A protein complex (E3, e.g., SCF and APC) then facilitates the substrate-specific attachment of multiple ubiquitins to the protein. The polyubiquitinated protein is then recognized by the proteosome. This protein-degradation machine is a large protein complex, which exists as multiple copies in the cytoplasm and nucleus. They are approximately cylindrical in shape with a chemically active inner surface. The ubiquitinated substrates are recognized at the end, and selected proteins enter the proteasome. Once inside the proteosome, enzymes degrade the proteins to short peptides that are released from the proteosome.

7.4 | Cell-Cycle Check Points

The molecular devices that drive cell division must overcome a series of hurdles put forward on their track by cell division inhibitory proteins. Cell-cycle networks process an array of external and internal signals before deciding to embark on the journey through the cell cycle. In addition to the controls that regulate the choice of whether to divide or not, there exists a hierarchy of molecular regulators that check replication and repair DNA and couple these processes to progression through the cell cycle.

Constructing an analogy to a relay race may be useful in understanding the roles of various types of protein complexes in cell division. Consider a relay race in which every member of the relay team must run along a track decorated with hurdles. In order to complete the circuit, the runners must pass the baton to each successive athlete in a controlled manner. Runner 2 cannot begin running until runner 1 is close, but cannot go too far into the next leg without receiving the baton. In this analogy, the passing zone is the checkpoint, failure of runner 1 to appear, or fumbling of the baton triggers the checkpoint and prevents runner 2 from entering leg 2 until the baton is secured. If the baton is dropped or passed on after the second runner has left the passing zone the team is disqualified even though the runners are physically capable of completing the circuit. Runner 2 serves as a sensor that must judge both the incoming runner's speed and his or her take-off speed. If the baton transfer takes place correctly, runner 2 then acts as a transducer until the baton is passed to runner 3 in the execution step. One can imagine many variations on this theme, such as slow or injured runners or heavy batons, to account for the molecular scenarios affecting the cell cycle. We will review the molecular basis of cell-cycle checkpoints in the next section.

7.5 | Phases of the Cell Cycle

7.5.1 G1 Phase

Commitment of a cell to either divide or stay quiescent occurs in the G1 phase. An animal cell may remain quiescent in G0 (G1) for weeks or even years. Before making the decision to begin the energy-expensive process of DNA replication, a eukaryotic cell must make sure that the conditions are appropriate for cell division. The first cyclin to appear in a cell stimulated by growth factors belongs to the cyclin D family. The synthesis of cyclin D and its subsequent assembly with the catalytic partner Cdks is dependent on growth factor stimulation. Withdrawal of growth factor from the medium leads to an abrupt cessation of cyclin D synthesis and to the rapid degradation of the preformed pool of protein. The activity of cyclin–cdk complexes is also regulated by members of a family of inhibitory proteins (p16, p21, and p27) whose levels are also regulated by targeted protein degradation.

Cyclin D–Cdk complexes drive the cell cycle through their impact on gene transcription. Once activated, these kinase complexes regulate the activity of other protein complexes through phosphorylation. One target of cyclin D–Cdk complex is the Retinoblastoma protein (pRb). In normal cells, the Rb protein is present in the nucleus through all phases of cell cycle. In its dephosphorylated state, pRb is found in association with transcription factor E2F (Fig. 7.8). Association of pRb with E2F hinders the transcriptional activity of E2F. Active Cdk–cyclin D complexes phosphorylate pRb on multiple serines and threonines. pRb detaches from E2F upon phosphorylation, and the untethered E2F binds DNA and activates genes that will be required to transit into the DNA-replication phase. One of these genes encodes cyclin E. Its product continues to phosphorylate pRb, eliminating its inhibitory effect on cell division.

The expression of cyclin E in the late G1 phase signals commitment of the cell to divide. At this point, progression through the cell cycle is no longer dependent

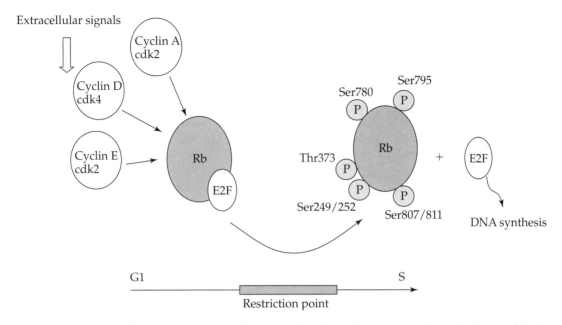

FIGURE 7.8   Role of the retinoblastoma protein (pRb) in the cell cycle. Dephosphorylated pRb binds to and holds inactive gene-regulatory proteins such as E2F. Phosphorylated pRb detaches from the regulatory protein, freeing it to activate proliferation. Rb becomes dephosphorylated as the cell exits mitosis and is then phosphorylated late in G1 phase as the cell prepares to go past START.

on external signals. Cells at this stage are said to have passed through the START or restriction point. From this point on, cells revert to a cell-autonomous program that largely ignores extracellular signals until mitosis is complete. The ultimate target for the drive from G1 to S phase is a protein complex called the DNA Replication Complex (RC). This protein complex must be in its activated state to initiate DNA replication. In both human and yeast cells, the complex is not phosphorylated in G1 and therefore is not active. The Cdk2–cyclin A complex activates the DNA-replication complex by phosphorylating it and triggers the actual onset of DNA replication. At this point, the cell has passed through the G1/S checkpoint.

The G1 phase of the cell cycle is subject to the greatest variation in duration. Almost all the variation in proliferation rates in the adult body of an animal is due to the time cells spend in the G1 phase. The average duration of the cell cycle is between 12 and 24 hours for most animal cells in culture. Nerve and muscle cells in the adult "rest" in G1 for years before they commit to DNA replication and are said to be in G0. Other cells, such as those in the early embryo, have virtually no G1 phase and a very short cell-cycle time. This abbreviated cell-cycle time in the early embryo coincides with the subdivision of the fertilized egg into hundreds of individual nucleated cells without the requirement for individual cell growth. Early embryonic cells are either in S or M phases and can complete the cell cycle in as little as 8 min.

7.5.2 S Phase

In animal cells cultured in vitro, the S phase takes about 8 hours. At the end of the S phase, both the genetic and somatic components of the cell are ready to divide.

DNA Replication: Once committed to undergo DNA replication, cells enter S phase and begin the process of incorporating nucleotides into DNA and duplicating the genome. DNA replication yields two new duplexes, with each duplex consisting of one of the two original strands plus its copy (Fig. 7.9). DNA synthesis begins at special regions of DNA called replication origins or *replicons*. The circular DNA of bacterial cells has a single point of origin for DNA duplication, whereas eukaryotic chromosomes initiate replication at multiple points. Recent estimates put the number of *replicons* in the human genome between 10,000 and 100,000. The activated DNA-replication complex binds to DNA at replication forks and recruits DNA polymerase, an aggregate of several different subunits that is largely responsible for DNA replication. Production of new DNA chains involves multiple steps. These steps include unwinding the double helix, introducing a primer to initiate the replication process, and using the existing DNA as a template for the sequential assembly of nucleotide bases into the new strand. The unwound DNA helix, with each strand serving as template for its complement, is called the *replication fork* (Fig. 7.9). DNA polymerase proceeds along a single-stranded molecule of DNA at the replication fork recruiting free nucleotide bases to hydrogen bond with their complements on existing strands and to form covalent bonds with the previous nucleotide of the same (growing) strand. DNA polymerase needs a small RNA primer onto which it can begin to attach nucleotides. The enzyme *primase* attaches the RNA primer to appropriate locations on the fork. Another enzyme eventually removes the primer, and DNA polymerase fills the resulting gap. DNA polymerase moves in the 5′ to 3′ direction in both the leading strand (5′ to 3′) and lagging strand (3′ to 5′). The only way the enzyme complex can move in 5′ to 3′ direction in the lagging strand is by making a copy of this strand in spurts (called Okazaki fragments).

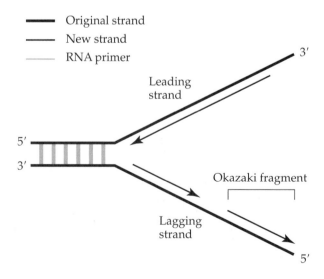

FIGURE 7.9 DNA synthesis in eukaryotes. DNA synthesis occurs by bidirectional growth of both strands from a single origin in prokaryotes and multiple origins in eukaryotes. DNA replication is semiconservative; that is, each daughter DNA molecule contains one old and one new strand. Arrows indicate direction of DNA replication.

Not all replication origins are initiated and active at the same time; some begin early in S phase and others initiate later. If the cell were to go through mitosis without DNA replication being completed, the results would be disastrous. The DNA-replication checkpoint ensures that mitosis is dependent upon completion of DNA replication. The protein complex involved in this checkpoint is called *DNA polymerase epsilon*. As would be expected, the replication checkpoint sensor machine is associated with the replication forks. If DNA synthesis is inhibited with the use of hydroxyurea (HU) in early S phase, replication origins that normally initiate later in S phase are kept dormant by a mechanism that prevents the association of DNA polymerase to the replication origin. The sensor components also activate a kinase cascade that couples DNA replication with chromatid cohesion and spindle elongation.

The process of nucleotide addition during DNA replication is rapid (50–500 nucleotides/s); even under normal conditions, DNA damage can occur as a result of faulty replication that is not repaired. Exposure to ionizing radiation results in more extensive damage. When normal cells are exposed to these agents, entry into mitosis is delayed or prevented by the DNA-damage checkpoint. The cdk1 inhibitor p21 is required for a sustained S-phase arrest after activation of the DNA-damage checkpoint. When damage is irreparable, many cells are diverted to a cell-suicide pathway called *apoptosis*. Like the cell cycle, the events of apoptosis are very similar in most organisms. The cell becomes isolated from its neighbors, chops up its chromatin, and then fragments itself. The surrounding cells ingest the remains of dead cells. In contrast, some tumor cells continue to cycle even after exposure to DNA-damaging agents. The failure of some tumor cells to undergo apoptosis when their genome is damaged is an important aspect of cancer. The discovery of mutations in checkpoint genes in cancer cells and the demonstration that certain yeast mutants continue into anaphase even after DNA damage led to the elucidation of DNA-damage checkpoint pathways. The DNA-damage sensor machine is activated by the presence of single-stranded DNA, probably in the context of repair of the lesion. The sensor components in turn activate a cascade of kinases that ultimately prevents the initiation of anaphase by inhibiting the anaphase-promoting complex (APC), which targets proteins for degradation.

Several transducing kinases are shared in the replication and DNA-damage checkpoint pathways pointing to a dependency of the two pathways. A number of important "tumor suppressor genes" such as the breast cancer gene BRCA1, the ataxia telangiectasia gene, and p53 function in DNA-damage check-point pathways. Loss of function of these genes results in cells that divide without checking for DNA damage. Consequently, these cells accumulate mutations in the genome that ultimately lead to cancer.

Duplication of Somatic Components: The number of ribosomes and mitochondria increases dramatically in S phase. Chromosome-associated proteins (*histones*) that combine with DNA to form chromatin are synthesized so that they are available for association with newly formed DNA. Otherwise, single-stranded DNA is highly unstable; proteins that bind to it during replication keep it from being degraded.

Another important event is the duplication of the centrosome, the organelle that directs the separation of chromosomes (Fig. 7.10). The centrosome consists of

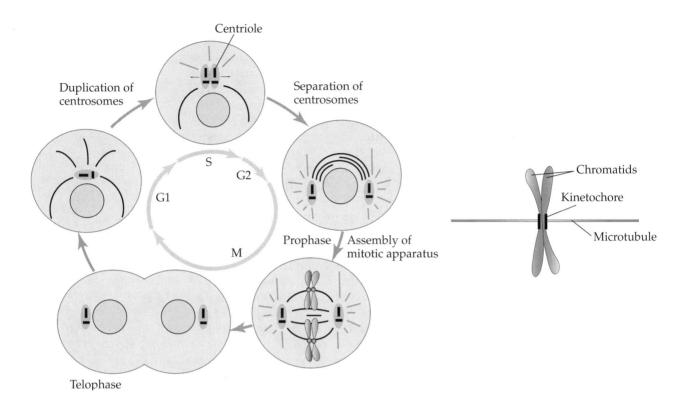

FIGURE 7.10   Centrosomes play important roles in polarization and division of cells. The barrel-like structures at the center of the centrosome are called centrioles. These are oriented at right angles to each other and are connected by thin fibrils. Late in the G1 phase, the distance between the two centrioles increase and the centrosome duplicates during S phase, with one old and one new centriole present in both copies. The two centrosomes remain paired until prophase of the M phase, during which they migrate to opposite sides of the nucleus. A set of microtubules connects the two centrosomes at the polar ends of the cell. Another set of microtubules extending from centrosomes attach to the sites called kinetochores on the centromeres of chromatid pairs, eventually pulling the two chromatids of a single chromosome to opposing polar ends of the cell.

two barrel-like centrioles. The centrioles are oriented at right angles and are connected by thin fibrils. During G1, the two centrioles in a pair separate. During S phase, a daughter centriole grows from the base of each parent centriole, and the two pairs of centrioles move away from one another at the beginning of M phase. The duplicated (daughter) centrosomes contain one old and one new centriole each.

Originating from paired centrioles are microtubules, another important component of the cell cytoskeleton. As discussed in Chapter 3, these are long cylindrical tubes, about 25 nm in diameter and made up of monomers of the proteins, α and β tubulin (Fig. 7.11). The tubulin monomers assemble first as a sheet, which then folds into a tube. $\alpha\beta$-tubulin dimers bind side by side and end to end with each other to form a microtubule. Microtubule assembly depends upon GTP and Mg, and their length depends on the concentration of free tubulin in the cytoplasm as well as on a variety of microtubule-associated proteins. Like actin filaments, microtubules are polarized with a + and a − end.

In cell division, the microtubule–centriole system plays an important role in guiding chromosomes to the ends of a dividing cell. As a result of this process, both daughter cells contain the right number and type of chromosomes.

Microtubules also function as tracks for transport of cellular organelles. The motor protein that drives the organelle transport along microtubules is a member of the kinesin family, which, like myosin, couples ATP hydrolysis to mechanical work. Kinesin along with myosin are called protein motors because they are both enzymes that convert chemical energy into mechanical work. In in-vitro systems, beads decorated with kinesin move toward the plus end of microtubules (Fig. 7.12). Kinesin and dynein, another mictotubule-based motor, also plays a role in the movement of chromosomes and in the sliding of microtubules past one another during chromosome segregation.

7.5.3 G2 Phase

The completion of the DNA replication signals the end of the S phase and marks the beginning of G2 phase. The transition from S to G2 is propelled by cdk1–cyclin A. In this phase, replicated chromosomes are untangled and begin the process of condensation. A key enzyme in both of these processes is DNA topoisomerase II, which functions to remove tangles and allow accurate condensation to occur. Chemical

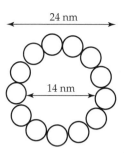

Transverse section microtubule

FIGURE 7.11
Microtubules are hollow cylindrical fibers that are composed of 13 parallel, but staggered protofilaments containing alternating α and β-tubulin subunits. They form tracks for the transport of organelles and movement of chromatin toward opposing poles in a dividing cell.

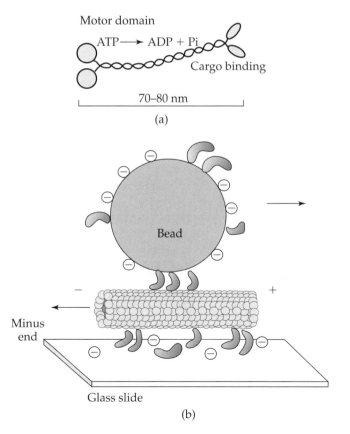

FIGURE 7.12 Schematic showing in-vitro motility involving the ATP-driven protein motor kinesin and microtubule tracks (a). The globular motor head regions of kinesin interact with microtubular tracks, whereas the tail domains bind to the organelles to be transported (b). Kinesin dimers can walk on the microtubule without losing contact for several microns.

inhibition of topoisomerase II blocks cells in G2. The sensor components of this pathway are not yet understood. The duration of the G2 phase in human cells is about 3 hours.

7.5.4 M Phase

The ultimate goal of mitosis is to separate the chromosomes equally into two (not three or six) identical daughter cells. The M phase of the cell cycle is a series of separate events identified as prophase, metaphase, anaphase, and telophase (Fig. 7.13).

1. *Prophase* is the subphase during which the diffuse chromosomes condense into structures visible with the light microscope. At the end the G2 phase, the duplicated chromosomes are in an extended intertwined partially condensed thread-like form. These threads must then be packed in a way that allows for error-free segregation to the opposite poles of the cell. Separation of 46 long, tangled, and fragile threads efficiently without breaking them within the small confines of a human cell is a true engineering tour de force. To facilitate this process, chromosomes undergo condensation in which each of the long, fine threads, already wrapped around nucleosomes, is further packed into a well-defined and separate chromosomal rope (Fig. 7.14). Nucleosomes pack into a coil that twists into another larger coil, and so forth, to produce condensed supercoiled chromatin fibers. The coils fold to form loops and the loops in turn coil further, forming two identical sister chromatids each containing a single DNA molecule.

FIGURE 7.13 A schematic of mitotic division in eukaryotic cells. During interphase, the cell integrates external signals for growth and adhesion and replicates its chromosomes and centrosome. After DNA replication, each chromosome consists of identical, paired chromatids. At the beginning of the M phase, chromosomes condense. The nuclear envelope breaks down, the duplicated centrosomes move to opposite poles, and the paired chromosomes become aligned in a plane at the equator of the cell. Chromatids separate from each other and begin to move toward the poles, the nuclear envelope reforms, and the chromosomes decondense and are no longer visible.

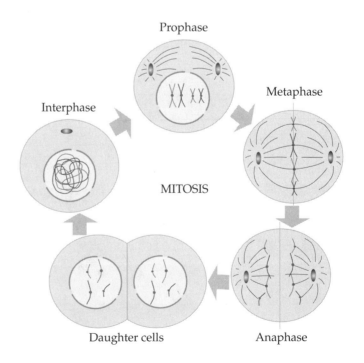

Interphase

Prophase

Metaphase

MITOSIS

Anaphase

Daughter cells

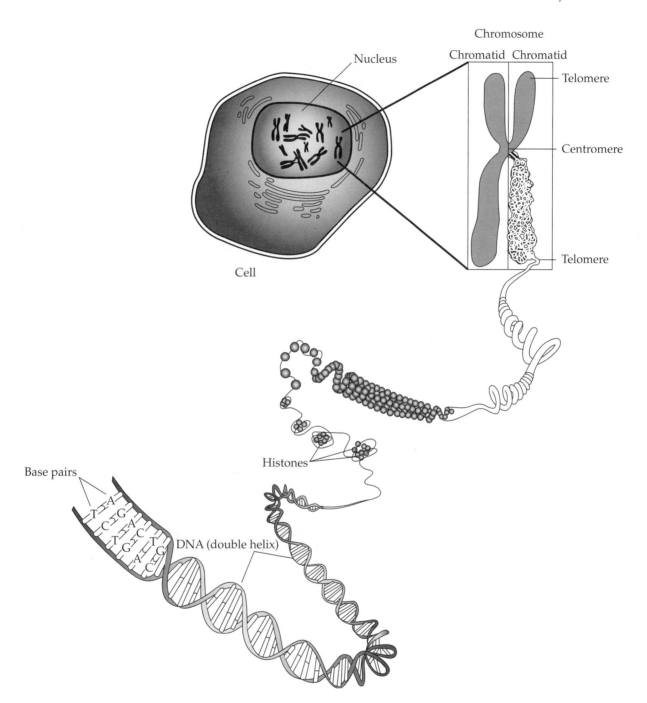

FIGURE 7.14   The complex of DNA and associated proteins in a eukaryotic cell is referred to as chromatin. The DNA carries the genetic information, and the associated proteins organize the chromosome physically and regulate the activities of DNA. The DNA double helix whose diameter is about 2 nm wraps around bead-like structures called nucleosomes. Nucleosomes are composed of proteins of the histone family and have a diameter of about 11 nm, still not visible under the light microscope. Histone H1 clamps DNA on to the surface of nucleosome. During M phase, nucleosomes pack into coils and loops, eventually forming supercoiled chromatin fibers. (Modified from www.nhgri.nih.gov/DIR/VIP/Glossary/illustration/.)

The sister chromatids are joined together in a region called the centromere in the condensed chromosome (Fig. 7.14). DNA-binding proteins called *Securins* provide the cohesion between the two chromatids. The sites for attachment of the microtubule cytoskeleton to the outer surface of centromere are called *kinetochores*. DNA condensation is inititated by phosphorylation of DNA-packaging proteins including *histone* H1. M phase is the only stage of the cell cycle at which condensed chromosomes become visible under the light microscope.

2. *Metaphase* is the subphase of M during which the nuclear envelope is disassembled and broken down to membrane vesicles. The microtubule-organizing structures (centrosomes) move to the opposite poles of the cell (Fig. 7.15). An array of microtubules connects the two centrosomes to form the mitotic spindle. Once the two spindle poles have been established, the cell has two clearly defined ends and an axis along which chromosomal segregation can take place. Microtubules that connect kinetochores to the cell poles exert tension on the chromosomes and align them along an equatorial plane (Fig. 7.16). At this point, the stage is set for the alignment and separation of chromatids.

It has been known for many years that eukaryotic cells arrest in metaphase if treated with microtubule polymerization inhibitors such as nocodazole or

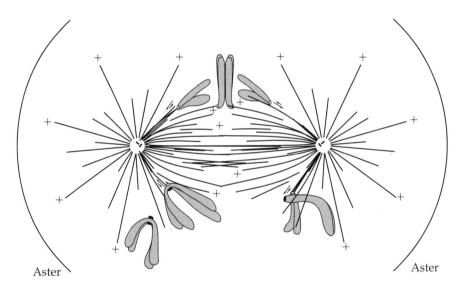

Aster   Aster

FIGURE 7.15   Spindle structure and chromosome behavior. In vertebrate cells, the mitotic spindle consists of two overlapping arrays of microtubules oriented with the + ends distal to and the − ends proximal to the poles. (a) Initially, one kinetochore (of the two kinetochores) becomes attached to a single microtubule and moves rapidly to the pole (long arrow). (b) During this movement, additional MTs become attached to the outer plate of the same kinetochore. (c) The chromosome oscillates to and from the pole until another MT from the opposite spindle pole attaches to the remaining kinetochore. (d) Opposing MT-generated tension on the two kinetochores results in the chromosome adopting an average position around the equator. (e) Assuming that all checkpoints are passed the action of the APC during anaphase allows the two chromatids to separate and there is a net movement toward the spindle poles. (Modified from Reider and Salmon, *Trends in Cell Biology* 1998, 8:310.)

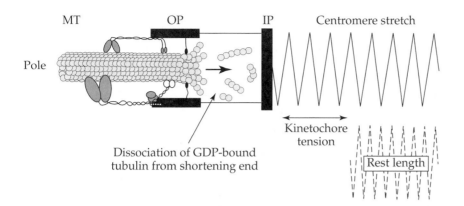

FIGURE 7.16 Model of the role of tension in kineto-chore–MT interactions. An attached kinetochore moves poleward at a rate of approximately 2 µM/min-1. During this process, it pulls on its associated MTs to stretch the centromere region. The force for poleward motion is provided by MT motors (a dynein family member), which are attached to the kinetochore. During poleward movement, the kinetochore MTs shorten by disassembly at the kineto-chore. (Modified from Reider and Salmon, *Trends in Cell Biology* 1998, 8: 310.)

colchicine. These drugs are common chemotherapeutic agents in the treatment of cancer because of their ability to inhibit cell division. Metaphase arrest also occurs if even one chromosome fails to orient appropriately at the metaphase plate. Abnormalities in the spindle pole body, centromere, or kinetochores all result in activation of the checkpoint and metaphase arrest. The sensor component of the checkpoint detects the presence of unattached kinetochores and is thought to be sensitive to the tension exerted by the spindle microtubules. (See Fig. 7.16.) A single unattached kinetochore prevents the cell from entering anaphase and inhibits spindle elongation. Tension exerted on the kinetochore by the kinetochore microtubules results in the dephosphorylation of one component of the sensor machinery. When all of these sensors are dephosphorylat-ed, the checkpoint is inactivated and the cell enters anaphase.

3. The major event in *anaphase* is the separation and movement of paired chro-matids to opposite poles of the dividing cell (Fig. 7.15). The chromatids of each chromosome separate from one another, and the associated microtubules pull them to opposite poles. Shortly after the chromosomes arrive at the poles, the polar microtubules elongate by polymerization and push the poles apart in preparation for reformation of the nuclear envelope and division of the cytoplasm.

4. *Telophase* is the last stage of M phase. In this stage, the cell actually divides into two cells. The shrinking waist between the two poles of the cell is a sig-naling event for the actual cell division. The forces for cleavage are provided by an overlapping contractile ring of actin and myosin filaments in a manner similar to the mechanism of muscle contraction.

The mitotic cyclins are degraded sequentially as the cell goes through various sub-phases of the M phase. Cells cannot exit mitosis and enter G1 if the activity of these cyclins persists. Because cyclin-dependent kinases play such important roles in the control of the cell cycle, they are potential targets for therapeutic intervention. Genetic evidence supports a strong link between aberrant Cdk control and the molecular pathology of cancer, providing the rationale for developing small-molecule Cdk in-hibitors as anticancer drugs. The inherent complexity of Cdk regulation offers a num-ber of possible routes to their inhibition. The development of peptides that mimic cdk function is being explored. Another promising strategy for cancer therapy appears to be the design and use of cdk-selective ATP competitive ligands that block cdk activity.

Both mechanical and control aspects of cell division utilize a number of principles that are familiar to engineers. The control systems that regulate the various phases of the cell cycle are well suited to engineering analyses. The use of positive and negative feedback loops, checkpoint controls, and parallel circuitry are all familiar concepts to engineers. Cell biologists are now turning toward investigations of the cell cycle in a more systems-based way. Central to the regulation of the cell cycle are multicomponent protein machines with identifiable tasks. For example, the anaphase-promoting complex functions to regulate mitosis and exit from mitosis and the SCF regulates the transition into the S phase. These machines were not constructed de novo; rather, they evolved over millions of years. Consequently, their component parts have not been directly designed for optimum function, but have been selected from preexisting structures/genes by a process of random mutation and selection.

7.6 | Cell Culture

The conditions for the cultivation of yeast cells have been refined over the centuries by the baking and brewing industries. Adapting these techniques for the laboratory was a relatively simple matter, and many scientists routinely grow and manipulate yeast in vitro. With their rapid rate of reproduction, relatively simple culture conditions, and ease of genetic manipulation, yeast is the system of choice for basic scientists interested in the eukaryotic cell cycle. Much of our present day knowledge regarding cell division comes from the study of yeast and Dr. Lee Hartwell, a yeast geneticist, recently awarded the Nobel prize for medicine for his studies on the yeast cell cycle.

Like bacteria, yeast can be grown to extremely high densities as a suspension in a nutrient medium and do not need attachment to a culture vessel in order to propagate. If conditions are good, bacterial and yeast cells will continue to divide for many generations and are essentially immortal.

In contrast, as discussed earlier, many animal cells cannot divide unless attached to a substratum and will only undergo a limited number of cell-division cycles before they senesce. The ability to grow animal cells for extended periods in vitro revolutionized the study of cell biology. The culture of animal cells directly from a tissue is called primary cell culture; these cells have a limited life span, but retain many of the characteristics of the tissues from which they were derived. In some cases, primary animal-cell cultures became spontaneously immortalized and could be propagated for many generations in vitro. Because of their extended life span, these "cell lines" were more amenable to experimentation; however, in many cases, they lost some of the differentiated characteristics of the original tissue.

7.6.1 Primary Cell Culture of Animal Cells

Most tissues are made up of multiple cell types, and one goal of cell culture is to be able to cultivate specific cell types. To do this, the scientist first isolates the tissue of interest under sterile conditions and minces it into small fragments. To separate the cells from one another, the scientist incubates the tissue fragments with a cocktail of proteolytic enzymes (e.g., trypsin and collagenase). Depending on the concentration and type of enzyme, the various populations of cells (e.g., epithelial and fibroblast) can be separated from one another.

In most situations, it is impossible to get a completely pure population of, for example, epithelial cells, and the culture will eventually be overgrown with fibroblasts. Isolated cells are generally cultivated at 37°C as attached cells in specially treated petri dishes or tissue culture flasks. Cells in the dishes are covered with a defined nutrient medium that contains many different components, including salts, amino acids, vitamins, and trace elements.

However, for the most part, cells will not divide unless this medium also contains an exogenous source of growth factors. This is usually provided by the inclusion of fetal calf serum (5–20 percent) in the medium. For most nutrient media, the buffering capacity depends upon an atmosphere of 5 percent CO_2. Consequently, mammalian cell culture is carried out in humidified incubators in an atmosphere of 5 percent CO_2 at 37°C.

Even under the best conditions, with the exception of some embryonic cells, most primary cell cultures undergo less than 10 rounds of replication before they senesce. Consequently, if experiments require primary cell cultures, the scientist has to regularly sacrifice animals and isolate the cells.

Over the past 50 years, some cells in primary cell cultures spontaneously continued to grow when the rest of the cells had senesced. These cells were cloned and have been propagated ever since as cell lines.

7.6.2 Cell Lines

Cell lines can originate following a spontaneous immortalization event or can be deliberately created by mutagens or by insertion of an immortalization gene such as the SV40 virus T antigen. Many cell lines exist and can be obtained from the American Type Culture Collection (ATCC). The simple fact that these cells are immortal shows that they differ from their tissue of origin. However, the ease with which they can be grown, together with the ability to follow the cells for many generations and genetically manipulate them has led to their adoption for many different studies.

Although cell lines have an extended life span, most are not transformed into cancer cells and will not form tumors or grow in the absence of attachment to a substratum. In contrast, cell lines derived from tumors or immortalized cells transfected with certain oncogenes are transformed.

Cell lines are grown in a nutrient medium containing serum or in an enriched defined medium containing specific growth factors, but no serum. The conditions for growth are similar to those described earlier for primary cell cultures.

Because the cells grow quite quickly, they must be passaged at regular intervals. To do this, cells are removed from one petri dish using a mixture of trypsin and the calcium chelator EDTA. The trypsin/EDTA is removed, and the detached cells can then be "split" into two or more dishes and cultivation continued.

Although the use of immortalized cell lines has led to enormous advances in cell biology, one must always bear in mind that these cells are not "normal." In many cases, adaptation to the culture conditions results in genetic drift and chromosomal abnormalities. For example, many cell lines are no longer diploid, some exhibit more than 46 chromosomes and some have fewer. Spontaneously immortalized cells must by definition have a genetic change that allows them to grow indefinitely. Many tumor cell lines can only be reliably used for a limited number of passages because their genetic makeup changes dramatically during the course of cell culture.

7.7 ASSIGNMENTS

7.1 The length of a typical bacterial cell is 700-fold smaller than the length of the DNA it contains. Search the literature to learn about the packing of DNA in prokaryotic cells. Which proteins play a role in the organization of DNA in bacteria?

7.2 Eukaryotic DNA and associated proteins are embedded in a viscous fluid-like medium contained in the nucleus. For a DNA-binding protein with radius $R = 45$ nm and velocity $V = 200$ nm/s, determine the drag force exerted by the surrounding viscous medium. The fluid force F (dyn) acting on a spherical particle of radius R (μm) moving at a creeping speed V (μm/s) in a viscous fluid with viscosity μ (dyn-s/μm^2) is given by the relation $F = 7\pi\mu RU$.

7.3 The fluorescence-activated cell sorter (FACS) can analyze cell DNA content and separate cells into the various phases of the cell cycle. Cells are stained with a DNA-binding dye such as propidium iodide or Hoechst. The FACS detects the amount of light signal from each cell and relates them to DNA content. Cells with unreplicated DNA (G1) are assigned one arbitrary unit. This corresponds to $2n$ in diploid cells. Cells that have completely replicated their DNA (G2/M) are assigned two arbitrary units corresponding to $4n$ in a diploid cell. Cells that are in the process of DNA synthesis (S phase) have one–two arbitrary units. Draw a typical schematic plot of population distribution of cultured cells with respect to DNA content.

7.4 Kinesins constitute a family of 100 eukaryotic motor proteins that interact directly with microtubules. Transport of motile organelles, such as mitochondria, along microtubules is powered by the motor proteins kinesin and dynein. Conduct a literature search to elucidate the role of kinesins and dyneins in cell division and, in particular, in movement of chromatids to the polar ends of a dividing cell.

7.5 Cyclin-dependent kinases were first recognized in the yeast cell cycle and were identified as cell-division-cycle proteins or Cdcs. These proteins are homologous to the Cdks in mammals. Yeast genes that act at each checkpoint of the cell cycle have been identified, leading to a working model of checkpoint pathways. The KEGG database presents a wiring diagram of the cell cycle for the budding yeast (saccharomyces cerevisiae) in their Web page. Go to the Web page http://www.genome.ad.jp/kegg/pathway/sce/sce04110.html, and identify all the components of the pathway by double-clicking the boxes in green. What is the MAPK signaling pathway? Indicate the points of interconnection of the pathways beginning with different input signals.

7.6 Budding yeast has been used for the isolation of mutants that are blocked at specific steps in the cell cycle. Temperature-sensitive *cdc* mutants are particularly useful because they grow normally at one temperature (permissive temperature), but express their growth-arrested phenotype when grown at a higher temperature (restrictive temperature). Cells can also be treated with the chemotherapy drug hydroxyurea (which inhibits the enzyme ribonucleotide reductase and blocks DNA synthesis). The effect of hydroxyurea can be reversed by washing. Suppose that you have discovered a new *cdc* mutant in yeast and have observed the following: You incubate a culture of the cells at their restrictive temperature (37°C) for 2 hours (the approximate length of the complete cell cycle in yeast). Then you transfer the cells to medium, which is at the permissive temperature (20°C) and contains hydroxyurea. Even after many hours, none of the cells divide. With a fresh culture of cells, you reverse the order of treatment: You incubate the cells at 20°C for 2 hours in medium containing hydroxyurea and then transfer them to medium at 37°C without hydroxyurea.

The cells undergo one round of division. In what phase of the cell cycle does your *cdc* mutation block growth at the restrictive temperature? Briefly explain your answer.

7.7 In the yeast (S. cerevisiae), the gene cdc2 encodes a protein required for the transition from G1 to S phase. Which protein in *Xenopus* (frog) is homologous to cdc2? For the answer, use National Library of Medicine's BLAST software, and search for homologs.

7.8 Explain the changes that occur in the phosphorylated state of the cell-cycle machinery as cells traverse G_2 to M phase.

7.9 Microtubules connecting to the kinetochores of chromatids pull these objects toward the polar ends by shortening at a rate V. Assume that the fluid drag force F (dyn) acting on each chromatid is given by the expression

$$F = 7\pi\mu VD,$$

where μ (ι10 dyn-s/cm^2) is the viscosity of the cytoplasm and D is the largest radius of the chromatid. The motion is slow, the mass of chromatid is small, and the balance of forces dictate that the microtubule system pulling on the chromatid is under tension and that this tensile force is equal to the fluid drag force. Search the literature, and find out how such a microtubule system can carry tension while its length decreases with time. Does polymerization or depolymerization at the tips of the microtubules play a role in this process?

7.10 The tumor suppressor protein p53 has defined roles in a number of cell-cycle checkpoints in response to stresses such as DNA damage, viral infection, and oncogene expression. p53 is also known to regulate duplication of the centrosome. Using the NCBI PubMed database, search the literature for information on p53. What role does this protein play in the response to mitotic spindle damage? Describe what is meant by "mitotic slippage." (*Hint*: See Meek DW. *Pathol Biol* 2000, 48(3): 247–254.)

7.11 Upon stimulation by growth factors, the first cyclin to appear early in the G1 phase is cyclin D. Cyclin D binds to and activates Cdk4 and Cdk7. There is significant evidence that mitogen-activated protein kinases (MAP) facilitate the accumulation of cyclin D and the degradation of Cdk inhibitors. Conduct a literature search to explain how cyclin D-dependent kinases integrate growth factor-induced signal transduction into the cell cycle.

7.12 The retinoblastoma gene is the prototypical example of a tumor suppressor gene. If its function is lost, cells become cancerous. For example, retinoblastoma is a childhood cancer that is associated with pRB. Conduct a literature search, and discuss the role of pRB gene in this disease.

7.13 Ubiquitin-mediated proteolysis of cell-cycle regulators is a crucial process during the cell cycle. The anaphase-promoting complex (APC) is a large multi-protein complex with ubiquitin ligase activity. Conduct a search to determine which cyclins and other regulatory proteins are subject to APC-mediated proteolysis.

7.14 The Cdk2–cyclin A complex can form stable complexes that contain the ZRXL motif (where the capital letters denote single-letter representation of amino acids and X can be any amino acid). Determine the identity of some of these proteins using the National Library of Medicine software BLAST.

Ans: Transcriptionally active E2F-1, the pRb-related proteins p107 and p130, and all members of the Cip/Kip CKI family.

7.15 Cyclin D contains an LXCXE sequence (single-letter amino-acid code, where X represents any amino acid). This motif is also found in a number of pRb-associated proteins and may serve a targeting function. Conduct a sequence homology search using BLAST to identify candidates for pRb-binding proteins in the human.

7.16 What is peptidomimetics technology (search for "synthetic tumor suppressor molecule")? Which industry is most likely to utilize this technology? Cite examples of CDK-selective ATP-competitive ligands (search for "CDK inhibition and cancer therapy").

7.17 Conduct a literature search to determine the proteins found in yeast SCF. Which protein has the substrate-recognition site?

7.18 *Mathematical models of cell cycle dynamics* have begun to emerge in scientific journals, some of which are listed in the accompanying list of references. Perhaps the simplest of these models is that of Goldbeter (1991), which focuses on G2/M phase transition in rapidly dividing cells of the early embryo. The model reduced the mechanics of the cell-division cycle to the expression of three representative proteins: cyclin, cyclin-dependent kinase (Cdk), and a cyclin protease. In the model, cyclin is synthesized constitutively. Cyclin activates cyclin-dependent kinase, which activates the cyclin protease, which in turn degrades cyclin. Differential equations governing the concentrations of these three proteins produce limited cycle oscillations under biological conditions. More recently, Gardner et al. (1998) considered a mathematical model for controlling cell-cycle dynamics using a reversible binding inhibitor. The model is composed of a series of rate equations for the concentrations of proteins involved in the cell cycle. One of these proteins is the target protein of the reversible binding inhibitor (a drug). Such control models and models of the cell-division cycle, which represent more accurately the biochemical complexity

of cell-division cycle (Novak and Tyson, 1997; Chen, et al., 2000), may bring much needed quantitative analysis to this subject. Conduct a literature search, and discuss the experimental validity of the various cell-cycle-division models that exist in the literature.

7.19 *Apoptosis:* Point mutations leading to ras gene activation occur spontaneously in thousands of cells per day in a human. The vast majority of the mutation events lead to growth arrest and subsequent cell death. An oncogenic mutation that leads to a strong and persistent activation of MAPK is associated with apoptosis, whereas relatively transient and weak activation of MAPK promotes survival. Loss of adhesion to ECM is also a promoter of apoptosis. Conduct a literature search to uncover the modes with which some cancer cells pass short-circuit checkpoints and avoid apoptosis.

REFERENCES

Chen KC, Csikasz-Nagy A, Gyorffy B, Val J, Novak B, and Tyson JJ. "Kinetic analysis of a molecular model of the budding yeast cell cycle." *Mol Biol Cell* 2000, 11: 379–391.

Gardner TS, Dolnik M, and Collins JJ. "A theory for controlling cell cycle dynamics using a reversibly binding inhibitor." *Proc Natl Acad Sci USA* 1998, 95: 14190–14195.

Goldbeter A. "A minimal cascade model for the mitotic oscillator involving cyclin and cdc2 kinase." *Proc Natl Acad Sci USA* 1991, 88: 9107–9111.

Halloy J, Bernard BA, Loussouarn G, and Goldbeter A. "Modeling the dynamics of human hair cycles by a follicular automation." *Proc Natl Acad Sci USA* 2000, 97: 8328–8333.

Novak B and Tyson JJ. "Regulation of the eukaryotic cell cycle: molecular antagonism, hysteresis, and irreversible transitions." *J Theor Biol* 2001, 210: 249–273.

Rieder CL and Salmon ED. "The vertebrate cell kinetochore and its roles during mitosis." *Trends in Cell Biology* 1998, 8: 310–318.

Development of Multicellular Organisms

CHAPTER

8

8.1 | Introduction

Development is a fascinating process during which an organism takes on successive forms that characterize its life cycle. Organisms grow according to detailed body plans described by their genomes and which have remarkable commonality across species. For example, rats, cats, humans, and horses have the same number of bones (with the exception of the tail) and muscles. In all of these mammals, the thigh contains a single long bone whereas the lower leg contains two. It is as if different mammalian species are reflections of the same shape viewed through different types of distorting mirrors. In keeping with this commonality, the genes that govern animal development are well conserved in evolution.

Development begins with a fertilized egg and continues into adulthood through millions of cycles of cell division. The forms that animals and plants take during their life cycles are asymmetric and diverse. How does a single cell, by dividing over and over again, create such extraordinary shapes and forms? The answer lies in the fact that living cells are not homogenous spherical balls, but have considerable molecular and structural asymmetry. Although all somatic cells of an organism have identical genomes, they do not have identical protein and mRNA contents, and successive rounds of cell division accentuate the differences in cell contents.

Animal body patterns appear to have evolved around an ancient genetic toolkit. Gene-specific transcription factors play crucial roles in executing the body plan encoded by DNA. It is now well established that early in embryogenesis the body plan is laid down by the activity of a number of "master regulator" genes. The products of these genes are all transcription factors. Many of the genes that are involved in the regulation of cell division are also involved in the development of the embryo into a multicellular organism. Evolutionary changes in developmental gene regulation have shaped large-scale changes in animal body plans. In the following, we briefly describe the main stages of the development of sexually producing organisms and discuss the principles governing this complex process.

8.2 | From Unfertilized Egg to Zygote

8.2.1 Meiosis

Sperm and eggs are single cells called gametes. They are haploid cells in that they carry a single set of chromosomes. If gametes were to carry two sets of chromosomes as diploid cells do, the resulting fertilized egg would have four sets of chromosomes, leading to a doubling of the number of chromosomes with each new generation. A special form of cell division called meiosis ensures that mature gametes only contain 1 set of chromosomes.

The precursors of sperm and eggs are called germ cells. They are diploid cells in that they contain two sets of chromosomes, one from each parent. In each pair of homologous chromosomes one is from the mother and the other is from the father. Germ cells must divide in a way that mixes these homologous pairs and at the same time reduces the number of chromosomes by half. A germ cell divides twice to produce gametes.

The sum total of these divisions is called meiosis (Fig. 8.1). It is the form of cell division that leads to daughter cells with only one set of chromosomes. The mixing of maternal and paternal chromosomes during fertilization as well as DNA recombination that occurs during meiosis imparts the genetic differences that distinguish the offspring from both of its parents. Meiosis therefore serves two primary functions: It maintains diploidy generation after generation and stimulates genetic variation.

The primary steps that occur during meiosis are illustrated in Fig. 8.1. The chromosomes of the germ cell duplicate before the first cell division (stages 1 and 2). The duplicated chromosomes remain as one structural unit and each such chromosome consists of two identical sister chromatids. After chromosome duplication,

FIGURE 8.1 Meiotic cell division produces gametes. During interphase the parent cell has $2n$ chromosomes. The chromosomes duplicate and form sister chromatids. Homolog pairs of sister chromatids then synapse and exchange genetic material. Sister chromatids remain together, but homologs separate to opposing poles. The nuclear membrane forms around the chromosomes, and the equatorial plane contracts and separates the newly formed cells. The next cell division occurs without DNA synthesis. Chromatid pairs align on the equatorial plane. The chromatids of each pair separate and move to opposing poles, and from then on the division proceeds as described for mitotic division.

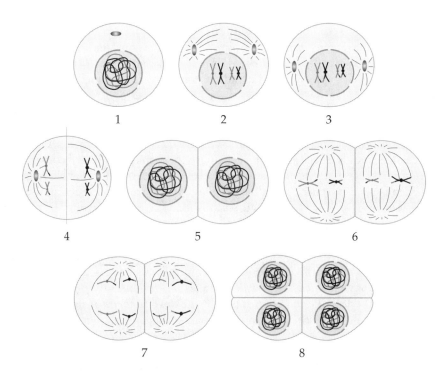

homologous pairs of chromosomes are positioned side by side and they exchange equivalent segments of DNA in an act called *synapsis* (stages 2 and 3). This results in mixing of the DNA from the mother and father and the overall event is called *crossover*. Initial analysis of the human genome indicates that crossovers occur much more frequently on the distal regions of the chromosomes and on shorter chromosome arms in general. The data also indicate that at least one crossover per chromosome arm occurs in each meiosis.

Following synapsis, sister chromaids remain bound to each other, but they are no longer identical. Each homolog of a chromosome travels to opposing poles of the dividing cell (stage 4 and 5). Thus, each daughter cell has either the maternal or the paternal chromosome from each homologous pair. Allocation of maternal and paternal chromosomes to daughter cells is a random process. Note that at this stage in cell division a paternal chromosome carries maternal DNA and maternal chromosome carries paternal DNA due to the crossover. In the second cell division, sister chromatids separate to generate daughter cells (stages 7–8). These cells have a $1n$ complement of chromosomes and are called *haploid cells*.

The random allocation of chromosomes to daughter cells during meiosis results in extensive diversity. To illustrate the point, consider the meiotic division of a cell with n homologous pairs of chromosomes. The number of possible varieties of chromosome allocations is then equal to 2^n. Humans have 23 pairs of chromosomes ($n = 23$) and therefore each individual can produce $2^{23} \approx 8,000,000$ distinct human gametes ($n = 23$). This is a low estimate, because swapping of genetic material during *crossover* is an additional source of new combinations of genes in interbreeding populations.

The human germ cells that divide to produce unfertilized eggs and sperm are called, respectively, *primary oocytes* and *primary spermatocytes* (Fig. 8.2). Spermatocytes are found in the testes and have 22 pairs of homologous chromosomes, plus an X chromosome from the mother and a Y chromosome from the father,

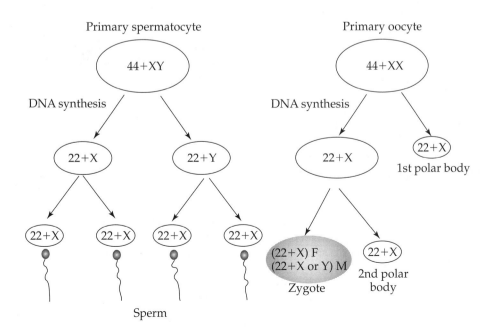

FIGURE 8.2 Meiosis leading to sperm and oocytes.

which carries the male sex-determining genes. Meiosis produces four haploid sperm, each with single set of chromosomes, two of these sperm carry the X chromosome and the other two carry the Y chromosome. In the normal male, the production of sperm is called spermatogenesis and takes place throughout life.

Oocytes are also diploid cells containing 22 pairs of homologous chromosomes, with one copy inherited from each parent. The female sex-determining genes lie on the X chromosome and oocytes receive one X chromosome from the father and one from the mother. The primary oocytes produce only one fertilizable egg. The first division occurs in the ovarium. The cytoplasm is not divided equally, but one daughter cell gets a nucleus and most of the cytoplasm whereas the other is composed of a nucleus and very little cytoplasm. This smaller cell is called the 1st polar body. The second meiotic division occurs after fertilization producing a haploid fertilizable egg and a 2nd polar body. Completion of meiosis may take as long as fifty years in the human female because the oocytes arrest in the meiotic prophase until fertilization. Also, a female germ cell produces a single functional egg and three cells with practically no cytoplasm (polar bodies). In human females the ovaries produce a finite number of eggs that are released at monthly intervals. These eggs do not divide through mitosis and when all the eggs have been released the female ceases to be fertile and enters menopause. Human genome studies show that DNA mutations are found much more frequently on sperm than on eggs, consistent with the fact that spermatocyes undergo many mitotic cell divisions before meiosis whereas oocytes do not.

8.2.2 Radial Symmetry of the Unfertilized Egg

The cytoplasm of an egg is a storage house for nutrients and molecules that play a role in development. In many species, these cytoplasmic components are not homogeneously distributed. Consider for example the structure of the unfertilized egg of the South African clawed frog *Xenopus laevis*. This is a large cell of about 1 mm in diameter and is enclosed in a transparent extracellular capsule called the jelly coat (Fig. 8.3). Most of the volume of the cell is occupied by so-called *yolk platelets*, which are membrane-bound aggregates of lipid and protein. The yolk platelets and other nutrient molecules in an unfertilized egg are relatively dense and therefore concentrate by gravity in the lower hemisphere of the egg. This lower hemisphere is called the *vegetal hemisphere*. The haploid nucleus of the unfertilized egg is located in the upper, less dense hemisphere called the *animal hemisphere*. The animal and vegetal regions of the egg contain different selections of mRNA molecules and other molecules involved in protein synthesis. The vertical axis of the spherical cell passes through the cell centrosome and nucleus toward the vegetal hemisphere. This axis is called the *animal vegetal axis* and its poles (points of intersection with the cell surface) are called the *animal pole* and *vegetal pole*, respectively. The animal pole coincides with the point of detachment of polar bodies from the egg. The animal–vegetal axis of the unfertilized egg defines the anteroposterior axis of the future embryo. The unfertilized egg appears to have radial symmetry about the animal–vegetal axis.

A female egg and a male sperm fuse to produce a *zygote* (fertilized egg) (Fig. 8.3). In some species that use external fertilization such as fish, the eggs release an attractant peptide to attract the sperm of the same species, causing them to swim vigorously toward the site of maximum attractant concentration. In mammals, sperm–egg surface interactions are mediated by several different families of adhesion molecules, including integrins. Integrins on the surface of the egg interact with

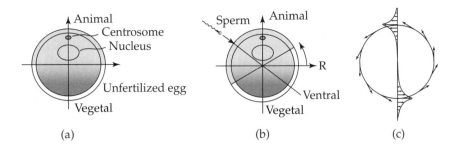

FIGURE 8.3 Polarization of the fertilized frog egg. The unfertilized egg has radial symmetry around an axis that passes through the cell centrosome and nucleus (a). The hemisphere that contains the centrosome and nucleus is called the *animal* hemisphere, whereas the lower hemisphere contains large numbers of yolk platelets full of nutrients. The point of contact between the egg and the second polar body coincides with the animal pole of the egg. The entry of sperm into the egg induces rotation of the cell cortex around the cell center (b). This movement is initiated by molecular motors such as myosin and involves the relative sliding movement of the cell cortex on the underlying cytoskeleton. Cortical rotation is resisted by viscous forces exerted by the surrounding fluid (c). The resultant moment exerted by the fluid is proportional to the rate of rotation of the cortex as well as the cell radius squared. This moment has the opposite sense of direction to the angular velocity of the cortex. Since this is the only external moment acting on the cell, Newton's laws of motion dictate that the inner region of the cell must move in the opposite direction of the cortex (c). As a result of this complex movement, radial symmetry is destroyed.

a sperm protein called *fertilin*. The final step in fertilization is sperm–egg fusion. The process is facilitated by a chemical process called acrosome reaction. The tip of the head of the sperm is capped by a bag of enzymes called the *acrosome*. These enzymes enable the sperm to digest its way into an egg.

Once the membranes of the sperm and egg fuse, the egg is metabolically activated and undergoes structural changes to its cortex and cytoplasm. Activation of the egg is triggered by a wave of calcium release that passes transiently across the egg. Activation also relieves prophase arrest and leads to the second meiotic division of the egg nucleus. The nucleus of the egg fuses with the nucleus of the sperm to create the diploid nucleus of the zygote.

8.2.3 Fertilization Introduces Polarization

The outer cortex of an unfertilized egg is rich in actin. In the frog egg, the part of the cortex surrounding the animal hemisphere is heavily pigmented, and the underlying cytoplasm has more diffuse pigmentation. The vegetal hemisphere is not pigmented, and its outer cortex is also optically clear. Thus, the cortex of the cell is different in composition in the two regions. The point of entry of the sperm in the animal hemisphere and the animal–vegetal axis create a plane that is called the *dorsal–ventral plane*. Fertilization causes the cell cortex to rotate about 30° (Fig. 8.3). Cortical rotation induces viscous resistance forces in the surrounding fluid; the resulting fluid torque acts on the egg in the direction opposite to the direction of rotation. Since this is the only external torque acting on the cell, the bulk of the cell interior must move in the opposite direction to cortical rotation (Fig. 8.3). As a result, the clear cortex associated with the vegetal hemisphere overlies the lightly pigmented marginal zone of the animal hemisphere opposite to the point of entry of the sperm, creating a *gray crescent*. The axis stemming from the gray crescent to the entry point of the sperm is called the *dorsoventral* axis. This becomes the principal axis of the embryo (animal) that goes from back (dorsal) to the belly (ventral). Treatments that block cortical rotation produce an embryo with a central gut, but no dorsoventral asymmetry.

8.3 | Cleavage: First Stage of Development

8.3.1 Morula

Cleavage refers to the first stage of embryonic development during which the cells divide rapidly without significantly increasing the overall volume of the zygote. The resulting ball of cells after two to three cell divisions is known as a morula (Fig. 8.4). These cell divisions are mitotic, with each daughter cell having identical copies of the DNA of the dividing cell. Thus, all cells of an embryo have the same genome, but through cell division, movement, and differentiation they develop into different tissues and organs. The volume of the *morula* is about the same as the fertilized egg, so that each cell division must lead to new cells each with a smaller volume. At this stage of development, cell-cycle time is very short and the cells have a very short G1 phase. The plane of the first two divisions is parallel to the animal–vegetal axis and perpendicular to each other. Subsequent divisions also occur in perpendicular planes. At this stage of development, DNA synthesis and cell division proceed with no growth and little gene expression.

As noted in the previous section, the chemical composition of the fertilized frog or sea urchin egg is asymmetric. Both of these species have large round eggs that are amenable to biological manipulation. When an eight-cell sea urchin embryo is bisected horizontally along the midplane, the resulting four-cell clusters do not develop into normal embryos whereas a vertical cut results in two small, but normal, larvae (Fig. 8.4).

This shows that by the time the sea urchin embryos reach the eight-cell stage, its cells exhibit differences in their development potential. These differences can at least be partially explained by the inhomogeneous distribution and therefore unequal allocation of maternal mRNA encoding development-related proteins (Fig. 8.5). Some

FIGURE 8.4 Bisection of eight-cell sea urchin embryos in two different midplanes. A horizontal plane in the figure separates each cell into chemically distinct upper hemisphere (animal hemisphere) and lower hemisphere (vegetal hemisphere). When the embryo is bisected vertically, the resulting four-cell clusters develop into small, but normal, larva (a). Bisection along the middle horizontal plane leads to a different result (b). Cells from the animal pole remain embryonic, whereas the cells from the vegetal plane develop into a small, but abnormal, embryo.

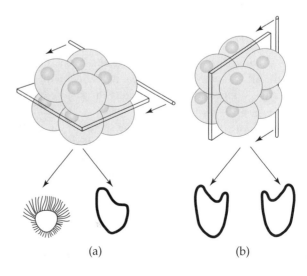

(a)   (b)

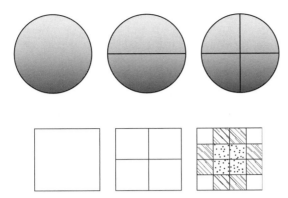

FIGURE 8.5 Division of an asymmetric fertilized egg results in cells with differing cell contents represented by different shading.

FIGURE 8.6 Three cycles of divisions of a fertilized egg give rise to three different cell types.

of these proteins are gene-specific transcription factors and are called the master regulators of body plan.

Similar experiments using early mammalian embryos give strikingly different results. In mammalian eggs, removal of a few cells from the morula does not affect the developing embryo. Cells taken and cultured from this stage of embryogenesis are called embryonic stem cells. Each stem cell is capable of forming a complete viable organism. A possible mechanism for this process is shown in Fig. 8.6 in which a fertilized egg gives rise to an eight-cell morula through three cycles of cell division. These eight cells can be grouped into three types according to the number of neighboring cells that each has. As discussed later in the chapter, cell–cell interactions have an effect on cell protein content. Recent research with the fruit fly demonstrates the actual concentration gradients of key proteins involved in the development of the nervous system in early embryo.

The fruitfly experiments illustrate an important element of development called cell differentiation—that is, the generation of cellular specificity. Differentiation defines the specific structure and function of the cell. For example, the human body has about 10^{14} cells, but only about 200 functionally distinct cell types. Cell types differ from one another in the portfolio of expressed proteins. How does a neuron end up having a different protein portfolio than a muscle cell or intestinal epithelial cell (given that the nuclei of all cells contain the same DNA)? The simplest explanation for differences in phenotype amongst different cell types with the same genome is that they selectively express different sets of genes. Determining how this occurs in a reproducible and predictable manner continues to be a major challenge for biologists in the genome age. There are several mechanisms that regulate differential gene expression. For example, we know that certain DNA-binding proteins affect the folding of DNA, preventing genes in certain parts of the DNA from interacting with transcription factors. This is illustrated in Fig. 8.7.

Cells at the early morula stage have the same set of transcription factors. The set of transcription factors begins to vary as the number of cell divisions increases, resulting in different DNA configurations in the daughter cells. Major transitions in chromosome and chromatin structure occur during development and make significant contributions to establishing and maintaining different cell types.

Cells that have the capacity to give rise to every type of cell in the adult body are called *totipotent* cells. Because embryonic stem cells are totipotent, they can be used to generate an adult with specific properties. Indeed, this is the basis of one form of transgenic animal. As discussed later, embryonic stem cells can be genetically

FIGURE 8.7 DNA-binding proteins can block parts of DNA from the action of transcription machinery. The idealized DNA molecules shown in the figure belong to stem cells (top row) and cells that differentiated into cell types 1 and 2 (bottom row).

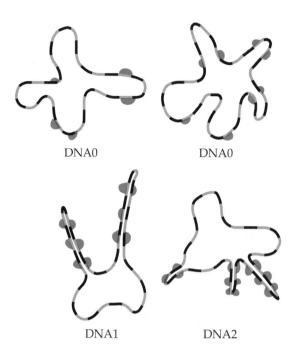

manipulated in vitro, by the knockout of an endogenous gene, then implanted into the uterus of an adult female. In many instances, a complete animal develops from the implanted embryo stem cells, illustrating the totipotent nature of cells from this early stage of embryo.

Many plant cells (but not animal cells) remain *totipotent* well after the plant has grown to adult size. For example, a carrot root cell in charge of food storage can divide and give rise to a complete plant, when provided with suitable nutrient medium and appropriate cues (sunlight). In animals, as cells in the embryo differentiate into different cell types, totipotency is reduced. Generally speaking, the more differentiated the cell, the less likely it is to retain totipotency. A potential mechanism for the decreasing cell capacity to transform from one cell type to another was illustrated in Fig. 8.7. Recent success in cloning sheep and pigs shows that given the right signals, even the adult cells can be made to alter the geometry of their DNA–protein complex with appropriate external signals. It is becoming clear that even the nuclei of fully differentiated cells have all the genetic machinery required to direct development of an animal from a single cell. Cloning animals is not easy, however, with the existing technology. Dolly was the only embryo out of the 272 embryos developed from *enucleated* eggs containing adult nuclei. (See figure in Chapter 5.)

8.3.2 Blastula

During subsequent cell divisions, the morula changes morphology to become a mass with a spherical cavity (Fig. 8.8). At this stage, the embryo is called a *blastula*, and the central cavity is called a *blastocoel*. The process of blastocoel formation (blastulation) occurs in mammals at the 32-cell stage, in frogs and sea urchins at about the 128-cell stage, and in birds much later, when the number of cells reaches about 80,000. The morphology and progress of cleavage depend on the egg contents (the

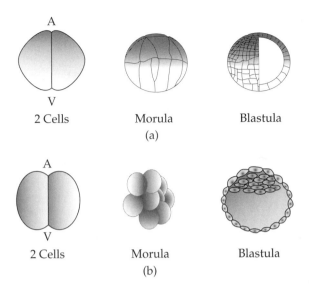

A

V

2 Cells   Morula   Blastula

(a)

A

V

2 Cells   Morula   Blastula

(b)

FIGURE 8.8 Cleavage refers to the first stage of embryonic development during which the cells divide rapidly without significantly increasing the overall mass volume of the zygote. Amphibian and mammary embryos exhibit different courses of development. The frog zygote develops into a spherical shell of cells called the *blastula* at the late stage of cleavage (a), whereas the human zygote develops into what is called *blastocyst* (b). The outer layer of cells is called *trophoblast*. Embryonic cells massed at the top of the fluid-filled sack (*blastocoel*) give rise to the human embryo.

proteins and the mRNA stored in the egg by the mother). In sea urchin eggs, the yolk content is sparse, and cell division produces daughter cells of roughly equal size. On the other hand, frog zygotes contain abundant amounts of yolk, and in these zygotes, cell division progresses more rapidly in the animal hemisphere than in vegetal hemisphere where the presence of large amounts of yolk impedes division. A complex set of movements in the frog blastula then results in invagination of the vegetal pole to form the gut and other interior tissues. Cells in the animal pole go on to form the skin and other external tissues.

The process of cleavage in mammalian eggs is somewhat different since the mammalian zygote must produce both an embryo and extra embryonic structures that provide a connection with the mother. Early cell divisions in mammalian embryos produce a loosely associated ball of cells. At about the eight-cell stage (morula), the adhesion between cells becomes stronger, increasing the surface area of contact and leading to the formation of tight junctions, the accumulation of fluid, and the formation of the blastocoel. During the next two cell divisions, the blastula undergoes a change of morphology, and the compact inner cells that eventually form the embryo are positioned on top of the *blastocoel*. Both units are surrounded by a layer of extraembryonic cells called the trophoblast. This shell of outer cells eventually becomes part of the placenta. The fluid contained in the blastocoel is actually secreted by the cells of trophoblast. At this stage, the mammalian embryo is called a *blastocyst*. The outermost cells of the resulting blastocyst attach to the cells lining the uterus, initiating implantation of the embryo into the uterine wall. The blastocoel prevents cells from different regions of the blastula from interacting.

8.4 | Gastrulation

Gastrulation is a phase of embryonic development when cells change their shape and move within the embryo. At this stage of embryogenesis, cells separate from each other and move in precisely determined patterns to create the basic tissue

FIGURE 8.9 Gastrulation phase of embryonic development, during which cells change their shape and move within the embryo. Cells move as individual cells or as a sheet of cells. The figure shows some of the modes of cell movement observed during gastrulation. *Invagination* involves a layer of cells moving inward toward the center of the blastula. Inward migration of single cells from the outer layer of embryo is known as *ingression*. The term *delamination* is used to characterize the inward movement of newly formed daughter cells while keeping the shell of blastula intact. *Involution* refers to folding of cell layers, leading to new cell–cell contacts and opportunities for induction. A compact domain of embryo extending to cover a larger surface area through cell movements is characterized as *convergent extension*.

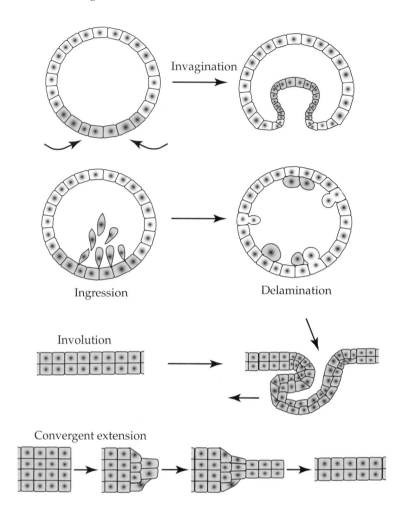

framework of the organism. Gastrulation brings cells to new positions, enabling them to contact cells that were initially distant from them. Cells move either as individual cells or as sheets of cells (Fig. 8.9). The stage of embryonic development when gastrulation occurs is called the *gastrula* and it follows the blastula stage of cleavage. One mode of cell movement entails the movement of single cells away from a layer (ingression). Another mode of cell movement is the deformation of epithelial sheets into tubes or baglike structures. The specific combinations of forces that lead to such complex patterns of movement are yet to be identified.

Gastrulation brings together groups of cells from different parts of the embryo and leads to the formation of three primary germ layers: ectoderm, mesoderm, and endoderm (Fig. 8.10). The external layer *ectoderm* ultimately gives rise to external epithelia and to the nervous system; the middle layer *mesoderm* develops into muscle, blood, and skeletal elements; and the internal layer the *endoderm* produces internal organs of the body (Fig. 8.10).

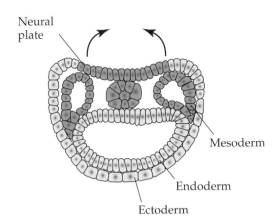

FIGURE 8.10 Three germ layers developed during gastrulation: ectoderm, mesoderm, and endoderm. Through a series of cell movements and inductive interactions, these three layers of cells give rise to all the tissues in the body.

8.4.1 Induction

Different cell types in an organism selectively express different sets of genes. Some genes such as actin and tubulin are always on; others such as intestinal specific enzymes are only expressed in the intestine. One important mechanism of cell differentiation is *induction*; the determination of cell type as a result of the influence of an adjacent group of cells. Induction also refers to the switching of cells from one cell type to another by the influence of an adjacent group of cells.

The process of induction relies on two important features: the capacity to produce a signal (ligand) by the inducing cells and the ability of the responding cells to receive and interpret the signal and respond to it accordingly. Inductive signals can originate through cell–cell contact or alternatively, the signals take the form of secreted peptides (*morphogens*) (Fig. 8.11). The most common mechanism of induction is the release of morphogens from the inducing tissue. Provided that the responding tissue has the appropriate response machinery, it can differentiate in response to these signals. The differential response to an inducing signal leads to the

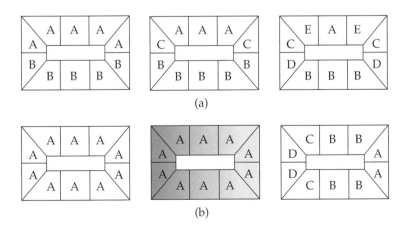

FIGURE 8.11 Induction by direct cell–cell contact (a) and by diffusion of a morphogen through the embryo (b).

formation of characteristic tissue-differentiation patterns. Two mechanisms are thought to mediate pattern formation. Firstly, a chemical gradient of the inducing molecule(s) could be established. In this case, cells further from the source would be exposed to less inducer. Secondly, cells could respond differently to the same inducer, either quantitatively (e.g., by being insensitive) or qualitatively (e.g., by differentiating into another cell type). During development, all of these mechanisms operate.

In addition to induction by secreted molecules, there are many examples of neighboring cells influencing each other's fate by direct contact. In one scenario, a cell expressing a transmembrane growth factor receptor contacts another cell that expresses the transmembrane growth factor ligand. The ligand–receptor interaction then stimulates the signal transduction pathway linked to the receptor and leads to changes in the fate of the cell expressing the receptor.

Another aspect of cell–cell contact that is essential for the development of tissues and organs is the higher order assembly of multicellular aggregates to form organ-specific structures. This important aspect of differentiation is governed by the expression of cell-adhesion molecules that link adjacent cells to one another and by the interaction of cells with the extracellular matrix. Remarkably, the behavior and differentiation state of like cells is profoundly influenced by these cell–cell and cell–matrix contacts. In this instance, not only do these cells have the identical genome, they also express the same genes. For example, epithelial cells lining the ducts of the mammary gland are essentially identical to one another. However, preventing cell–cell or cell–matrix interactions profoundly alters not only cell morphology, but also the profile of expressed genes. In other words, organization of like cells into linked collectives affords an additional level of differentiation control and allows for the coordinated regulation of function such as the secretion of milk proteins during lactation.

Consequently, inductive signals alter the mode of gene expression in the responding cell. In embryogenesis, inductive signals result in changes in the expression of transcription factors involved in development and patterning.

Induction is *instructive* when cells respond to a signal by differentiating into one cell type (e.g., mesoderm), but in the absence of the signal, it differentiates into another cell type (ectoderm). During normal development, inductive interactions may occur between cells that have been adjacent from the outset or between cells that are brought together during *gastrulation*. A series of successive inductions can generate many different kinds of cells from a few cell types (Fig. 8.12).

Growth factors act as inducing agents (morphogens) in much of development. Addition of an appropriate growth factor to the medium induces mesoderm from ectoderm even in the absence of contacting endoderm cells. The cell type induced depends on the concentration and class of the growth factor. For example, *Activin*, a member of the *TGF-β* growth factor family, induces ectodermal cells that would otherwise grow into epidermis to develop onto *notochord* at high concentrations. Ectodermal cells differentiate into muscle cells at lower concentrations of *activin*. *Wingless (Wnts)* proteins are another family of secreted factors that regulate embryonic patterning, cell proliferation, and determination of cell type. *Wnts* interact with members of a seven-transmembrane-containing surface receptor family called *frizzled* to activate downstream signaling events. These and other signal transduction pathways are discussed in more detail in Chapter 6.

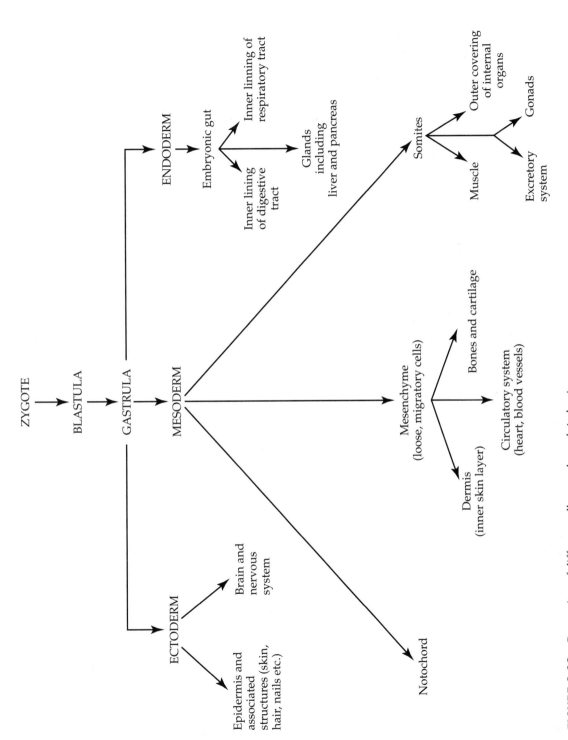

FIGURE 8.12 Generation of different cell types through induction.

8.5 | Pattern-Generating Genes

The previous sections showed that embryonic cells contain different sets of proteins and mRNA molecules and that the cell content is region specific. In species such as the frog and sea urchin, the fertilized egg is inhomogeneous and its division leads to cells with different protein and mRNA contents. In insects such as the fruit fly, the nucleus of the fertilized egg divides thousands of times before division of the cytoplasm takes place (Fig. 8.13). Because there is a significant variation of the cell content within the embryo, the DNA in these nuclei are subject to the action of different sets of proteins.

In mammals, the cells at the early stage of embryogenesis have nearly identical contents and are therefore totipotent. Subsequent divisions and inductive interactions lead to spatial heterogeneity in protein content. Proteins that drive development are sets of transcription factors that are called master regulators. These transcription factors act sequentially on DNA. In the following, we illustrate the mechanism of their action by considering examples from the development of the fruit fly, an insect whose genome has been decoded, and the mouse, a mammal used as a model for human in biological studies.

8.5.1 Fruit Fly

The fruit fly is an inexpensive model for the study of development because its growth cycle takes only about 10 days and its growth needs are modest (Fig. 8.13).

FIGURE 8.13 Fruit fly embryo. Nuclear division is not accompanied by cell division until about 2000–3000 nuclei form. Maternal mRNA of the bicoid gene is localized at the anterior end, and the mRNA of nanos gene is concentrated at the posterior end. The products of these genes play roles in the activation or repression of gene transcription involved in development. About six hours after the fertilization, the midsection of the fruit fly embryo exhibits 14 stripes, each a few cells thick, with alternative expression of even-skipped (Eve) and fushi tarazu (Ftz) proteins, both of which are involved in the generation of development of smaller-scale patterns.

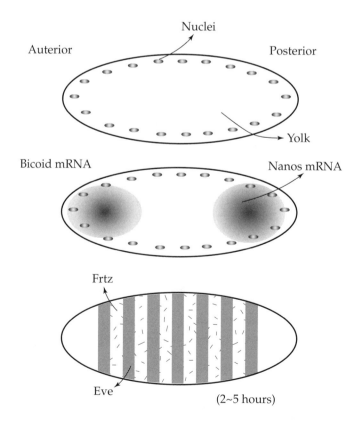

The fruit fly, like other insects, develops as highly modular segments. The transcription factors that induce modular body segments have largely been identified. Let us illustrate the mechanism of action of these transcription factors by focusing on pattern development in a fruit fly larva in the anterior/posterior axis. The cytoplasm of the fruit fly embryo does not divide with the initial nuclear divisions. A fruit fly larva stained with an antibody for a transcription factor called *even-skipped* (*Eve*) exhibits seven stripes, each three or four cells wide, on the surface of the embryo (Fig. 8.13). These stripes are separated by segments of roughly equal thickness that contain no Eve protein, but express the protein fushi tarazu (Ftz). How do such regular stripelike repeating patterns evolve from the initial spatial variation of cell content in the fertilized egg?

The maternal mRNA molecules in the fertilized egg of the fruit fly are not free to diffuse within the cytoplasm, but are distributed highly asymmetrically along the axis of symmetry of the egg. The mRNA molecules for a protein called *bicoid* are localized to the anterior pole of the embryo, whereas the mRNA for the *nanos* protein is localized at the posterior end. Once bicoid mRNA is translated by the growing embryo, the resulting bicoid protein diffuses posteriorly, forming a protein gradient along the anterior/posterior axis. The bicoid protein is a transcription factor that binds to the regulatory segment of the *hunchback* gene to initiate transcription of the gene. The effect of bicoid on expression of the hunchback gene depends strongly on its concentration in the nucleus. The hunchback promoter just upstream of the transcription site has three low-affinity and three high-affinity binding sites for the bicoid protein (Fig. 8.14a). The number of binding sites occupied by bicoid determines the frequency of gene transcription. At high concentrations, all six sites on the promoter will be occupied and the rate of transcription will be highest. At lower concentrations, fewer sites will be occupied and hunchback transcription will not be as high. Because the concentration of *bicoid* protein is high at the anterior end,

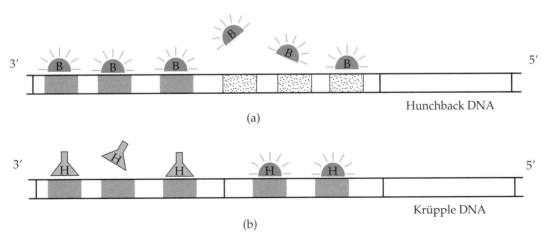

FIGURE 8.14 Activator and repressor sites for the hunchback (a) and krüpple genes (b). The transcription factors that bind to activator sites are shown as half circles, whereas those that bind to repressors are shown as triangles. The high-affinity regulator sites are darker than the low-affinity sites. The capital letters B and H stand for bicoid and hunchback proteins, respectively.

mRNA produced by the transcription of the hunchback gene is largely localized in the anterior region.

The hunchback protein is a transcription factor that controls the expression of a number of transcription factors such as the *krüpple* gene involved in pattern formation. The hunchback protein binds to appropriate sites on the enhancer domain of DNA regulating the expression of the krüpple gene (Fig. 8.14b). The hunchback protein has low affinity for a repressor site and high affinity for an activator site in the krüpple promoter. At high concentrations, the hunchback protein is able to interact with the repressor site and repress transcription. At moderate concentrations, the low affinity of hunchback to the repressor site ensures that it remains unoccupied. The high affinity of the activator site enables it to interact with lower levels of hunchback, allowing the transcription of the krüpple gene.

The resulting *krüpple* protein distribution is shown in Fig. 8.15. The expression of *even-skipped* in each stripe is controlled independently by the action of different sets of transcription factors. The expression of *even-skipped* in stripe 2 is regulated by the *bicoid* protein and the proteins *hunchback*, *krüpple*, and *giant*. These proteins bind to upstream regulatory sites controlling the transcription of the *eve* gene.

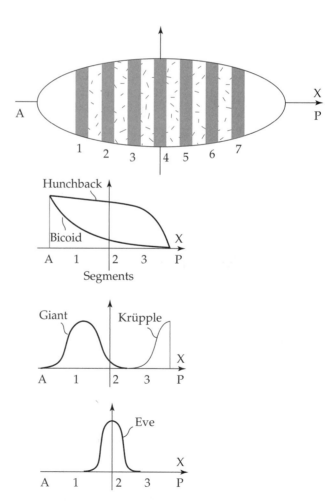

FIGURE 8.15 Head-to-tail pattern formation in the fruit fly. The figure shows the spatial distribution of transcription factors *bicoid*, *hunchback*, *giant*, and *krüpple* along the axis of a fruit fly larva between segments 1 and 3. The expression of the even-skipped (Eve) protein in stripe 2 is regulated by these transcription factors.

The coordinated effects of these transcriptional activators and repressors depend strongly on their concentrations. Since these concentrations vary along the anterior–posterior axis, transcription of the eve gene is dependent on the spatial position along the main axis of the embryo, resulting in the striped patterns in the midsection of the embryo.

The brief discussion of the body stripes of the fruit fly larva presented here illustrates a basic mechanism of pattern generation during development. Axial region formation in the fruit fly embryo begins with the diffusion of various transcription factors through the common cytoplasm to influence transcription from thousands of nuclei. These transcription factors activate different sets of genes at different threshold concentrations. The action of one patterning gene induces the action of others, which in turn regulate the transcription of yet other transcription factors involved in pattern formation.

8.5.2 Mammals

The same set of homologous genes marks out the head-to-tail axis of mammals and insects. Patterning along the head-to-tail axis of animals is controlled by *homedomain* genes. These genes carry a common DNA-binding sequence motif called the homeobox. In animals, there are a large number of homedomain or hox genes. The complex regulatory circuit formed by these genes regulates pattern formation along the main body axis. The circuit receives inputs from many transcription factors.

Let us illustrate the decision-making process of the hox gene circuit by considering an idealized circuit composed of only four transcription factors (Fig. 8.16). Let us assume that these transcription factors exist in either high concentration (1) or low concentration (0). The outcome of patterns formed will depend on the states of the four nodes in the circuit. Since each node can take one of only two different values, the circuit could at most yield $2^4 = 18$ different patterns.

State 1   0 - off, 1 - on   Fate 1 - Thorax

	A	B	C	D
Hox 1	0	0	0	1
Hox 2	1	0	1	0
Hox 3	1	1	0	0
Hox 4	0	1	1	1

State 2   Fate 2 - Abdomen

	A	B	C	D
Hox 1	1	0	1	0
Hox 2	0	1	0	1
Hox 3	1	1	1	0
Hox 4	1	0	0	0

FIGURE 8.16 A hypothetical homeobox gene language. Hox genes can be either on (1) or off (0). The expression of Hox d2 and d3 in the context of no expression of other Hox genes specifies that those cells will go on to form tissue in the tail of the mouse. However, if other Hox genes are inappropriately expressed or the timing of expression is wrong, then the ability of Hox d2 and Hox d3 to specify a tail fate is lost.

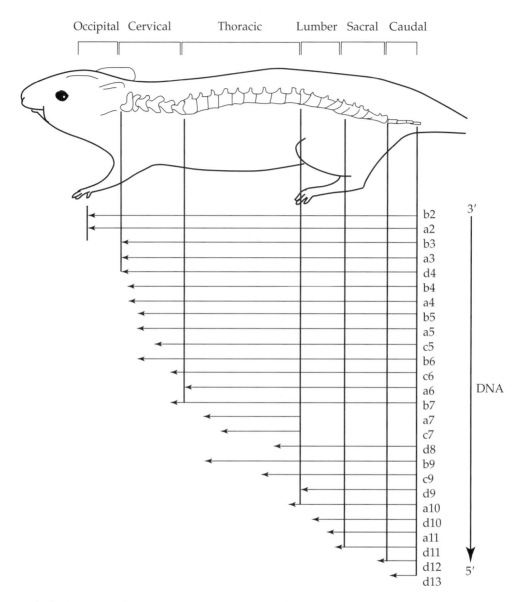

FIGURE 8.17 The body pattern of a mouse and the spatial range of activation of hox genes that mediate mouse development. (Modified from Hunt and Krumlaut, *Ann Rev Cell Biol* 1992, 8:227.)

The number of hox genes that regulate the development of mammals is large (Fig. 8.17). Unique combinations of on and off Hox switches create codes that specify thoracic, cranial, and lumbar segments of the body. Genetic manipulation of these and other similar codes early in development can transform one body part to another. In the hypothetical example shown in Figure 8.16, expression of Hox d2 and d3 in the context of no expression of other Hox genes specifies that those cells will go on to form tissue in the tail of the mouse. However, if other Hox genes are inappropriately expressed or the timing of expression is wrong, then the ability of Hox d2 and Hox d3 to specify a tail fate is lost.

Remarkably, the spatial and temporal order of Hox gene expression is also specified by the position of the genes on the chromosome. Those genes most near 3′ on the chromosome are expressed earliest and are associated with the more cranial (head) segments, whereas genes toward the 5′ end of the chromosome are expressed later and specify a caudal (tail) fate. This fact indicates that the genes on DNA are not distributed randomly and that their spatial organization may be reflected in the development plans of animals. In fact, the initial sequencing and analysis of the human genome showed that the HOX gene clusters on DNA are in the most repeat-poor (noncoding) regions of the human genome, probably reflecting the very complex coordinate regulation of the genes in clusters.

Recent research on comparative genomics suggests that the diversity of body plans is not necessarily due to gene number or protein function, but more likely due to changes in gene regulation by upstream promoters or enhancers. Most animals, no matter how different in overall shape, share several families of genes that regulate major aspects of body pattern. This common genetic toolkit for animal development contains many families of transcription factors including the Hox genes. The number and function of genes within the development gene toolkit has remained nearly constant in the course of evolution, with the exception of two rapid expansion points. The toolkit expanded significantly early in animal evolution and then again at the base of the vertebrate tree. The diversity in animal shapes is largely due to the changes in the function of regulatory elements in controlling the level, pattern, or timing of gene expression. Thus, regulatory DNA is the primary source of the genetic diversity that underlies morphological variation and evolution. That is, the same gene expressed at a different time may result in a different body plan. As noted previously, development-patterning genes have cis-regulatory regions composed of many independent elements. Each of these elements may evolve independently during the course of evolution. Regulatory DNA sequences have greater tolerance to all varieties of mutational change than coding DNA.

One cause of diversity of body plans in animals may be spatial differences in the level of expression of Hox genes along the head–tail axis. Comparative gene mapping indicates that Hox genes are expressed at different relative positions along the head–tail axis in mammals. The evolution of Hox gene regulation correlates with the evolution of axial diversity. As discussed by Carroll (2000), the relative positions of Hoxc8 gene expression shifted along the primary body axis in part through evolutionary changes in cis-regulatory elements.

Many other developmentally important genes are expressed in different combinations along the head–tail axis of the body. Quantitatively, this spatial variation may have arisen as a result of the asymmetric chemistry of the fertilized egg or early embryo (initial conditions) as well as the flux of development-affecting signals from the other parts of the embryo (boundary conditions). For example, in some species, the initial conditions are established by the uneven distribution of maternal mRNA and proteins involved in development, in the zygote. In mammals, boundary conditions also incorporate development-related signals produced by the mother in addition to those produced by other regions of the embryo. In these circumstances, pattern formation in one region of the embryo is not only dependent on the initial conditions (e.g., mRNA and protein content at an earlier time) in that region, but also on the boundary conditions specifying the flux of signals from the rest of the embryo. In other words, cells and tissues can affect each other's gene expression by releasing secreted proteins or by cell–cell contact signaling (induction).

8.6 | Stem Cells and Tissue Engineering

The recent political debate regarding stem-cell research and human cloning has brought this previously arcane area of biology to the fore. The idea that some cells retain the capacity to differentiate into every organ in the body and to even regenerate an entire organism is clearly very provocative. Such totipotent stem cells can theoretically be used to develop artificial organs to replace damaged or diseased tissues. The commercial prospects of this technology are enormous and have spawned a number of for-profit companies solely devoted to the translation of basic stem-cell research into commodities of financial value. The present debate surrounds the use of embryonic verses adult stem cells, and it is this distinction that we discuss first.

8.6.1 Embryonic Stem Cells

The generation and use of embryo stem (ES) cells in transgenic mouse technology has been described in preceding sections. In this case, the ES cells are manipulated in vitro (genes inserted or deleted) prior to their implantation into the uterus. ES cells taken from early embryos are not differentiated and retain the capacity to contribute to all organs in the body of the resulting transgenic animals. Under certain conditions, ES cells can be induced to differentiate into specific tissues (e.g., nervous tissue) in vitro. A great deal of effort is being exerted into defining the conditions required to make ES cells turn into predetermined tissues. If this is achieved with human ES cells, it is conceivable that this sort of technology could be used to provide replacement hearts, kidneys, nervous tissue, etc.

ES cells have two fundamental advantages over adult stem cells for this purpose. Firstly, stem cells taken from early embryos are truly totipotent; the same ES source could potentially be used to generate many different tissues suitable for use in any individual. Secondly, ES cells have not yet developed the cell-surface markers that mark them as specific for one particular individual. These surface markers are the basis of organ rejection. A continuing problem in the field of organ transplantation is the recipient's treatment of the transplanted organ as foreign and a target for immunological rejection. When transplanted into a recipient with significantly different surface properties, the transplanted organ is attacked by the host immune system unless the immune system is suppressed by treatment with powerful immunosuppressants. The great promise of ES cells lies in their potential for providing a universal and unexpendable source of new organs.

Clearly, many scientific, ethical, and political hurdles need to be overcome before this becomes a reality. In the United States, these concerns, together with the small number of human ES cells that are available, have accelerated research of stem cells isolated from adults.

8.6.2 Adult Stem Cells

Tissues such as the mammary gland, gastrointestinal tract, testis, hematopoietic system, and skin are continually making new cells in the adult. The source of these new cells is a small population of "organ specific" stem cells that exists within adult organs. Until recently, it was assumed that these stem cells had a very limited repertoire of fates; a stem cell from the mammary gland was not likely to become anything other than a mammary cell. However, it is now clear that under certain conditions, but then only rarely, stem cells from these tissues can be induced to differentiate into other organs. The cloned sheep "Dolly" was almost certainly derived from a stem cell isolated from the udder (mammary gland) of an adult sheep. Recent work has shown that cells isolated from the bone marrow can be used to generate nervous tissue. However, because of the immunological rejection problems involved, the utility of adult stem cells for organ replacement is likely to be restricted to cells isolated from the individual destined to become the recipient.

In a unique attempt to combine the flexibility of ES cells, but avoid the ethical issues surrounding them, CellTech has created embryos parthenogenically. In this classical embryological manipulation, an egg can be induced to divide and undergo embryogenesis without fertilization or injection of DNA. Simply pricking the egg or exposing it to a small electrical shock tricks the embryo into beginning embryogenesis. However, although such an embryo does undergo the early stages of organogenesis it cannot give rise to a viable embryo. Cells isolated from the developing kidney of parthenogenically derived cow embryos developed into viable kidney-like structures when implanted under the skin of the donor cow. At this time, it is not clear whether these cells would be rejected if implanted into an animal that was not the donor. It is also not clear that this procedure would alleviate the concerns of groups opposed to any research on human embryos created for the purpose of harvesting organs, even though these embryos do not have the potential to create human life.

8.6.3 Tissue Engineering

Fundamental research on cell division and differentiation has major impacts in the areas of cancer treatment and birth abnormalities. The number of known proteins and protein complexes that are involved in cell-cycle control is rapidly increasing, and the more we know about the functions of these proteins, the better is our ability to design drugs that can alter their state of activation. The three-dimensional structures of many proteins have already been identified, and extensive databases for their ligands (small compounds) also exist. New tools of computer graphics enable assessment of the binding capacity of such ligands to a variety of proteins. Chemical engineers working in drug design have developed automated systems that can test the effectiveness of large numbers of compounds on cell-cycle control.

The gene products involved in cell division and differentiation form complex circuits (networks) with self-regulatory properties. The regulation systems used in living systems are of great interest to control engineers in this age of modern technology when minute sensors, molecular circuits, and motors are becoming essential components of the manufacturing industry. Biologists have begun utilizing the present knowledge of structure and function of electrical circuits in creating wiring diagrams for various cell-regulation systems. Design engineers can benefit from the examples of biological design that has been perfected by billions of years of the evolutionary process.

The biology of cell differentiation is at the very heart of the technology for regenerating tissue lost to injury, aging, disease, and genetic abnormality. The potential of early embryonic cells to differentiate into any one of the 260 human cell types has been well established. As detailed in this chapter, although all somatic cells of an individual have the same genome, switching of cells from one type to another becomes more difficult at later stages of embryonic development and adulthood. This is probably because DNA-binding molecules can modulate chromatin structure, enabling the expression of certain genes and hiding other genes from the transcription machinery. This raises the following question: Given the right induction signals and an appropriate biocompatible scaffolding to grow on, can a cell be converted from one cell type to another?

Could cells be directed to form tissues closely mimicking those that exist in vivo (cartilage tissue, muscle tissue, and so on)? Scientists working in the field of tissue engineering are addressing these questions. Their overall aim is to develop partial or whole replacement of organs, bones, cartilage, and connective tissue (Fig. 8.18). With the development of a number of three-dimensional tissue constructs, tissue engineers have already begun to meet some of the clinical needs. Dermal tissue equivalents such as

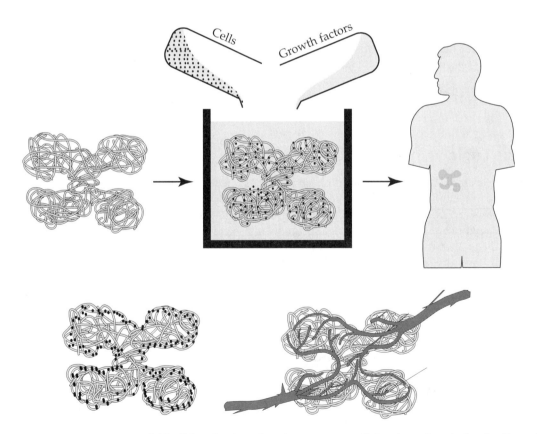

FIGURE 8.18 Tissue engineering is a field of biotechnology that aims to generate living tissue from isolated cells, scaffolds, matrix proteins, and growth factors. (Modified from http://www.cs.cmu.edu/People/tissue/tutorial.html.)

Dermagraft and TransCyte are used to cover chronic skin ulcers, burns, and wounds. Significant progress has been made to repair articular cartilage using cartilage constructs grown in vitro. These constructs undergo remodeling in vivo and adapt to the surrounding tissue. Tissue-engineered blood vessels have also been formed and are being used clinically. Whole joint structures have been developed de novo by combining specific cells and biodegradable polymers. Research efforts are also underway to use embryonic as well as adult stem cells in brain-tissue repair and cardiac-muscle regeneration. Clearly, the field of tissue engineering has much promise. Rapid progress in tissue engineering is dependent on adapting a multidisciplinary approach including genetic, clinical, cell biological, device, and bioengineering approaches.

There are more than 50 known growth factors and hormones that impact cell differentiation and cell-cycle regulation. The numbers of possible combinations and dosages are enormous, and cells respond differently to different combinations of these growth factors. Close-range induction processes can also be manipulated by altering substrate adhesion and scaffold structure, and again, there are a large number of possible inductive configurations. The full potential of tissue engineering will be realized with the automation of large-scale experiments and database construction for tissue-engineering informatics. In addition, there is a need to develop effective routes of administration of peptide growth factors for blood-vessel development following myocardial ischemia. Progress has already being made in creating a database for cell-type-specific microstructures and their function.

8.7 ASSIGNMENTS

8.1 In *mosaic development*, if a few of the blastomeres are eliminated from the blastula, a particular portion of the embryo will not form. Which species undergo mosaic development? Illustrate mosaic development by providing specific examples.

8.2 Consider a four-cell stage hypothetical embryo modeled as a planar square divided into four squares. Initially, all four cells have two sides contacting other cells and two sides contacting the boundary. These cells divide without altering the overall size of the embryo. Let *A*, *B*, and *C* denote, respectively, the cell types having two sides, one side, or no side contacting the boundary of the embryo. Determine the distribution of cell types in an 8-cell and an 18-cell embryo.

8.3 A 16-cell hypothetical embryo undergoes a sudden shape change as shown in Fig. 8.6. If *A*, *B*, *C*, and *D* refer, respectively, to the cell types with 1, 2, 3, and 4 sides in contact with other cells, determine the cell-type distribution before and after embryonic shape change.

8.4 Search the literature for the definition of Spemann's organizer. What role do homeobox genes play in specifying the formation of Spemann's organizer?

8.5 Give the addresses of three Web sites that present movies of gastrulation of the frog egg or sea urchin egg. Describe the differences in gastrulation in these two species.

8.6 In the progression of colon cancer, mutations inactivating the tumor suppressor gene APC gene typically appear first; these mutations can be detected in small benign polyps as well as large malignant tumors. Mutations activating the *ras* oncogene occur later and are associated with the ability to divide without anchorage to a substrate. The loss of another tumor suppressor gene p53 seems to relieve the few remaining breaks on cell division control. Loss of p53 allows cells to divide and accumulate further mutations at a rapid rate because these cells progress through the cell cycle when they are in no state to do so. Search the literature to construct a Boolean network to model cancer progression through gene mutation and deletion.

8.7 *Activin* is a morphogen with the capacity to switch ectoderm to mesoderm. Discuss the surface receptors for activin and the so-called Smad intercellular signaling cascade associated with activin.

8.8 *Wingless (Wnts)* proteins are a family of secreted factors that regulate embryonic patterning, cell proliferation, and determination of cell type. *Wnts* interact with members of a seven-transmembrane-containing surface receptor family called *frizzled*. *Wnt* binding to its receptor activates gene transcription through a pathway involving the protein β-*catenin*. This multifunctional protein plays an important role in axis specification (polarization) of the embryo. In epithelial cells, this protein is found at the plasma membrane in association with cell–cell adhesion molecules called *cadherins*. It also associates with the tumor suppressor protein APC. In early embryogenesis, β-catenin accumulates opposite the sperm entry point by the end of the first cell division in cleavage, and it continues to accumulate in the cytoplasm of cells in the dorsal, but not ventral region through the early cleavage stage. When cells taken from the dorsal region at this stage of cleavage are implanted in host embryos, they induce a new dorsoventral axis. These results suggest that β-catenin is the first signaling molecule to indicate the emergence of back-to-belly polarity in the growing embryo. The concentration of β-catenin protein, like cyclins, is tightly controlled by protein degradation. In the absence of a wnt signal, phosphorylation of β-catenin by a serine/threonine kinase targets it for rapid degradation in the embryo. When the activity of the kinase is inhibited, β-catenin accumulates in both the cytoplasm and the nucleus, where it can bind to at least one transcription factor to activate gene expression. Identify two kinases that can phosphorylate β-catenin and at least one transcription factor whose function is activated by β-catenin. What genes are regulated by β-catenin?

8.9 Discuss the advantages of studying *Drosophila* as a model system in developmental biology. In your answer, you should consider the following points: generation time, the size of the chromosomes, the significance of polytene chromosomes, chromosome walking (what is this?), the mutants, and relevance of developmental mechanisms of *Drosophila* to that of vertebrates.

8.10 During development, the frequency of the cell cycle must be tightly controlled. Describe the mathematics of the control mechanism described by Gardner, et al. (1998) in an article entitled "A theory for controlling cell cycle dynamics using a reversibly binding inhibitor" (PNAS 95: 14190–14195).

8.11 In the 1950s, the mathematician Alan Turing put forward the hypothesis that chemicals generated incrementally during development might lead to varying concentrations within the embryo, depending on the distance from the source and the concentration gradient of key proteins involved in development may lead to the differentiation of cells during early embryonic development. Conduct a literature search, and discuss the emergence of a concentration gradient for the morphogen SOG in the fruit fly embryo. (SOG stands for short gastrulation and is a protein that promotes the early development of the nervous system in animals as diverse as humans and insects.)

8.12 Comparative genome analysis (such as chromosome painting) proved to be a powerful technique in providing information on the rates and patterns of chromosomal evolution on a whole genome scale. It appears that large tracts of mammalian genomes are remarkably conserved. Therefore, transferring information from species such as human and mouse (which already have gene-rich maps) to the gene maps of other mammals could be feasible. Conduct a literature search, and discuss the evolution of mammalian genome organization as inferred from comparative gene mapping. Explain the process of gene mapping using somatic cell hybrids?

REFERENCES

Bodnar JW. "Programming the Drosophila embryo." *J Theor Biol* 1997, 188: 391–445.

Carroll SB. "Endless forms: the evolution of gene regulation and morphological diversity." *Cell* 2000, 101: 577–580.

Castelli-Gair J. "Implications of the spatial and temporal regulation of Hox genes on development and evolution." *Int J Dev Biol* 1998, 42: 437–444. Review.

Ciliberto A and Tyson JJ. "Mathematical model of early development of the sea urchin embryo." *Bull Math Biol* 2000, 82: 37–59.

Darnell DK and Schoenwolf GC "The chick embryo as a model system for analyszing the mechanisms of development." *Methods Mol Biol* 2000, 135: 25–29.

Dillion R and Othmer HG. "A mathematical model for outgrowth and spatial patterning of vertebrate limb bud." *J Theor Biol* 1999, 197: 295–330.

Ferrandiz C, Liljegren SJ, and Yanofsky MF. "Negative regulation of the SHATTERPROOF genes by FRUITFULL during Arabidopsis fruit development." *Science* 2000, 21: 436–438.

Gardner TS, Dolnik M, and Collins JJ. "A theory controlling cell cycle dynamics using a reversibly binding inhibitor." *Proc Natl Acad Sci USA* 1998, 95: 14190–14195.

Heanue TA, Reshef R, Davis RJ, Mardon G, Oliver G, Tomarev S, Lassar AB, and Tabin CJ. "Synergistic regulation of vertebrate muscle development by Dach2, Eya2, and Six1, homologs of genes required for Drosophila eye formation." *Genes Dev* 1999, 13(24): 3231–3243.

Jenik PD and Irish VF. "Regulation of cell proliferation patterns by homeotic genes during Arabidopsis floral development." *Development* 2000, 127(8): 1267–1276.

Lander ES et al. "Initial sequencing and analysis of the human genome." *Nature* 2001, 409: 880–928.

Mercader N, Leonardo E, Piedra ME, Martinez-A C, Ros MA, and Torres M. "Opposing RA and FGF signals control proximodistal vertebrate limb development through regulation of Meis genes." *Development* 2000, 127(18): 3961–3970.

Murphy WJ, Stanyon R, and O'Brien SJ. "Evolution of mammalian genome organization inferred from comparative gene mapping." *Genome Biology* 2001, 2: 0005.1–0005.7.

Murray P and Edgard D. "Regulation of programmed cell death by basement membranes in embryonic development." *J Cell Biol* 2000, 150: 1215–1221.

Nakatsuka A, Murachi S, Okunishi H, Shiomi S, Nakano R, Kubo Y, and Inaba A. "Differential expression and internal feedback regulation of 1-aminocyclopropane-1-carboxylate synthase, 1-aminocyclopropane-1-carboxylate oxidase, and ethylene receptor genes in tomato fruit during development and ripening." *Plant Physiol* 1998, 118(4): 1295–1305.

Packer AI, Mailutha KG, Ambrozewicz LA, and Wolgemuth DJ. "Regulation of the Hoxa4 and Hoxa5 genes in the embryonic mouse lung by retinoic acid and TGFbeta1: implications for lung development and patterning." *Dev Dyn* 2000, 217(1): 62–74.

Srinivascan S, Rashka KE, and Bier E. "Creation of a SOG morphogen gradient in the Dropophila embryo." *Dev Cell* 2002, 2: 91–101.

Wolffe AP. "Chromatin and gene regulation at the onset of embryonic development." *Reprod Nutr Dev* 1998, 38: 581–808.

Wu L, Hansen D, Franke J, Kessin RH, and Podgorski GJ. "Regulation of Dictyostelium early development genes." *Immunol* 1994, 8(2): 231–237.

CHAPTER 9

Large-Scale Biology

9.1 | Introduction

The use of molecular perturbation techniques, imaging, and homology bioinformatics elucidated the identity and function of many human genes long before the completion of the Human Genome Project. In addition, complete signaling and metabolic pathways had been identified in bacteria and the findings projected to the human domain. Nevertheless, the current knowledge base of living systems is poised to increase dramatically with the decoding of the human genome and the recent developments of high-throughput tools for biological research. In fact, new bioassays are transforming biology from a science that was largely focused on gene identity and function discovery, one gene at a time, to one where the activities of thousands of genes are measured in a single experiment. The availability of new large-scale biological tools has challenged biologists to take a global view of the connectivity of the gene and protein networks underlying cell communication.

In this chapter, we introduce the emerging large-scale biological tools to the reader. One of these tools, gene microarrays, provides information about the gene-expression patterns of as many as 30,000 genes at one instant of time. The array technique is also being applied to study protein–DNA binding related to cancer research and protein–small-molecule binding for new drug design. The second global assay provides protein expression and interaction maps of cells and tissues and is known as proteomics.

These new biology tools are of particular interest to engineers and computer scientists for two reasons. These techniques require microscale and often nanoscale machining and sophisticated robotics. Secondly, analysis of the data produced by these tools requires mathematical analysis involving signal processing, pattern recognition, supervised machine training, and hierarchical clustering. Engineers and computer scientists are essential parts of the interdisciplinary teams pushing New Biology forward.

9.2 | Microarrays

Microarrays are an important new tool for large-scale biological investigations and have been used extensively in recent years in the study of cell response to changes in the biochemical and electromechanical environment. Microarrays have great promise in reducing the costs of development of new pharmaceutical agents. Using microscope slides, precision robots, and other off-the-shelf equipment, researchers created microarrays of circular spots each containing one type of molecule for systematic biological research (Fig. 9.1). Depending on the type of the microarray, the circular spots may contain DNA segments, proteins, or small molecules. DNA fragments remain functional when dried on glass or silicon chips. Protein chips are based on wet chemistry at present in order to prevent proteins from denaturation. They are in the form of disposable arrays of microwells in silicon elastomer sheets placed on top of microscope slides. These arrays are exposed to cell contents or other biological specimens. The interactions of biomolecules with arrays of spots are quantified by using fluorescence and radioisotope marking. In the following, we review briefly the various forms of this promising instrument of large-scale biology.

9.2.1 DNA Microchips

Mutations or alterations in genes are the causes of many forms of diseases. It is often difficult to identify and characterize these mutations because most large genes have many regions where a mutation could occur and cause disease. Examples of mutated genes that cause cancer are BRCA1 and BRCA2, which are believed to cause as

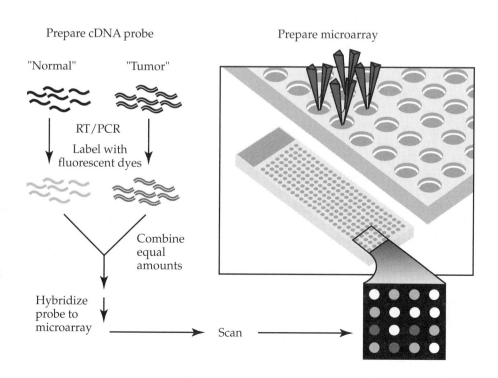

FIGURE 9.1 DNA microarrays are used to quantify the expression of large arrays of genes at discrete time intervals. The microarray technology utilizes dry chemistry and automation to decorate a planar substrate (a glass coverslide or a silicon chip) with a rectangular array of thousands of circular spots, with each spot presenting a unique nucleotide sequence complementary to an mRNA. When the contents of a cell or cells are poured onto microarrays, fluorescent-labeled mRNAs attach to their corresponding sticky spots, leading to a map of gene expression in accordance with the intensity of fluorescence bound to each spot. (Modified from http://www.nhgri.nih.gov/DIR/VIP/Glossary/Illustration/microarray_technology.html.)

many as 60 percent of all cases of hereditary breast and ovarian cancers. In BRCA1 alone, over 500 different mutations have already been discovered.

The DNA chip is used to identify the type of mutations that exist in a certain gene. DNA chips contain an array of synthetic single-stranded DNA sequences identical to those found on a normal targeted gene such as BRCA1.

To determine whether an individual possesses a mutation for BRCA1, a scientist first obtains a sample of DNA from her blood, as well as a sample that does not contain a mutation in either gene. After separating the samples of DNA into single strands and cutting them into smaller, more manageable fragments, the researcher labels the fragments with fluorescent dyes. The individual's DNA is labeled with green dye and the normal DNA is labeled with red dye. Both sets of labeled DNA are then inserted into the chip and allowed to hybridize, or bind, to the synthetic BRCA1 DNA fragments on the chip. If the individual does not have a mutation for the gene, both the red and green samples will hybridize with the sequences on the chip. If the individual does possess a mutation, the red (normal) DNA will still hybridize perfectly with the DNA on the chip, but the green (individual's) DNA will not hybridize properly in the region where the mutation is located. The scientist can then examine this area more closely to confirm that a mutation is present. DNA chips may find potential use in assessing individual risk for diseases such as cancer, heart disease, and diabetes.

9.2.2 DNA Microarrays

Although all of the cells in the human body contain the same genetic material, the same genes are not active in all of those cells. Studying which genes are active and which are inactive in different kinds of cells helps scientists understand more about how these cells function. DNA microarrays allow researchers to examine the expression of tens of thousands of genes simultaneously in a cell population (Fig. 9.1).

Microarrays are formed on glass slides or gene chips by placing a rectangular array of thousands of spots containing functional DNA (Fig. 9.1). Each spot may contain an oligonucleotide or cDNA fragment whose sequence associates with a distinct mRNA. mRNA from a particular cell or cell population is labeled with fluorescent tags and allowed to hybridize to the DNA on the slide whose sequences are complementary to the those of the mRNA. After a scanner measures the fluorescence of each sample on the slide, scientists can determine how active the genes represented on the microarray are in the cell. Strong fluorescence indicates that high levels of mRNA hybridized to a particular microarray spot and, therefore, that the gene is very active in the cell. Conversely, no fluorescence indicates that none of the mRNA hybridized to the slide and that the gene is inactive in the cell.

Microarray technology has found increasing use in the study of cancer. The typing and staging of cancer today is largely based on histological studies. Pathologists examine tissue slides using light microscopy and assign cancer to various categories. A more accurate and detailed classification of cancer based on molecular profiling may lead to more sophisticated classification and the more restricted use of toxic chemotherapy. For example, researchers have found success in associating microarray gene expression patterns obtained from tumor tissue with well-defined clinical outcomes. Classification based on clinically relevant clustering of microarray profiles is a crucial step in the design of treatment strategies targeted directly to each specific type of cancer. Additionally, by examining the differences in gene activity between untreated and treated-radiated or oxygen-starved cells, scientists can better understand how different types of cancer therapies affect tumors and can develop more effective treatments.

9.2.3 Can Microarrays Provide Information about the Gene Network Structure?

DNA microarray experiments are easy to perform and the data processing is largely automated with the range of gene expression presented numerically as well as in various shades of greens and reds. With microarrays, one can generate immense data sets indicating the gene product level response of cells to directed perturbations. However, at present, there are considerable difficulties associated with the separation of noise from signal in the actual data. Moreover, levels of expression of many of the genes involved in cancer or other diseases may not be detected by a cDNA microarray because their regulation may be exerted posttranscriptionally; for example, by selected degradation (see cell-division-cycle checkpoints discussed in Chapter 7) or posttranslational modification (such as phosphorylation).

A comprehensive system-level model that can relate microarray data to an underlying gene network is yet to emerge. Computational studies based on Boolean networks indicate the difficulty of inferring network properties from microarray data. Ideker et al. (2000) addressed the problem of determining the structure of a hypothetical Boolean gene network from data resulting from the perturbation of the expression of one node in the network at a time. A Boolean gene network with $N = 5$ nodes (genes) and $K = 2$ arrows going to each node had a single inferred network. When the number of genes in the network (N) was increased from 5 to 10 (while K was kept equal to 2), the number of inferred networks increased to 60. For $N = 100$ and $K = 2$, there were 3×10^{26} inferred networks. Even the data involving double permutations were not sufficient to differentiate between inferred networks. These studies indicate the need to use multiple approaches to generate information related to cellular gene networks. The development of new and effective methods of inferring network structure from biological data is an exciting new challenge for engineers and computer scientists in the 21st century.

9.2.4 Protein Microarrays

Protein microarrays are under development for measuring the function of thousands of proteins simultaneously and creating protein snapshots of cells, profiling the massive number of enzymes and other proteins in various states. Presently available protein microarrays contain as many as 10,000 wet spots of protein Fig. 9.2. As demonstrated by the ability of deposited proteins to react with their target proteins and small molecules, the technique used in the fabrication of protein spots preserves the function of these proteins.

A variation of the protein chip is the kinase chip. This tests the ability of kinases to phosphorylate protein populations immobilized on planar substrates. These chips have been developed and used to evaluate the functions of protein kinases encoded by the yeast genome. There are currently 122 known yeast protein kinase genes. To test their ability for autophosphorylation, the kinases were directly adhered

FIGURE 9.2 Dimensions of a typical protein microarray chip.

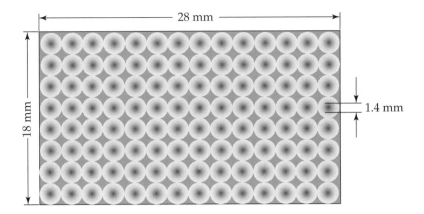

to microwells and ATP was added. After the reactions were completed, the kinase slides were washed and the phosphorylation signals quantified using a high-resolution phosphoimager. In other assays, microarrays were formed of known protein kinase substrates (targets), and their phosphorylation was monitored in the presence of ATP and one kinase agent at a time. Results indicate that most substrates are preferentially phosphorylated by a particular kinase or set of kinases. Similar methods are in development for the study of kinase activity in human cells.

The development, adaptation, and integration of different microelectronic and chemical sensors into miniaturized biochips are underway in several laboratories. Small molecule chips to be used in drug development are also in various stages of development. These new instruments of large-scale biology will potentially provide functional online analysis of living cells in physiologically relevant environments, with the direct impact of eliminating or economizing animal tests.

9.3 | Proteomics

Proteomics is the large-scale study of gene-expression products at the protein level. Such studies aim to gain insights into the active states of all relevant proteins. The field derives its name from "proteome," the protein content of a cell or tissue type. This approach has the potential to detect perturbations in the proteome of a population of cells due to disease and drug action. The two fundamental steps in traditional proteomics are (a) separation of proteins in a sample using two-dimensional gel electrophoresis (2-DE) and (b) their subsequent quantification and identification using mass spectroscopy. Proteomic tools are quite attractive for medical biology as they can be used to quantify the protein content of solid tissues (such as tumor) as well as biofluids such as blood serum and urine. The two-dimensional gel positions the proteins in a cell on a planar surface according to their molecular weight and electrical charge and, as such, creates a complex mathematical pattern that can be used for early detection and treatment of diseases. Moreover, the gel is more than just an image, but contains the actual proteins associated with disease. Parts of the gel can be cut and the proteins in the gel can be isolated for further studies using mass spectroscopy. Identification of disease-stage-specific proteins is an important step in designing small molecule drugs against these proteins. The principles behind these procedures are discussed briefly in the sections that follow.

9.3.1 Protein Electrophoresis

Electrophoresis is the core technology to separate complex protein mixtures from biological samples. Protein profiling in cell mixtures under normal and perturbed conditions is important for drug target identification. Electrophoresis refers to the migration of charged molecules in solution in response to an electric field. Figure 9.3 shows a typical electrophoresis procedure used to separate proteins. Proteins can be separated both on the basis of their mass and on the basis of the relative contents of acidic and basic residues.

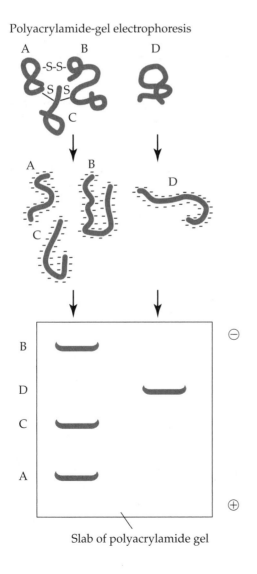

FIGURE 9.3 A schematic of a protein electrophoresis gel.

To separate proteins according to mass, the mixture of proteins is dissolved in a solution sodium dodecyl sulfate (SDS), a detergent that disrupts nearly all noncovalent interactions in native proteins. Other chemicals are introduced to break disulfide bonds so that the resulting denatured polypeptides become "rods" of a negative charge cloud with equal charge or charge densities per unit length. Anions of SDS bind to main chains at a ratio of one SDS for every two amino-acid residues, which gives a complex of SDS with a denatured protein a large net negative charge that is proportional to the mass of the protein. The charge of the native protein is negligible in comparison with the charge acquired on binding SDS. The SDS complexes with the denatured proteins are then electrophoresed on a polyacrylamide gel. Larger protein–SDS complexes, although they have large negative charges, move much slower toward the (+) end of the gel because of the dramatic increase in friction coefficient f with increasing mass and volume.

Thus, small proteins move rapidly through the gel, whereas large ones stay at the top, near the point of application of the mixture. A linear relationship exists between the logarithm of the molecular weight of an SDS-denatured polypeptide and the distance migrated by the molecule to that migrated by a marker dye-front. The proteins in the gel can be visualized by staining them with silver or a dye.

Proteins are also separated on the basis of their charge, and the resulting protein distribution on the gel is referred to as isoelectric focusing. The isoelectric point (pI) of a protein is the pH at which the net charge is zero. Proteins are amphoteric compounds; their net charge is determined by the pH of the medium in which they are suspended. In a solution with a pH above its isoelectric point, a protein has a net negative charge and migrates towards the anode in an electrical field. Below its isoelectric point, the protein is positively charged and migrates towards the cathode. At the isoelectric pH, the velocity of migration (mobility) is zero. When a mixture of proteins is electrophoresed in a pH gradient in a gel in the absence of SDS, each protein moves until it reaches a position in the gel at which the pH is equal to pI of the protein. Isoelectric focusing, when combined with SDS gel electrophoresis results in very high-resolution separations, and the method is called two-dimensional electrophoresis or 2DE. In this procedure, the sample is first subject to isoelectric focusing. This gel is then placed horizontally on top of an SDS gel and electrophoresed vertically to yield a two-dimensional pattern of spots. Automated systems can be used transferring selected protein spots on 2DE gels to microplates for mass spectropy identification. Difference gel electrophoresis (DIGE) enables the researchers to run in a single 2DE gel, several samples labeled with different fluorescence probes for the purpose of direct comparison while avoiding gel–gel variations.

Recent studies utilize two-dimensional gels in combination with microarrays in the discovery of genes associated with cancer metastases (Fig. 9.4). As discussed earlier, the expression of a gene product depends not only on the rate of its transcription, but also on the stability of its mRNA, the rate of translation, and the rate at which the protein is degraded. The combination of cDNA microarrays to discover transcriptional regulation and two-dimensional gel electrophoresis to monitor changes in protein concentration offers significant advantages to the use of either technique alone.

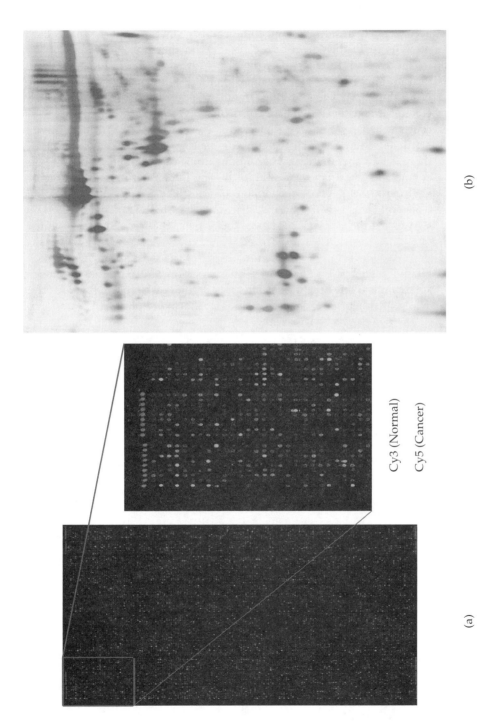

Cy3 (Normal)

Cy5 (Cancer)

(a)

(b)

FIGURE 9.4 Parallel analysis of transcriptional (a) and post-transcriptional (b) dysfunction associated with breast cancer. RNA and protein are isolated from a tumor and histologically normal tissue obtained from a single donor. (Courtesy of Dr. Richard Somiari, Windber Research Institute.)

9.3.2 Mass Spectroscopy

Mass spectroscopy is used to estimate the molecular weight of a protein with great accuracy. The resulting data can be compared with the molecular weight of known proteins to further identify the proteins in the spots on a two-dimensional electrophoresis gel. In mass spectrometry, a low concentration of sample is introduced into an ionization chamber where they are bombarded by a high-energy electron beam (Fig. 9.5). The molecules fragment, and the positive ions produced are accelerated through a charged array into an analyzing tube. The path of the charged molecules is bent by an applied magnetic field. Ions having low mass (low momentum) are deflected most extensively by this field; as a result, these particles collide with the walls of the analyzer. High-momentum ions are deflected least, and they also collide with the analyzer wall. Ions having the proper mass-to-charge ratio, however, follow the path of the analyzer, exit through the slit, and collide with the Collector. This generates an electric current, which is then amplified and detected. By varying the strength of the magnetic field, the frequency of fragments with different mass-to-charge ratio can be evaluated.

As shown in Figure 9.6, the output of the mass spectrometer shows a plot of relative intensity vs. the mass-to-charge ratio (*m/e*). The most intense peak in the spectrum is termed the **base peak**, and all others are reported relative to its intensity. The peaks themselves are typically very sharp and are often simply represented as

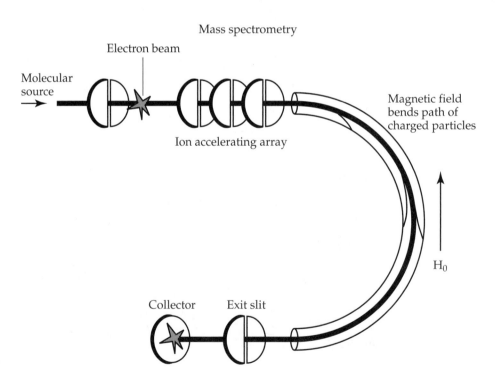

FIGURE 9.5 Schematic drawing of mass spectrometer.

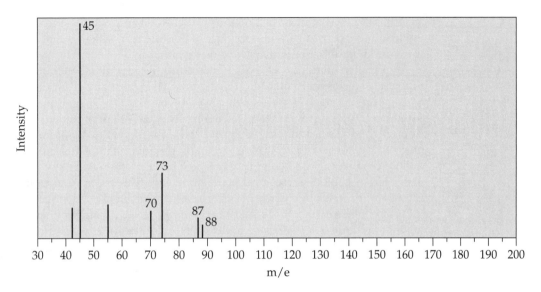

FIGURE 9.6 Mass spectrum of $C_5H_{12}O$. The compound has a molecular weight of 88.15.

vertical lines. Such spectrum graphs are being used in cancer research as mathematical patterns that can be used in the identification of cancers with high metastatic potential.

The process of fragmentation follows simple and predictable chemical pathways. The ions thus formed are the most stable cations that a molecule can form. The highest molecular weight peak observed in a spectrum typically represents the parent molecule, minus an electron, and is termed the **molecular ion** (M+). Fragments can be identified by their mass-to-charge ratio, but it is often more informative to identify them by the mass that has been lost. For example, loss of a methyl group generates a peak at *m*-15. Current models of mass spectroscopy are quite accurate in their estimates of molecular weight, which is then used in the identification of the protein.

Mass spectroscopy and two-dimensional electrophoresis are powerful techniques, but their application to protein mapping for all proteins is not straightforward. Some types of proteins—highly insoluble, phosphorylated, and membrane-bound proteins, as well as very small proteins—are typically lost when samples are analyzed by conventional MS analysis. The detection of phosphorylated proteins by MS is aided currently by chemical replacement of phosphate groups with biotin followed by avidin-based capture or by replacing phosphate groups with a thiol-based linker. These approaches now permit the identification of serine/threonine as well as tyrosine phosphorylation. Antibodies against phosphotyrosine (but not phosphoserine) are suitable for affinity purification of target proteins. Alternatively, in order to detect phosphorylated proteins and large, peripheral and integral membrane proteins, the two-dimensional gel electrophoresis of samples is replaced by two-dimensional liquid chromotography. This new method identified 1484 different proteins in yeast whole-cell lysate, far more than have previously identified in a single study. Separation of hydrophobic proteins in immobilized pH gradient (IPG) strips has been improved by combining more effective reducing agents.

Another limitation of proteomics concerns low-abundance proteins. Only a fraction of genes are switched on in a given cell type at any given time. The protein concentrations may vary by as much as eight orders of magnitude in cell lysates. Routine 2D analysis is biased toward long-lived abundant proteins. Proteins that are subject to structural modification and those that have low concentrations may be masked by housekeeping proteins. Sample prefractionation techniques are used to reduce the diversity and complexity of protein mixtures so that low-abundance proteins can also be detected. Prefractionation increases the concentration of distinct subsets of proteins in narrow pH range two-dimensional gels. Sequential extraction of proteins from a cell or tissue can be used to prefractionate proteins based on their relative solubility in a series of buffers. Prefractionation can also be executed using other physiochemical properties such as net charge, mobility, size, and hydrophobicity. Subcellular fractionation is used to enrich functionally related proteins based on their localization within the cell because the location of proteins within different compartments of the cell may vary without significant variation in the overall cellular concentration.

Finally, unlike the genome, the proteome is cell specific and constantly changing; therefore, a human proteome project documenting the protein content of different cell types is a challenging task. The dynamic range of protein expression and protein modification by phosphorylation makes characterization of the proteome even more difficult. Sensitivity of proteomics in detecting and quantifying low-abundance proteins suffers from a lack of an amplification method such as the use of PCR in genomics. Nevertheless, this new technology has advanced within a short period of time and will likely remain as an important arsenal in large-scale biological research.

9.3.3 Development of a Protein–Protein Interaction Database

Identifying the connections between various signaling and transcriptional regulatory pathways is an important aspect of characterizing cellular networks. Identifying which proteins physically interact with one another in a cell would go some way in elucidating the connectivity of these networks. To approach this, several groups have taken advantage of the principle of the yeast two-hybrid screen used to identify interacting proteins from a library. In this method, the gene for a protein of interest is fused with the DNA-binding domain of a transcriptional activator. Because this does not contain an activating region, it cannot activate transcription even though it can bind DNA.

This fused gene is then used as "bait" to probe a library of prey cDNAs. The "prey" genes are fused with the transactivating domain of the activator. Although these fusion proteins can bind activators, they cannot bind DNA in the absence of an interaction with the DNA-binding region. Only when the proteins encoded by bait and prey genes physically interact with one another, are the DNA binding and transactivating regions brought together and transcription of a reporter gene takes place. The activity of the reporter gene allows the investigators to identify which yeast colonies express the fusion partners (Fig. 9.7).

Consequently, these colonies express the genes of proteins that can interact with one another. By expanding and automating this method, several groups have created yeast protein–protein interaction databases that include over 4000 potential interacting partners. In array screening, 6000 yeast colonies, each expressing a different "prey" molecule were mated with yeast strains containing "bait" molecules. In the case of Uetz et al., 192 bait genes were used and 48 possible partners were identified. Some of these partners were already known to interact in cells thus confirming the validity of the approach. Remarkably, most proteins only interacted with a small number of partners implying that cellular networks exhibit limited connectivity. Although this approach can reveal potential protein-protein interactions it cannot put them in cellular context. For example, protein A and protein B may interact with one another in a two-hybrid screen but are normally found in separate cellular compartments. Consequently, each new interaction needs to be rigorously confirmed using other methods to rule out such false positives.

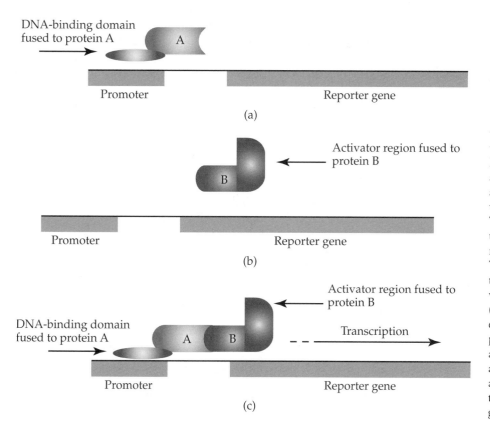

FIGURE 9.7 Detection of protein–protein interactions using the two-hybrid system. (a) The gene for a protein to be used as "bait" is fused to the DNA-binding region of a transcriptional activator. Because this does not contain an activating region, it cannot activate transcription even though it can bind DNA. (b) The genes for the "prey" protein are fused to the activating region of the transactivator. These cannot bind DNA in the absence of an interaction with the DNA-binding region. (c) When the proteins encoded by bait and prey genes physically interact with one another, the DNA-binding and -transactivating regions are brought together, and transcription of a reporter gene takes place.

9.4 | Protein Structure

Most proteins are in the form of loose strings (with short coiled sections) as they are manufactured by ribosomes. These molecules contort quickly into various partially folded states before attaining their completely folded final form. Proteins in living cells assume correct folding in one tenth of a second. As was discussed in Chapter 2, the three-dimensional structure of a polypeptide is a direct consequence of its sequence of amino acids. Proteins spontaneously fold into their unique three-dimensional form under physiological conditions. Because protein molecules are collections of electrically charged particles, the interactions of atoms in a protein are largely determined by the laws of electrostatics. Despite the knowledge of the physical principles involved, we are far from predicting three-dimensional protein folding from amino-acid sequence. Today, only experimental methods such as X-ray crystallography and nuclear magnetic resonance (NMR) yield detailed structural information on proteins. The three-dimensional protein structures obtained by X-ray techniques constitute about 80 percent of the three-dimensional protein structure data stored in the Protein Data Bank (PDB). The rest is made up of NMR data ($\sim$16 percent) and theoretical modeling. However, the current state of computational proteomics allows estimates of three-dimensional structures of proteins based on the known structure of a highly similar protein.

Knowing the structure of a protein sequence allows us to probe the function of the protein, discover the mechanism of substrate ligand binding, and design novel proteins. The field of computational proteomics explores the potential use of protein structure in protein therapeutics and protein-based designs. Large-scale three- and four-dimensional protein structural information will enable a revolution in rational design, having a particular impact on drug design and optimization. Additionally, understanding structure has potential applications in mapping the functions of proteins in metabolic pathways and deducing evolutionary relationships.

9.4.1 Three-Dimensional Structures Obtained by X-Ray Diffraction

X-ray diffraction is a technique that directly images molecules (Fig. 9.6). In optics, the uncertainty in locating an object increases with the wavelength of the radiation used to observe it. Both X-ray wavelengths and covalent bond length are in the range of angstroms. The type of X rays used in three-dimensional structure determination have wavelengths ranging from approximately 0.5 Å to 1.5 Å whereas the wavelength of visible light ranges from 4000 Å to 7000 Å. This is why individual molecules cannot be seen using light microscopy, but can be detected using X rays. X-ray lenses and microscopes do not exist, and for a structure to be visualized, the protein first needs to be crystallized. X rays collide with the electrons of the molecules in the crystal and as a consequence are diffracted as shown in Fig. 9.8. The diffraction pattern is recorded on a photographic beam and digitally stored in a computer. In the crystal form of a molecule, its atoms are arranged in regularly repeating three-dimensional lattices. Refraction of X rays from this lattice structure yields patterns providing information on the structure itself. The intensities of the diffraction maxima (darkness of the spots on film) are then used to construct mathematically the three-dimensional structure of the molecule. The X-ray structure can

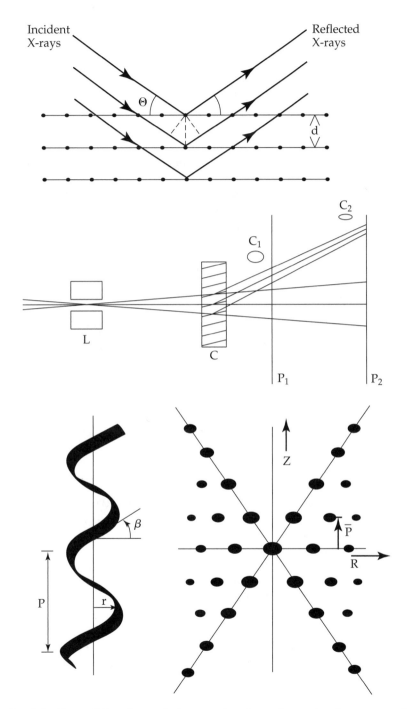

FIGURE 9.8 Schematic of the X-ray diffraction technique used to determine molecular structure from a protein crystal.

be considered as an image of the electron density of the object under study. Electron density is represented by counters of constant intensity either in planar sections cutting the molecule or in three dimensions.

Remarkably, crystalline proteins assume very nearly the same structures that they have in solution. Many enzymes continue to have identical catalytic activity in the crystalline state, and structure based on X-ray crystallography is highly relevant to biology. However, protein crystallization is as much an art as a science. Proteins do not exist in crystal form in living systems, and it is a challenge to obtain the crystal form of a protein. As a result, many important molecules are missing from the current protein-structure databases. In general, molecules with highly hydrophobic portions such as transmembrane proteins have proved particularly hard to crystallize. Much success has been had, however, in crystallizing extracellular domains of membrane-anchored receptors, lacking the hydrophobic membrane-anchoring regions.

Protein crystals differ from those of smaller molecules in crystal form in that protein crystals are highly hydrated (50 percent by volume water). The high water content of protein crystals reduces the rigid order, characteristic of crystals of smaller molecules such as NaCl. Due to the water content, molecules in a protein crystal are disordered in the length scale of a few angstroms, and as a result, the electron-density map of a protein may not reveal clearly the positions of individual atoms. The method can neither resolve the positions of hydrogen atoms nor distinguish among nitrogen, oxygen, and carbon. Despite the presence of fuzziness at the angstrom scale, the distinctive shape of the polypeptide backbone can be recognized. Tracing of the backbone together with knowledge of the amino-acid residue sequence of the protein allows the positions and orientations of its side chains to be deduced. The resulting three-dimensional structure is a collection of Cartesian coordinates for every nonhydrogen atom in a protein molecule. In 2000, the National Institutes of Health in Bethesda, MD began a protein-structure initiative aimed at producing 10,000 X-ray crystal structures, to be made available on the Internet.

9.4.2 NMR and 3-D Protein Structures

Even though membrane proteins represent 30 percent of the proteome, relatively little is known about the structure of these proteins, because of their resistance to crystallization. NMR utilizes the tiny magnetlike properties of atomic nuclei in which at least one proton or one neutron is unpaired. The data obtained by NMR are used in determining distances between various atoms in a molecule. The distance determinations in return yield the three-dimensional structure of the protein.

In the NMR method, a small volume of concentrated protein solution is subjected to the field of a powerful magnet. The natural magnets in the sample orient themselves to line up with the NMR. The researchers then expose the sample to a series of split-second radio-wave pulses (wavelength much greater than visible light). As the RF pulse continues, some of the protons on lower energy levels absorb energy from the RF field and make a transition into the higher energy state. This has the effect of tipping the net magnetization toward the transverse plane. When the RF pulse is discontinued, the net magnetization vector goes back toward its original direction. Reaction of nuclei to the radio waves (the time rate of change of net magnetization vector) comprises a signal called the chemical shift. Patterns of these chemical shift signals are analyzed with Fourier transform methods in order to deduce the distances between neighboring atoms. A set of algorithmic tools such as *Distance Geometry* is then used to predict three-dimensional structure.

NMR can yield time-averaged structures of proteins in the fluid-membrane environment and therefore can be used to determine the structure of membrane proteins. NMR is generally used to determine the structure of small proteins in solution. It is the method of choice for proteins or protein domains up to 30 kDa (<300 amino-acid residues). The largest protein whose structure was determined by NMR is about 40 kDa. X-ray techniques have already revealed the rough three-dimensional structures of proteins that are 60 times as large. Unlike the X-ray methodology, NMR yields the positions of at least some of the hydrogen atoms in a molecule. Because it uses molecules in solution, the method is not limited to those protein molecules that crystallize. The Institute of Physical and Chemical Research in Yokohoma, Japan (RIKEN) is using an assembly of 16 state-of-the-art high-resolution NMR spectrometers in its "Protein Folds Project" initiative to determine the three-dimensional structures of smaller proteins on small atomic length scales. Some molecules (such as thioredoxin from *E. coli*) have been studied by both NMR and X-ray techniques and the resulting structures are practically identical.

9.4.3 Homology Modeling

The small molecule drugs available today on the market work only on about 500 unique targets. Considering that there may be as many as 30,000 different types of proteins in humans, there is a need for computational prediction of three-dimensional protein structure.

Computational proteomics seeks to determine protein structure through computational means. Protein modeling is quite literally orders of magnitude faster and much less costly than X-ray crystallography. This new field benefits from the X-ray crystallography data available on more than 10,000 crystal structures in the publicly accessible protein databank (PDB). Homology modeling utilizes these known crystal structures to predict the structure of the polypeptides with similar primary sequence or protein fold class. How good are these predictions of protein structures? For what purposes are they useful?

Homology modeling (Fig. 9.9) is also known as comparative modeling or knowledge-based modeling. Homology refers to proteins having a common evolutionary origin. The three-dimensional structures of homologous proteins are conserved to a greater extent than their primary structures (sequences). In homology modeling, scientists rely on the structural knowledge of proteins for which three-dimensional structures have already been determined to infer the structures of homologous proteins for which only the sequence is known. The output of the homology model is a three-dimensional-space model indicating the spatial positions of all main-chain and side-chain nonhydrogen atoms.

The known three-dimensional structure of a protein is used in the initial positioning of various amino-acid residues of a homologous protein. The computation minimizes the elastic strain energy of the collection of atoms in the protein subject to constraints on distances between various carbon atoms and atom groups. There is an effective low-end cut-off point for effective homology modeling of ∼30 percent of sequence identity. But even at this lower limit, the computed structure describes accurately the arrangements of secondary structure elements comprising the protein

FIGURE 9.9 Schematic illustration of homology modeling.

1. Align sequence with structures:

3D   GRISFFEDAGF--GBCYECSSDC--NLQP

3D   GKITFYEDRGFQGHCYECSSDC--NLQP

SEQ   GKITFYEDRG-------RCYECSSDCPNLQP

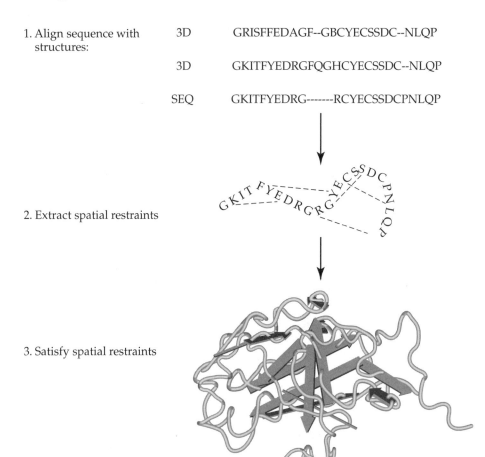

2. Extract spatial restraints

3. Satisfy spatial restraints

in 3D. Therefore, the method is useful in elucidating the overall function of the protein by identification of common three-dimensional structural elements that can be correlated with particular enzymatic activity.

The computations at low degrees of homology yield relatively inaccurate surface structures. This is a disadvantage for drug design purposes since most chemical activities of proteins reside on the surfaces of the protein. But even at the low end of sequence identity, some portions of the protein's structure (such as the sites for enzyme action) might be highly conserved, providing the potential for computational library screening used in computational drug design. Databases that contain large numbers of protein structures obtained through homology modeling include ModBase (http://guitar.rockefeller.edu/modbase/). This database currently contains three-dimensional models for substantial portions of approximately 17,000 proteins from 10 complete genomes including those of yeast and *E. coli*. The database also includes the fault assignments and alignments on which the models were based. The Web-based tool www.SWISS-MODEL.com is one of the more popular programs to generate models of protein three-dimensional structures using homology modeling.

When no homologue of known structure from which to make a three-dimensional model exists, a logical step would be to predict secondary structure, with the aim of providing the location of helices and sheets within a protein. There are a number of automated secondary structure predictors (PHD, SOPMA, and SSPRED) already available on the Web. If a sequence has more than 500 amino-acid residues, it is likely composed of discrete functional domains. Faster computers and new rapid algorithms are lowering the barrier of computation speed in protein-structure modeling.

9.5 | Systems Biology

Until the late 1990s, biological research had to be based on a clearly expressed specific hypothesis, before funding by the National Institutes of Health, or other funding agencies. More recently, larger grants are being awarded to interdisciplinary discovery teams working under more general hypotheses. An emerging trend in high-throughput biology today is to establish a comprehensive laboratory that contains functional genomic tools, microarrays, and robotic-enhanced proteomics. These high-throughput experimental systems generate immense amounts of noisy data; therefore, their usefulness will depend on the extent they are integrated to bioinformatics and other computational tools. The concept of homology has proved particularly useful in the identification of protein (gene) functions. Scientists are interrogating the proteins of exotic species living in extreme conditions for their potential use in biotechnology. Existing knowledge on metabolic and signaling pathways serve as a useful guide for large-scale biology. However, we do not yet have effective analytical tools to study complex gene and protein networks. The significance of spatial patterns on DNA concerning the positioning of development-related genes is under investigation, but mathematical models that account for the spatial positioning of proteins (membrane bound versus cytoplasmic) in signal transduction pathways are yet to emerge.

Most modern biologists still devote a great deal of effort to demonstrating the selectivity and precise outcome of a particular linear pathway in model cell systems. The general idea is that specific manipulation of certain pathways implicated in disease or development-related processes will lead to cures or a profound understanding of ontogeny. Whilst this may be true for certain poorly connected parts of the cell "network," if, as we emphasize throughout this book, most pathways are parallel and exhibit some degree of connectivity, then interpretation of results obtained in such experiments will be difficult. Exceptions are diseases that result from defined monogenic lesions.

These exceptions are rare; about 2 percent of human diseases can be attributed to monogenic disorders. Most cancers and heart diseases are the result of complex interactions among many different gene products, in other words, they are polygenic. However, these disease processes are not simply the result of autonomous alterations in gene structure and function, since the onset and progression of most solid tumors and heart diseases can be attributed to environmental influences such as diet and exposure to carcinogens such as cigarette smoke.

Given the preceding discussion, which points to great flexibility in the circuitry used to generate a given output phenotype, should we not explore the possibility that advantage could be taken of the alternative routes that must inevitably lead to the same phenotype even in the face of apparently irreversible defects in some pathways. Rather than searching for more and more specific agents, perhaps equal effort should be devoted to the discovery of agents with broad effects and to those that influence alternative, compensatory pathways.

As indicated, nature has given us some indications of the sorts of agents that may be in this category. The importance of dietary agents such as vitamins A and D in the onset and control of such diseases as cancer and rickets points to a remarkable degree of cooperation between an individual and his or her surroundings. Clearly, we have evolved to coexist and be dependent upon our environment. One major unanswered, even unaddressed question, which remains in New Biology today is how the environment interacts with and influences the genetic and epigenetic circuitry that regulates cell and tissue homeostasis.

9.6 ASSIGNMENTS

9.1 Conduct a literature search, and discuss the application of microarrays in identifying those genes that may play important roles in breast cancer metastasis.

9.2 Tumor tissue is composed of cancer cells as well as normal differentiated cells and immune cells and therefore is highly heterogeneous. Thus, mRNA collected from tissue for microarrays contain mRNA from normal cells as well as cancer cells. Moreover, the cancer cells in a tumor differ considerably in metastatic potential. Conduct a literature search, and determine how researchers are addressing tissue heterogeneity in microarray experiments.

9.3 Currently available microarrays used in human biological studies have as many as 20,000 spots. Considering the fact that there will be some stochastic manufacturing variability in the stickiness of the spots to their target mRNAs, the data produced contain error. Discuss the other sources of error in microarray data and the methods used by scientists to reduce microarray data noise in relation to the signal.

9.4 Two-dimensional gel electrophoresis produces accurate protein-expression maps of tissue. Until recently, the interpretation of the gel experiments was very difficult because, due to manufacturing errors, a certain protein would occupy a slightly different spatial position in different gels under identical experimental conditions. Conduct a literature search, and determine how scientists have addressed the problems associated with comparison of protein expression maps.

9.5 Nuclear magnetic resonance imaging of protein structure is based on the magnetic spin properties of atoms. The processional frequency of a small magnet in a magnetic field Bo is given by the *Larmor Equation*: $\omega = Bo\ \gamma$ where ω is the frequency (1/s), Bo is the magnetic field, and γ is the gyromagnetic ratio, the ratio between the magnetic moment and angular momentum. Conduct a physics literature search about this equation, and find out the gyromagnetic ratios of magnetically active isotopes of carbon, fluorine, phosphorus, and sodium.

9.6 *Threading modeling* is a technique for predicting the three-dimensional structure of a protein. It is also called inverse folding, sequence for structure fitness, or FOSFOS. Threading is employed when insufficient structural information is available to permit homology modeling. Describe the principal features of threading using the Web site www.mbi.ucla.edu/people/frsvr/frsvr/.

9.7 Fluorescence In Situ Hybridization (FISH) is a technique used to identify the location of a particular sequence (gene) on a chromosome. Conduct a literature search to determine what makes FISH a more versatile technique than others used to study chromosomes.

9.8 Spectral karyotyping (SKY) is a laboratory technique that allows scientists to visualize all 23 pairs of human chromosomes at one time, with each pair of chromosomes painted in a different fluorescent color. Many diseases are associated with particular chromosomal abnormalities. For example, chromosomes in tumor cells frequently exhibit aberrations called translocations, where a piece of one chromosome breaks off and attaches to another chromosome. Identifying such chromosome abnormalities and determining their role in disease is an important step in developing new methods for diagnosing many genetic disorders. Conduct a literature search to determine how SKY works.

9.9 One of the most inefficient steps in drug discovery process is determining which small molecule compounds to target as new drug candidates. Cell-based assays provide biologically relevant information in the discovery of new therapies. Conduct a literature search on cell-based screening in drug discovery.

9.10 The MEMS is an engineering technology that has recently been adapted to monitor blood-pressure levels in the organs or blood clots of patients with heart failure or with abdominal aortic aneurysm, an abnormal widening of the aorta. If it works, it could provide doctors with an easier way to detect serious problems.

Conduct a literature search on the emerging BioMEMS technology used to design microchip implants.

9.11 Use the software available at portal.curagen.com to identify proteins that interact with the yeast protein Msh5. What does this information tell us about the mechanisms involved in meiotic recombination?

9.12 SNP Analysis. Studies involving single nucleotide polymorphisms (SNPs) have gained increased attention recently because of the interest in individual differences in disease susceptibility as well as individual responses to drug treatment. The review by R. Tolle ("Information technology tools for efficient SNP analysis." *Am J Pharmacogenomics* 2001, 1:303–14) provides a guide to the SNP sources and tools. Write a report discussing the numerical procedures used in SNP analysis.

9.13 Conduct a literature search to discuss the proteome of mouse and the evolution of gene families in the mouse genome in relation to the human genome.

REFERENCES

Chun HM et al. "MBO(N)D: A multibody method for long-time molecular dynamics simulations." *J Comp Chem* 2000, 21: 205–213.

Dudek MJ et al. "Protein structure function prediction using a combination of sequence homology and global energy minimization: II. Energy functions." *J Comp Chem* 1998, 19: 548–573.

Eisenmenger F et al. "A method to configure protein side chains from the main-chain trace in homology modeling." *J Mol Biol* 1993, 231: 849–860.

Ito T, Chiba T, Ozawa R, Yoshida M, Hattori M, and Sakaki Y. "A comprehensive two-hybrid analysis to explore the yeast protein interactions." *PNAS* 2001, 4569–4754.

Ito T, Tashiro K, Muta S, Ozawa R, Chiba T, Nishizawa M, Yamamato K, Kuhara S, and Sakaki Y. "Toward a protein-protein interaction map of the budding yeast: A comprehensive system to examine two-hybrid interactions in all possible combinations between the yeast proteins." *PNAS* 2002, 97: 1143–1147.

Lemer CM et al. "Protein structure prediction by threading methods: evaluation of current techniques." *Proteins: Struct Funct Genet* 1995, 23: 337–355.

Maggio ET and Ramnarayan K. "Recent developments in computational proteomics." *Trends in Biotechnology* 2001, 19: 266–272.

Norwell JC and Machalek AZ. "Structural genomics programs at the US National Institute of General Medical Sciences." *Nat Struct Biol* 2000, 7: 935–939.

Origene Technologies Inc for two-hybrid kits, specialized plasmids, libraries, and manuals.

Peterson TN et al. "Protein secondary structure prediction at 80% accuracy." *Proteins: Struct Funct Genet* 2000, 41: 17–20.

Russel RB and Sternberg MJE. "Protein structure prediction: how good are we?" *Current Biology* 1995, 5: 488–490.

Sanchez R, Pieper U, Mirkovic N, deBakker PIW, Wittenstein E, and Sali A. "ModBase, a database of annotated comparative protein structure models." *Nucleic Acid Research* 2000, 28: 250–253.

Uetz P, Giot L, Cagney G, Mansfield TA, Judson RS, Knight JR, Lockshon D, Narayan V, Srinivasan M, Pochart P, Qureshi–Emili A, Li Y, Godwin B, Conover D, Kalbfleisch T, Vijayadamodar G, Yang M, Johnston M, Fields S, and Rothberg JM. "A comprehensive analysis of protein-protein interactions in Saccharomyces cerevisiae." *Nature* 2000, 403: 623–627.

Williams DM and Cole PA. "Kinase chips hit the proteomics ear." *Trends Biochem Sci* 2001, 26: 271–273.

Glossary

Adenine (A)

A nitrogenous base, one member of the base pair AT (adenine–thymine).

Allele

Alternative form of a genetic locus; a single allele for each locus is inherited from each parent (e.g., at a locus for eye color, the allele might result in blue or brown eyes).

Alternative splicing

Different ways of combining a gene's exons to make variants of the complete protein

Amino acid

Any of a class of 20 molecules that are combined to form proteins in living things. The sequence of amino acids in a protein and hence protein function are determined by the genetic code.

Amplification

An increase in the number of copies of a specific DNA fragment; can be in vivo or in vitro.

Annotation

Adding pertinent information, such as gene coded for amino-acid sequence, or other commentary, to the database entry of raw sequence of DNA bases.

Antisense

Nucleic acid that has a sequence exactly opposite to an mRNA molecule made by the body; binds to the mRNA molecule to prevent a protein from being made.

Apoptosis

Programmed cell death, the body's normal method of disposing of damaged, unwanted, or unneeded cells.

Arrayed library

Individual primary recombinant clones (hosted in phage, cosmid, YAC, or other vector) that are placed in two-dimensional arrays in microtiter dishes. Each primary clone can be identified by the identity of the plate and the clone location (row and column) on that plate. Arrayed libraries of clones can be used for many applications, including screening for a specific gene or genomic region of interest.

Autosomal dominant

A gene on one of the non-sex chromosomes that is always expressed, even if only one copy is present. The chance of passing the gene to offspring is 50 percent for each pregnancy.

Autosome

A chromosome not involved in sex determination. The diploid human genome consists of a total of 46 chromosomes: 22 pairs of autosomes and 1 pair of sex chromosomes (the X and Y chromosomes).

Bacterial artificial chromosome (BAC)

A vector used to clone DNA fragments (100- to 300-kb insert size; average, 150 kb) in *Escherichia coli* cells. Based on naturally occurring F-factor plasmid found in the bacterium *E. coli*.

Base

One of the molecules that form DNA and RNA molecules.

Base pair (bp)

Two nitrogenous bases (adenine and thymine or guanine and cytosine) held together by weak bonds. Two strands of DNA are held together in the shape of a double helix by the bonds between base pairs.

Base sequence

The order of nucleotide bases in a DNA molecule; determines structure of proteins encoded by that DNA.

Bioinformatics

The science of managing and analyzing biological data using advanced computing techniques; especially important in analyzing genomic research data.

Biotechnology

A set of biological techniques developed through basic research and now applied to research and product development. In particular, "biotechnology" refers to the use by industry of recombinant DNA, cell fusion, and new bioprocessing techniques.

BLAST

A computer program that identifies homologous (similar) genes in different organisms, such as human, fruit fly, or nematode.

Cancer

Diseases in which abnormal cells divide and grow unchecked. Cancer can spread from its original site to other parts of the body and can be fatal.

Capillary array

Gel-filled silica capillaries used to separate fragments for DNA sequencing. The small diameter of the capillaries permits the application of higher electric fields, providing high-speed, high-throughput separations that are significantly faster than traditional slab gels.

Carcinogen

Something that causes cancer to occur by causing changes in a cell's DNA.

cDNA library

A collection of DNA sequences that code for genes. The sequences are generated in the laboratory from mRNA sequences.

Cell

The basic unit of any living organism that carries on the biochemical processes of life.

Centromere

A specialized chromosome region to which spindle fibers attach during cell division.

Chimera (variant, chimaera)

An organism that contains cells or tissues with a different genotype. These can be mutated cells of the host organism or cells from a different organism or species.

Chloroplast chromosome

Circular DNA found in the photosynthesizing organelle (chloroplast) of plants instead of the cell nucleus where most genetic material is located.

Chromosome

The self-replicating genetic structure of cells containing the cellular DNA that bears in its nucleotide sequence the linear array of genes. In prokaryotes, chromosomal DNA is circular, and the entire genome is carried on one chromosome. Eukaryotic genomes consist of a number of chromosomes whose DNA is associated with different kinds of proteins.

Clone

An exact copy made of biological material such as a DNA segment (e.g., a gene or other region), a whole cell, or a complete organism.

Cloning

Using specialized DNA technology to produce multiple, exact copies of a single gene or other segment of DNA to obtain enough material for further study. This process, used by researchers in the Human Genome Project, is referred to as cloning DNA. The resulting cloned (copied) collections of DNA molecules are called clone libraries. A second type of cloning exploits the natural process of cell division to make many copies of an entire cell. The genetic makeup of these cloned cells, called a cell line, is identical to the original cell. A third type of cloning produces complete, genetically identical animals such as the famous Scottish sheep, Dolly.

Cloning vector

DNA molecule originating from a virus, a plasmid, or the cell of a higher organism into which another DNA fragment of appropriate size can be integrated without loss of the vector's capacity for self-replication. Vectors introduce foreign DNA into host cells, where the DNA can be reproduced in large quantities. Examples are plasmids, cosmids, and yeast artificial chromosomes; vectors are often recombinant molecules containing DNA sequences from several sources.

Codon

The three nucleotide words on DNA. Each codon may translate into a single amino acid.

Comparative genomics

The study of human genetics by comparisons with model organisms such as mice, the fruit fly, and the bacterium *E. coli*.

Complementary DNA (cDNA)

DNA that is synthesized in the laboratory from a messenger RNA template.

Complementary sequence

Nucleic acid–base sequence that can form a double-stranded structure with another DNA fragment by following base-pairing rules (A pairs with T and C with G). The complementary sequence to GTAC, for example, is CATG.

Conserved sequence

A base sequence in a DNA molecule (or an amino-acid sequence in a protein) that has remained essentially unchanged throughout evolution.

Contig

Group of cloned (copied) pieces of DNA representing overlapping regions of a particular chromosome.

Contig map

A map depicting the relative order of a linked library of overlapping clones representing a complete chromosomal segment.

Cosmid

Artificially constructed cloning vector containing the cos gene of phage lambda. Cosmids can be packaged in lambda phage particles for infection into *E. coli*; this permits cloning of larger DNA fragments (up to 45 kb) than can be introduced into bacterial hosts in plasmid vectors.

Crossing over

The breaking during meiosis of one maternal and one paternal chromosome, the exchange of corresponding sections of DNA, and the rejoining of the chromosomes. This process can result in an exchange of alleles between chromosomes.

Cytological band

An area of the chromosome that stains differently from areas around it.

Cytological map

A type of chromosome map whereby genes are located on the basis of cytological findings obtained with the aid of chromosome mutations.

Cytosine (C)
 A nitrogenous base; one member of the base pair GC (guanine and cytosine) in DNA.

Data warehouse
 A collection of databases, data tables, and mechanisms to access the data on a single subject.

Deletion
 A loss of part of the DNA from a chromosome; can lead to a disease or abnormality.

Deletion map
 A description of a specific chromosome that uses defined mutations—specific deleted areas in the genome—as 'biochemical signposts,' or markers for specific areas.

Deoxyribose
 A type of sugar that is one component of DNA (deoxyribonucleic acid).

Diploid
 A full set of genetic material consisting of paired chromosomes, one from each parental set. Most animal cells except the gametes have a diploid set of chromosomes. The diploid human genome has 46 chromosomes.

Disease-associated genes
 Alleles carrying particular DNA sequences associated with the presence of disease.

DNA (deoxyribonucleic acid)
 The molecule that encodes genetic information. DNA is a double-stranded molecule held together by weak bonds between base pairs of nucleotides. The four nucleotides in DNA contain the bases adenine (A), guanine (G), cytosine (C), and thymine (T). In nature, base pairs form only between A and T and between G and C; thus, the base sequence of each single strand can be deduced from that of its partner.

DNA bank
 A service that stores DNA extracted from blood samples or other human tissue.

DNA repair genes
 Genes encoding proteins that correct errors in DNA sequencing.

DNA replication
 The use of existing DNA as a template for the synthesis of new DNA strands. In humans and other eukaryotes, replication occurs in the cell nucleus.

DNA sequence
 The relative order of base pairs, whether in a DNA fragment, a gene, a chromosome, or an entire genome.

Domain
 A discrete portion of a protein with its own function. The combination of domains in a single protein determines its overall function.

Dominant
 An allele that is almost always expressed, even if only one copy is present.

Double helix
 The twisted-ladder shape that two linear strands of DNA assume when complementary nucleotides on opposing strands bond together.

Draft sequence
 The sequence generated by the HGP as of June 2000 that, while incomplete, offers a virtual road map to an estimated 95 percent of all human genes. Draft sequence data are mostly in the form of 10,000 base-pair-sized fragments whose approximate chromosomal locations are known.

Electrophoresis
 A method of separating large molecules (such as DNA fragments or proteins) from a mixture of similar molecules. An electric current is passed through a medium containing

the mixture, and each kind of molecule travels through the medium at a different rate, depending on its electrical charge and size. Agarose and acrylamide gels are the media commonly used for electrophoresis of proteins and nucleic acids.

Electroporation

A process using high-voltage current to make cell membranes permeable to allow the introduction of new DNA; commonly used in recombinant DNA technology.

Embryonic stem (ES) cells

An embryonic cell that can replicate indefinitely, transform into other types of cells, and serve as a continuous source of new cells.

Enzyme

A protein that acts as a catalyst, speeding the rate at which a biochemical reaction proceeds, but not altering the direction or nature of the reaction.

Escherichia coli (*E. coli*)

Common bacterium that has been studied intensively by geneticists because of its small genome size, normal lack of pathogenicity, and ease of growth in the laboratory.

Eukaryote

Cell or organism with membrane-bound, structurally discrete nucleus and other well-developed subcellular compartments. Eukaryotes include all organisms except viruses, bacteria, and blue–green algae.

Exogenous DNA

DNA originating outside an organism that has been introduced into the organism.

Exon

The protein-coding DNA sequence of a gene. *See also intron.*

Expressed sequence tag (EST)

A short strand of DNA that is a part of a cDNA molecule and can act as identifier of a gene; used in locating and mapping genes.

Fingerprinting

In genetics, the identification of multiple specific alleles on a person's DNA to produce a unique identifier for that person.

Finished DNA Sequence

High-quality, low-error, gap-free DNA sequence of the human genome. Achieving this ultimate 2003-HGP goal requires additional sequencing to close gaps, reduce ambiguities, and allow for only a single error every 10,000 bases, the agreed-upon standard for an HGP finished sequence.

Flow cytometry

Analysis of biological material by detection of the light-absorbing or fluorescing properties of cells or subcellular fractions (i.e., chromosomes) passing in a narrow stream through a laser beam. An absorbance or fluorescence profile of the sample is produced. Automated sorting devices, used to fractionate samples, sort successive droplets of the analyzed stream into different fractions depending on the fluorescence emitted by each droplet.

Flow karyotyping

Use of flow cytometry to analyze and separate chromosomes according to their DNA content.

Fluorescence in situ hybridization (FISH)

A physical mapping approach that uses fluorescein tags to detect hybridization of probes with metaphase chromosomes and with the less-condensed somatic interphase chromatin.

Forensics

The use of DNA for identification. Some examples of DNA use are to establish paternity in child-support cases, establish the presence of a suspect at a crime scene, and identify accident victims.

Fraternal twin

Siblings born at the same time as the result of fertilization of two ova by two sperm. They share the same genetic relationship to each other as any siblings born at different times.

Full gene sequence

The complete order of bases in a gene. This order determines which protein a gene will produce.

Functional genomics

The study of genes, their resulting proteins, and the role played by the proteins the body's biochemical processes.

Gamete

Mature male or female reproductive cell (sperm or ovum) with a haploid set of chromosomes (23 for humans).

GC-rich area

Long stretches of repeated G and C on a DNA sequence that indicate a gene-rich region.

Gene

The fundamental physical and functional unit of heredity. A gene is an ordered sequence of nucleotides located in a particular position on a particular chromosome that encodes a specific functional product (i.e., a protein or an RNA molecule).

Gene amplification

Repeated copying of a piece of DNA; a characteristic of tumor cells.

Gene chip technology

Development of cDNA microarrays from a large number of genes; used to monitor and measure changes in gene expression for each gene represented on the chip.

Gene expression

The process by which a gene's coded information is converted into the structures present and operating in the cell. Expressed genes include those that are transcribed into mRNA and then translated into protein and those that are transcribed into RNA, but not translated into protein (e.g., transfer and ribosomal RNAs).

Gene family

Group of closely related genes that make similar products.

Gene library

Chromosome-specific recombinant DNA libraries for each chromosomes.

Gene mapping

Determination of the relative positions of genes on a DNA molecule (chromosome or plasmid) and of the distance, in linkage units or physical units, between them.

Gene product

The biochemical material, either RNA or protein, resulting from expression of a gene. The amount of gene product is used to measure how active a gene is; abnormal amounts can be correlated with disease-causing alleles.

Gene therapy

An experimental procedure aimed at replacing, manipulating, or supplementing nonfunctional or misfunctioning genes with healthy genes.

Gene transfer

Incorporation of new DNA into and organism's cells, usually by a vector such as a modified virus; used in gene therapy.

Genetic code

The sequence of nucleotides, coded in triplets (codons) along the mRNA, that determines the sequence of amino acids in protein synthesis. A gene's DNA sequence can be used to predict the mRNA sequence, and the genetic code can in turn be used to predict the amino-acid sequence.

Genetic engineering

Altering the genetic material of cells or organisms to enable them to make new substances or perform new functions.

Genetic marker

A gene or other identifiable portion of DNA whose inheritance can be followed.

Genetic polymorphism

Difference in DNA sequence among individuals, groups, or populations (e.g., genes for blue eyes versus brown eyes).

Genetics

The study of inheritance patterns of specific traits.

Genome

All the genetic material in the chromosomes of a particular organism. Its size is generally given as its total number of base pairs.

Genomic library

A collection of clones made from a set of randomly generated, overlapping DNA fragments that represent the entire genome of an organism.

Genomics

The study of genes and their function.

Genotype

The genetic constitution of an organism, as distinguished from its physical appearance (its phenotype).

Germ cell

Sperm and egg cells and their precursors. Germ cells are haploid and have only one set of chromosomes (23 in all), while all other cells have two copies (46 in all).

Guanine (G)

A nitrogenous base, one member of the base pair GC (guanine and cytosine) in DNA.

Haploid

A single set of chromosomes (half the full set of genetic material) present in the egg and sperm cells of animals and in the egg and pollen cells of plants. Human beings have 23 chromosomes in their reproductive cells.

Hemizygous

Having only one copy of a particular gene. For example, in humans, males are hemizygous for genes found on the Y chromosome.

Hereditary cancer

Cancer that occurs due to the inheritance of an altered gene within a family.

Heterozygosity

The presence of different alleles at one or more loci on homologous chromosomes.

Highly conserved sequence

DNA sequence that is very similar across several different types of organisms.

High-throughput sequencing

A fast method of determining the order of bases in DNA.

Homeobox

A short stretch of nucleotides whose base sequence is virtually identical in all the genes that contain it. Homeoboxes have been found in many organisms from fruit flies to human beings. In the fruit fly, a homeobox appears to determine when particular groups of genes are expressed during development.

Homolog

A member of a chromosome pair in diploid organisms or a gene that has the same origin and functions in two or more species.

Homologous chromosome

Chromosome containing the same linear gene sequences as another, each derived from one parent.

Homologous recombination

Swapping of DNA fragments between paired chromosomes.

Homology

Similarity in DNA or protein sequences between individuals of the same species or among different species.

Homozygote

An organism that has two identical alleles of a gene.

Human artificial chromosome (HAC)

A vector used to hold large DNA fragments.

Hybridization

The process of joining two complementary strands of DNA or one each of DNA and RNA to form a double-stranded molecule.

In situ hybridization

Use of a DNA or an RNA probe to detect the presence of the complementary DNA sequence in cloned bacterial or cultured eukaryotic cells.

In vitro

Studies performed outside a living organism such as in a laboratory.

In vivo

Studies carried out in living organisms.

Interphase

The period in the cell cycle when DNA is replicated in the nucleus; followed by mitosis.

Intron

DNA sequence that interrupts the protein-coding sequence of a gene. An intron is transcribed into RNA, but is cut out of the message before it is translated into protein.

Junk DNA (noncoding DNA)

Stretches of DNA that do not code for genes. Most of the genome consists of so-called junk DNA, which may have regulatory and other functions.

Karyotype

A photomicrograph of an individual's chromosomes arranged in a standard format showing the number, size, and shape of each chromosome type; used in low-resolution physical mapping to correlate gross chromosomal abnormalities with the characteristics of specific diseases.

Kilobase (kb)

Unit of length for DNA fragments equal to 1000 nucleotides.

Knockout

Deactivation of specific genes; used in laboratory organisms to study gene function.

Library

An unordered collection of clones (i.e., cloned DNA from a particular organism) whose relationship to each other can be established by physical mapping.

Linkage

The proximity of two or more markers (e.g., genes and RFLP markers) on a chromosome; the closer the markers, the lower the probability that they will be separated during DNA repair or replication processes (binary fission in prokaryotes, mitosis, or meiosis in eukaryotes), and hence the greater the probability that they will be inherited together.

Linkage map

A map of the relative positions of genetic loci on a chromosome, determined on the basis of how often the loci are inherited together. Distance is measured in centimorgans (cM).

Locus (pl. loci)

The position on a chromosome of a gene or other chromosome marker; also, the DNA at that position. The use of locus is sometimes restricted to mean expressed DNA regions.

Long-Range Restriction Mapping

Map depicting the chromosomal positions of restriction-enzyme cutting sites. These are used as biochemical "signposts," or markers of specific areas along the chromosomes.

Macrorestriction map

Map depicting the order of and distance between sites at which restriction enzymes cleave chromosomes.

Mass spectrometer

An instrument used to identify chemicals in a substance by their mass and charge.

Megabase (Mb)

Unit of length for DNA fragments equal to 1 million nucleotides and roughly equal to 1 cm.

Meiosis

The process of two consecutive cell divisions in the diploid progenitors of sex cells. Meiosis results in four, rather than two, daughter cells, each with a haploid set of chromosomes.

Messenger RNA (mRNA)

RNA that serves as a template for protein synthesis. *See also genetic code.*

Metaphase

A stage in mitosis or meiosis during which the chromosomes are aligned along the equatorial plane of the cell.

Microarray

Sets of miniaturized chemical-reaction areas that may also be used to test DNA fragments, antibodies, or proteins.

Microinjection

A technique for introducing a solution of DNA into a cell using a fine microcapillary pipet.

Micronuclei

Chromosome fragments that are not incorporated into the nucleus at cell division.

Mitochondrial DNA

The genetic material found in mitochondria, the organelles that generate energy for the cell; not inherited in the same fashion as nucleic DNA.

Mitosis

The process of nuclear division in cells that produces daughter cells that are genetically identical to each other and to the parent cell.

Model organisms

A laboratory animal or other organism useful for research.

Modeling

The use of statistical analysis, computer analysis, or model organisms to predict outcomes of research.

Molecular biology

The study of the structure, function, and makeup of biologically important molecules.

Molecular farming

The development of transgenic animals to produce human proteins for medical use.

Molecular genetics

The study of macromolecules important in biological inheritance.

Molecular medicine

The treatment of injury or disease at the molecular level. Examples include the use of DNA-based diagnostic tests or medicine derived from DNA-sequence information.

Mutation

Any heritable change in DNA sequence.

Nitrogenous base

A nitrogen-containing molecule having the chemical properties of a base. DNA contains the nitrogenous bases adenine (A), guanine (G), cytosine (C), and thymine (T).

Northern blot

A gel-based laboratory procedure that locates mRNA sequences on a gel that are complementary to a piece of DNA used as a probe.

Nuclear transfer

A laboratory procedure in which a cell's nucleus is removed and placed into an oocyte with its own nucleus removed so that the genetic information from the donor nucleus controls the resulting cell. Such cells can be induced to form embryos. This process was used to create the cloned sheep Dolly.

Nucleic acid

A large molecule composed of nucleotide subunits.

Nucleolar organizing region

A part of the chromosome containing rRNA genes.

Nucleotide

A subunit of DNA or RNA consisting of a nitrogenous base (adenine, guanine, thymine, or cytosine in DNA; adenine, guanine, uracil, or cytosine in RNA), a phosphate molecule, and a sugar molecule (deoxyribose in DNA and ribose in RNA). Thousands of nucleotides are linked to form a DNA or an RNA molecule.

Nucleus

The cellular organelle in eukaryotes that contains most of the genetic material.

Oligonucleotide

A molecule usually composed of 25 or fewer nucleotides; used as a DNA synthesis primer.

Oncogene

A gene, one or more forms of which is associated with cancer. Many oncogenes are involved, directly or indirectly, in controlling the rate of cell growth.

Open reading frame (ORF)

The sequence of DNA or RNA located between the start-code sequence (initiation codon) and the stop-code sequence (termination codon).

Operon

A set of genes transcribed under the control of an operator gene.

Peptide

Two or more amino acids joined by a bond called a "peptide bond." *See also polypeptide.*

Phage

A virus for which the natural host is a bacterial cell.

Phenotype

The physical characteristics of an organism or the presence of a disease that may or may not be genetic. *See also genotype.*

Physical map

A map of the locations of identifiable landmarks on DNA (e.g., restriction-enzyme cutting sites, genes), regardless of inheritance. Distance is measured in base pairs. For the human genome, the lowest-resolution physical map is the banding patterns on the 24 different chromosomes; the highest-resolution map is the complete nucleotide sequence of the chromosomes.

Plasmid

Autonomously replicating extra-chromosomal circular DNA molecules, distinct from the normal bacterial genome and nonessential for cell survival under nonselective conditions. Some plasmids are capable of integrating into the host genome. A number of artificially constructed plasmids are used as cloning vectors.

Pleiotropy

One gene that causes many different physical traits such as multiple disease symptoms.

Pluripotency

The potential of a cell to develop into more than one type of mature cell, depending on environment.

Polygenic disorder

Genetic disorder resulting from the combined action of alleles of more than one gene (e.g., heart disease, diabetes, and some cancers). Although such disorders are inherited, they depend on the simultaneous presence of several alleles; thus, the hereditary patterns usually are more complex than those of single-gene disorders.

Polymerase chain reaction (PCR)

A method for amplifying a DNA base sequence using a heat-stable polymerase and two 20-base primers, one complementary to the (+) strand at one end of the sequence to be amplified and one complementary to the (−) strand at the other end. Because the newly synthesized DNA strands can subsequently serve as additional templates for the same primer sequences, successive rounds of primer annealing, strand elongation, and dissociation produce rapid and highly specific amplification of the desired sequence. PCR also can be used to detect the existence of the defined sequence in a DNA sample.

Polymerase, DNA or RNA

Enzyme that catalyzes the synthesis of nucleic acids on preexisting nucleic acid templates, assembling RNA from ribonucleotides or DNA from deoxyribonucleotides.

Polymorphism

Difference in DNA sequence among individuals that may underlie differences in health. Genetic variations occurring in more than 1 percent of a population would be considered useful polymorphisms for genetic linkage analysis.

Polypeptide

A protein or part of a protein made of a chain of amino acids joined by a peptide bond.

Primer

Short preexisting polynucleotide chain to which new deoxyribonucleotides can be added by DNA polymerase.

Probe

Single-stranded DNA or RNA molecules of specific base sequence, labeled either radioactively or immunologically, that are used to detect the complementary base sequence by hybridization.

Prokaryote

Cell or organism lacking a membrane-bound, structurally discrete nucleus and other subcellular compartments. Bacteria are examples of prokaryotes.

Promoter

A DNA site to which RNA polymerase will bind and initiate transcription.

Protein

A large molecule composed of one or more chains of amino acids in a specific order. The order is determined by the base sequence of nucleotides in the gene that codes for the protein. Proteins are required for the structure, function, and regulation of the body's cells, tissues, and organs; and each protein has unique functions. Examples are hormones, enzymes, and antibodies.

Proteome

Proteins expressed by a cell or organ at a particular time and under specific conditions.

Proteomics

The study of the full set of proteins encoded by a genome.

Pseudogene

A sequence of DNA similar to a gene, but nonfunctional; probably the remnant of a once-functional gene that accumulated mutations.

Purine

A nitrogen-containing, double-ring, basic compound that occurs in nucleic acids. The purines in DNA and RNA are adenine and guanine.

Pyrimidine

A nitrogen-containing, single-ring, basic compound that occurs in nucleic acids. The pyrimidines in DNA are cytosine and thymine; in RNA, cytosine and uracil.

Recessive gene

A gene that will be expressed only if there are two identical copies or, for a male, if one copy is present on the X chromosome.

Recombinant clone

Clone containing recombinant DNA molecules.

Recombinant DNA molecules

A combination of DNA molecules of different origin that are joined using recombinant DNA technologies.

Recombinant DNA technology

Procedure used to join together DNA segments in a cell-free system (an environment outside a cell or organism). Under appropriate conditions, a recombinant DNA molecule can enter a cell and replicate there, either autonomously or after it has become integrated into a cellular chromosome.

Recombination

The process by which progeny derive a combination of genes different from that of either parent. In higher organisms, this can occur by crossing over.

Regulatory region or sequence

A DNA base sequence that controls gene expression.

Repetitive DNA

Sequences of varying lengths that occur in multiple copies in the genome; represents much of the human genome.

Resolution

Degree of molecular detail on a physical map of DNA, ranging from low to high.

Restriction enzyme, endonuclease

A protein that recognizes specific, short nucleotide sequences and cuts DNA at those sites. Bacteria contain over 400 such enzymes that recognize and cut more than 100 different DNA sequences.

Restriction fragment length polymorphism (RFLP)

Variation between individuals in DNA fragment sizes cut by specific restriction enzymes; polymorphic sequences that result in RFLPs are used as markers on both physical maps and genetic linkage maps. RFLPs usually are caused by mutation at a cutting site.

Restriction-enzyme cutting site

A specific nucleotide sequence of DNA at which a particular restriction enzyme cuts the DNA. Some sites occur frequently in DNA (e.g., every several hundred base pairs); others much less frequently (rare-cutter; e.g., every 10,000 base pairs).

Reverse transcriptase

An enzyme used by retroviruses to form a complementary DNA sequence (cDNA) from their RNA. The resulting DNA is then inserted into the chromosome of the host cell.

Ribose

The five-carbon sugar that serves as a component of RNA. *See also ribonucleic acid; deoxyribose.*

Ribosomal RNA (rRNA)

A class of RNA found in the ribosomes of cells.

Ribosomes

Small cellular components composed of specialized ribosomal RNA and protein; site of protein synthesis.

RNA (Ribonucleic acid)

A chemical found in the nucleus and cytoplasm of cells; plays an important role in protein synthesis and other chemical activities of the cell. The structure of RNA is similar to that of DNA. There are several classes of RNA molecules, including messenger RNA, transfer RNA, ribosomal RNA, and other small RNAs, each serving a different purpose.

Sanger sequencing

A widely used method of determining the order of bases in DNA.

Satellite

A chromosomal segment that branches off from the rest of the chromosome, but is still connected by a thin filament or stalk.

Scaffold

In genomic mapping, a series of contigs that are in the right order, but not necessarily connected in one continuous stretch of sequence.

Segregation

The normal biological process whereby the two pieces of a chromosome pair are separated during meiosis and randomly distributed to the germ cells.

Sequence assembly

A process whereby the order of multiple sequenced DNA fragments is determined.

Sequence tagged site (STS)

Short (200 to 500 base pairs) DNA sequence that has a single occurrence in the human genome and whose location and base sequence are known. Detectable by polymerase chain reaction, STSs are useful for localizing and orienting the mapping and sequence data reported from many different laboratories and serve as landmarks on the developing physical map of the human genome. Expressed sequence tags (ESTs) are STSs derived from cDNAs.

Sequencing

Determination of the order of nucleotides (base sequences) in a DNA or an RNA molecule or the order of amino acids in a protein.

Sequencing technology

The instrumentation and procedures used to determine the order of nucleotides in DNA.

Sex chromosome

The X or Y chromosome in human beings that determines the sex of an individual. Females have two X chromosomes in diploid cells; males have an X and a Y chromosome. The sex chromosomes comprise the 23rd chromosome pair in a karyotype.

Shotgun method

Sequencing method that involves randomly sequenced cloned pieces of the genome, with no foreknowledge of where the piece originally came from. This can be contrasted with "directed" strategies, in which pieces of DNA from known chromosomal locations are sequenced. Because there are advantages to both strategies, researchers use both random (or shotgun) and directed strategies in combination to sequence the human genome.

Single nucleotide polymorphism (SNP)

DNA sequence variations that occur when a single nucleotide (A, T, C, or G) in the genome sequence is altered.

Single-gene disorder

Hereditary disorder caused by a mutant allele of a single gene (e.g., Duchenne muscular dystrophy, retinoblastoma, and sickle-cell disease).

Somatic cell

Any cell in the body except gametes and their precursors.

Somatic cell gene therapy

Incorporating new genetic material into cells for therapeutic purposes. The new genetic material cannot be passed to offspring.

Southern blotting

Transfer by absorption of DNA fragments separated in electrophoretic gels to membrane filters for detection of specific base sequences by radio-labeled complementary probes.

Spectral karyotype (SKY)

A graphic of all an organism's chromosomes, each labeled with a different color. Useful for identifying chromosomal abnormalities.

Splice site

Location in the DNA sequence where RNA removes the noncoding areas to form a continuous gene transcript for translation into a protein.

Sporadic cancer

Cancer that occurs randomly and is not inherited from parents; caused by DNA changes in one cell that grows and divides, spreading throughout the body.

Stem cell

Undifferentiated, primitive cells in the bone marrow that have the ability both to multiply and to differentiate into specific blood cells.

Structural genomics

The effort to determine the three-dimensional structures of large numbers of proteins using both experimental techniques and computer simulation

Suppressor gene

A gene that can suppress the action of another gene.

Synteny

Genes occurring in the same order on chromosomes of different species.

Tandem repeat sequences

Multiple copies of the same base sequence on a chromosome; used as markers in physical mapping.

Targeted mutagenesis

Deliberate change in the genetic structure directed at a specific site on the chromosome; used in research to determine the targeted region's function.

Technology transfer

The process of transferring scientific findings from research laboratories to the commercial sector.

Telomerase

The enzyme that directs the replication of telomeres.

Telomere

The end of a chromosome. This specialized structure is involved in the replication and stability of linear DNA molecules.

Teratogen

Substance such as a chemical or radiation that cause abnormal development of a embryo.

Thymine (T)

A nitrogenous base, one member of the base pair AT (adenine–thymine).

Transcription

The synthesis of an RNA copy from a sequence of DNA (a gene); the first step in gene expression.

Transcription factor

A protein that binds to regulatory regions and helps control gene expression.

Transcriptome

The full complement of activated genes, mRNAs, or transcripts in a particular tissue at a particular time

Transfection

The introduction of foreign DNA into a host cell.

Transfer RNA (tRNA)

A class of RNA having structures with triplet nucleotide sequences that are complementary to the triplet nucleotide-coding sequences of mRNA. The role of tRNAs in protein synthesis is to bond with amino acids and transfer them to the ribosomes, where proteins are assembled according to the genetic code carried by mRNA.

Transformation

A process by which the genetic material carried by an individual cell is altered by incorporation of exogenous DNA into its genome.

Transgenic

An experimentally produced organism in which DNA has been artificially introduced and incorporated into the organism's germ line.

Translation

The process in which the genetic code carried by mRNA directs the synthesis of proteins from amino acids.

Translocation

A mutation in which a large segment of one chromosome breaks off and attaches to another chromosome.

Transposable element

A class of DNA sequences that can move from one chromosomal site to another.

Uracil

A nitrogenous base normally found in RNA but not DNA; uracil is capable of forming a base pair with adenine.

Virus

A noncellular biological entity that can reproduce only within a host cell. Viruses consist of nucleic acid covered by protein; some animal viruses are also surrounded by membrane. Inside the infected cell, the virus uses the synthetic capability of the host to produce progeny virus.

Western blot

A technique used to identify and locate proteins based on their ability to bind to specific antibodies.

Wild type

The form of an organism that occurs most frequently in nature.

X chromosome

One of the two sex chromosomes, X and Y.

Xenograft

Tissue or organs from an individual of one species transplanted into or grafted onto an organism of another species, genus, or family. A common example is the use of pig heart valves in humans.

Y chromosome

One of the two sex chromosomes, X and Y.

Yeast artificial chromosome (YAC)

Constructed from yeast DNA, it is a vector used to clone large DNA fragments.

Zinc-finger protein

A secondary feature of some proteins containing a zinc atom; a DNA-binding protein.

*This glossary was compiled from the publicly accessible Web site http://www.ornl.gov/ TechResources/Human_Genome/glossary/glossary.html, funded by the <u>Human Genome Program</u> of the U.S. Department of Energy.

Index

A

Acids, 16–17
Acrosome, 219
ACT, 117
Actin, 99–100, 137
    microfilaments, formation of, 83–84
Activation energy, 24
Activator proteins, 120
Active membrane transport, 107–111
    defined, 108
    example of, 108–109
    facilitated transport, 108
    integral membrane proteins, 108
    membrane channels, 107–108
    membrane electrical potential, 110–111
Activin, 226, 237
Adenine, 115
ADP, 21, 97
    detachment from myosin, 101
    hydrolysis of, 21
ADP-actin, 83
Adrenaline, 88
Adult stem cells, 235
Agrobacterium tumefacians, 153
Alanine, 61
Alleles, 140
α-actinin, 84
α-glucose, 41, 43
α-helix sheet, 65
α-helix structure, 65
Alternative splicing, 116–117, 158
American Type Culture Collection (ATCC), 211
Amino acids, 32–33, 55, 57–58, 142
    aromatic, 61
    chemical properties found in, 58–62
    essential, 62
    glycine, 59–61
    with hydrocarbon chains, 59–61
    mapping between mRNA codons and, 63
    residue, 58
    sulfur-containing, 61
    water-loving, 61–62
Ammonia, 13–14, 29

Ammonia molecules, 7–8, 10–11
AMP, 21, 53
    hydrolysis of, 21
Anabolism, cells, 86–87
Anaphase-promoting complex (APC), 203
Animal cells, 73, 76
    cell culture, 210–211
    organelles of, 73–74
Animal cloning, 163–164
Animal hemisphere, 218–219
Animal pole, 218
Animal vegetal axis, 218
Anion, 13
Antisense strand, 51
APC, 198
Apoptosis, 203, 214
APT, 22–23
ARACHNE, 166
Archaea, 72
Arginine, 61
Aromatic amino acids, 61
Aromatic side chains, 61
Asparagine, 61
Aspartate, 61
Asymmetric carbon, 39
Asymmetric (chiral) carbon, 39
Atomic composition of living organisms and Earth's crust, 2
Atomic mass, 4
Atomic mass number, 3–4
Atomic mass unit, 3–4
Atomic number, 2–3
Atoms, 1, 2–6
    chemical reactivity, 5–6
    combining capacity of, 5
    and covalent bonds, 6
    electron configurations, 4
    number of electrons in, 2
    number of protons in, 2–3
    octet rule, 5–6
    physical properties of, 2–4
ATP, 1, 20–23, 89, 97–99, 244
    defined, 22
    hydrolysis of, 21, 22
    stability of, 22